Symmetry in Structural Health Monitoring

Symmetry in Structural Health Monitoring

Editors

Yang Yang
Ying Lei
Xiaolin Meng
Jun Li

MDPI • Basel • Beijing • Wuhan • Barcelona • Belgrade • Manchester • Tokyo • Cluj • Tianjin

Editors

Yang Yang
School of Civil Engineering,
Chongqing Univerity,
Chongqing 400044, China

Ying Lei
School of Architecture and
Civil Engineering,
Xiamen University
Xiamen 365001, China

Xiaolin Meng
Faculty of Architecture,
Civil and Transportation Engineering,
Beijing University of
Technology, 100 Pingleyuan,
Chaoyang District,
Beijing 100124, China

Jun Li
School of Civil and
Mechanical Engineering,
Curtin University,
Bentley, WA 6102, Australia

Editorial Office
MDPI
St. Alban-Anlage 66
4052 Basel, Switzerland

This is a reprint of articles from the Special Issue published online in the open access journal *Symmetry* (ISSN 2073-8994) (available at: https://www.mdpi.com/journal/symmetry/special_issues/Symmetry_Structural_Health_Monitoring).

For citation purposes, cite each article independently as indicated on the article page online and as indicated below:

LastName, A.A.; LastName, B.B.; LastName, C.C. Article Title. *Journal Name* **Year**, *Volume Number*, Page Range.

ISBN 978-3-0365-4237-9 (Hbk)
ISBN 978-3-0365-4238-6 (PDF)

Contents

Preface to "Symmetry in Structural Health Monitoring"

Structural health monitoring refers to the strategy and process of damage diagnosis and the characterization of engineering structures. With the development of urbanization, various types of infrastructure and mechanical equipment provide people with convenient life services. This also shows the importance of structural health monitoring where symmetry is widely used. By analyzing the symmetry of the structure and using sensors to collect data, it is possible to study the performance of the structure itself. This is a hot topic in the current research. The collected data contain the structural random vibration and environmental noise; good equipment can improve the efficiency of data collection, while effective denoising methods with the knowledge of structural dynamics can extract the characteristic parameters of the structure from the data. In this Special Issue on symmetry, we mainly discuss the application of symmetry in various aspects of structural health monitoring. For example, by considering the health monitoring of a known structure by obtaining the structure's static or dynamic response, using different signal processing methods, including some advanced filtering methods, to remove the influence of environmental noise and extract structural feature parameters to determine the structure's safety. These damage diagnosis methods can be effectively applied to various types of infrastructure and mechanical equipment. For this reason, the vibration control of various structures and the knowledge of random structure dynamics should be considered, which will promote the rapid development of the structural health monitoring. The signal extraction and evaluation methods are also worthy of study. The improvement in signal acquisition instruments and acquisition methods improves the accuracy of the data. A good evaluation method will help to correctly understand the performance with different types of infrastructure and mechanical equipment.

Yang Yang, Ying Lei, Xiaolin Meng, and Jun Li
Editors

symmetry

MDPI

Article

Installation Quality Inspection for High Formwork Using Terrestrial Laser Scanning Technology

Linlin Zhao [1,2,3,*], Jasper Mbachu [4], Bill Wang [5], Zhansheng Liu [1] and Huirong Zhang [1]

1 Faculty of Architecture, Civil and Transportation Engineering, Beijing University of Technology, Beijing 100124, China; liuzhansheng@bjut.edu.cn (Z.L.); zhanghuirong@bjut.edu.cn (H.Z.)
2 Beijing Key Laboratory of Earthquake Engineering and Structural Retrofit, Beijing University of Technology, Beijing 100124, China
3 Key Laboratory of Urban Security and Disaster Engineering of China Ministry of Education, Beijing University of Technology, Beijing 100124, China
4 Faculty of Society & Design, Bond University, Gold Coast, QLD 4226, Australia; jmbachu@bond.edu.au
5 Department of Business Information Systems, Auckland University of Technology, Auckland 1010, New Zealand; bill.wang@aut.ac.nz
* Correspondence: llzhao@bjut.edu.cn

Abstract: Current inspection for installation quality of high formwork is conducted by site managers based on personal experience and intuition. This non-systematic inspection is laborious and it is difficult to provide accurate dimension measurements for high formwork. The study proposed a method that uses terrestrial laser scanning (TLS) technology to collect the full range measurements of a high formwork and develop a genetic algorithm (GA) optimized artificial neutral network (ANN) model to improve measurement accuracy. First, a small-scale high formwork model set was established in the lab for scanning. Then, the collected multi-scan data were registered in a common reference system, and RGB value and symmetry of the structure were used to extract poles and tubes of the model set, removing all irrelevant data. Third, all the cross points of poles and tubes were generated. Next, the model set positioned on the moving equipment was scanned at different specified locations in order to collect sufficient data to develop an GA-ANN model that can generate accurate estimates of the point coordinates so that the accuracy of the dimension measurements can be achieved at the millimetre level. Validation experiments were conducted both on another model set and a real high formwork. The successful applications suggest that the proposed method is superior to other common techniques for obtaining the required data necessary for accurately measuring the overall structure dimensions, regarding data accuracy, cost and time. The study proposed an effective method for installation quality inspection for high formwork, especially when the inspection cannot be properly operated due to cost factors associated with common inspection methods.

Keywords: terrestrial laser scanning; high formwork; RGB; genetic algorithm; artificial neutral network

Citation: Zhao, L.; Mbachu, J.; Wang, B.; Liu, Z.; Zhang, H. Installation Quality Inspection for High Formwork Using Terrestrial Laser Scanning Technology. *Symmetry* **2022**, *14*, 377. https://doi.org/10.3390/sym14020377

Academic Editors: Yang Yang, Ying Lei, Xiaolin Meng and Jun Li

Received: 10 December 2021
Accepted: 1 February 2022
Published: 14 February 2022

1. Introduction

Formwork is a temporary structure on a construction site into which concrete or a similar substance is poured, while the falsework supports the shuttering moulds [1]. In building construction, formwork plays a major role in determining the duration and schedule of construction activities [2]. In fact, the cost of formwork construction (forming cost) accounts for 10% of the overall cost and erection and assembly of formwork systems (forming time) spend 50% of the overall construction time of the entire project [3]. Recently, with developments in the building industry, high formwork is used more widely. High formwork projects are more significant than general formwork projects due to their complex structures. Their large size and complex structure make them more vulnerable to potential safety risks.

In the installation of high formwork structures, errors are unavoidable, resulting in defects that differ from code requirements or design specifications. In general, small errors are permitted, but they must be within an acceptable error range. Otherwise, they can create potential safety risks. Any such safety risks can adversely affect subsequent processes, lower the efficiency of the construction work and increase construction costs, or worst of all, cause accidents [4]. Such accidents can lead to significant loss of life as well as financial losses. Hence, it is necessary to conduct proper, high-quality inspections for high formwork after installation. However, the inspection process for high formwork has not been dealt with systematically due to the lack of an appropriate inspection technology for such projects and a shortage of primary data necessary to establish a plan. In fact, inspecting these factors on construction site using common methods like tape measurements or plumb monitoring is exhausting and laborious.

The main concern of the inspection of installation quality for high formwork includes: (1) to check the position of the main components of the formwork such as vertical poles, horizontal tubes, bottom reinforcing tubes and the diagonal bracing; (2) to check whether the spacing of vertical pole and lift height (the distance between the two neighbour tubes) are in accordance with the requirements of the building code. Based on the investigations of [5–7], these factors are the main determinants of the proper function of a high formwork. According to China's building codes related to high formwork [8–10], they are the main inspection terms during the inspection process. Hence, the study proposed a method for effectively checking them.

Current installation quality assessment of high formwork is mainly based on manual inspection using traditional measurement instruments like measuring tapes, which require close-up access, and are not ideal for detecting defects or for accurately measuring the dimensions and recording the data. The inspection process is mainly based on human inspection and judgement, which is labour-intensive and inefficient, especially for large-scale structures.

Different approaches have been adopted, including theodolites, joint meters, and inclinometers, as well as tachymetry, global navigation satellite system (GNSS), and robotic total station [11,12]. They can offer accurate results, but they only provide discrete monitoring points and fail to attain full area coverage. Some measurement techniques are also too expensive. Moreover, not all available techniques are suitable for inspection of high formwork due to the size and complex shape of the structure, and level of required accuracy. When choosing a suitable measurement system, factors such as cost, time required for operation, and the required skills of the user should all be considered [13–15].

Inspection accuracy and efficiency can be improved by using terrestrial laser scanning (TLS) technology. Another benefit of TLS is its long range, which allows data collection in inaccessible fields. The TLS system can be regarded as the best monitoring technique for high formwork owing to its ability to provide high spatial resolution and accurate measurements. A major benefit of TLS is that it provides rapid and dense measurements in inaccessible areas [16]. TLS can produce point clouds with high level of details and accuracy, which is particularly good for detailed investigations on a large scale [17,18]. The large amount of accurately measured data has great potential to deliver measurements that are difficult to achieve by other methods in such a short period of time, with simple installation requirements. TLS measurements require only a laser scanner, a laptop computer with data processing software, and several reference targets.

Previous studies have provided sound evidence for the efficiency and effectiveness of inspection operation using TLS. Gordon et al. [19] and Akinci et al. [20], for example, demonstrated that manual inspection can miss important information such as site changes and defects. However, the adoption of TLS can improve efficiency in construction inspection via timely provision of comprehensive, as-built data. Although there are advantages to adopting TLS, some studies [21–23] also point out the challenges that adversely influence the effective use of TLS, including difficulties in effectively extracting the required information from scan data and in generating required accuracy results.

This study provides a new method for conducting quality inspection using TLS technology. The proposed method can effectively extract important elements such as poles and tubes in high formwork from the point cloud data. A GA-ANN model was developed to increase the accuracy of the TLS measurements. To achieve these, some techniques are first proposed to rapidly extract the important points from the dense data. Secondly, a small-scale high formwork model set with set dimensions and simulated defects was built to help in collecting sufficient data to train the ANN model. The obtained measurements based on the results from the ANN model were compared to the set dimensions to evaluate the performance of the ANN model. Moreover, the genetic algorithm was adopted to optimised ANN model. The performance of the GA-ANN model was compared with other ML algorithms such as support vector machine and random forest. Next, a laboratory test was conducted to validate the proposed method. Finally, the proposed method was used on a real high formwork to validate its effectiveness.

The uniqueness of the study lies in: (1) the consideration of TLS as an effective tool in quality inspection for high formwork after installation; (2) the rapid extraction of the standard poles and tubes from the point clouds; and (3) the development of a GA-ANN model that can generate the coordinates of important points to accurately calculate the distance between the important points and thus detect defects that fall out of the acceptable error range.

The organization of the study is as follows: the literature review and the highlights of the contribution of the current study are described in Section 2. The proposed method for installation quality inspection for high formwork are illustrated in Section 3. The validation of the proposed method on a real high formwork are shown in Section 4. Section 5 provides the conclusion and future research plans.

2. Literature Review

2.1. TLS System

Several studies have been carried out focusing on the use of terrestrial laser scanning (TLS) technology as an inspection approach including the application of TLS for health monitoring and the inspection of bridge structures [24,25]; condition monitoring and defect identification for historic structures and building facades [26,27]; checking concrete conditions [16,28]; and as-built modelling [25,29]. In a study of [24] TLS that was used in building damage inspection after an earthquake, the accuracy of the measurement was achieved at the millimetre level, a performance comparable with conventional methods like displacement transducers and inclinometers.

TLS has been widely adopted in the construction field for construction progress tracking [30,31], the reconstruction of buildings [32–35], and construction quality assurance [36–38]. However, limited studies have been performed using TLS to inspect the installation quality of high formwork.

2.2. Object Recognition

Extensive studies on the use of TLS in object recognition have been conducted. TLS data can generally be extracted based on features. Some studies also proposed to specify the point data based on the RGB value. For example, Pu and Vosselman [33] proposed a region-growing method to effectively extract planar objects from point clouds of a building façade, and a feature recognition method was used to classify the planar objects. In the study [39], the least square method was used to fit lines to extract important lines and points from the point clouds of structural elements on buildings damaged by an earthquake.

Some studies also focused on developing an automatic algorithm to recognize and segment required objects from point clouds. For example, Riveiro et al. [40] recommended an approach that can automatically extract masonry blocks from point clouds. In the study [41], machine learning methods were used to automatically classify the morphological segments of a hillslope affected by shallow landslides into seven classes (e.g., scarp, eroded area, deposit, rock outcrop and different classes of vegetation). Lee et al. [42] used a method that automatically extracts pipelines and their detailed parts, such as elbows and tees, from point clouds. Similarly, Czerniawski et al. [43] proposed as fully automatic approach for extracting pipe spools from point clouds. Some studies also developed methods that can recognize infrastructure objects based on point clouds. For example, Holgado-Barco et al. [44] extracted features from Light Detection and Ranging (LiDAR) data to model a road axis.

Most of the studies have focused on extracting data from buildings or infrastructure such as roads or bridges. However, few studies extract important elements from point cloud data based on the specific features of a high formwork. The combination of the RGB value of the point cloud data and the symmetry of the structure and special shape of the main components of the high formwork can help to rapidly extract important data from scan data.

2.3. Quality Assessment and Control

Traditionally, structural dimension checks are performed manually. Two main inspection methods are widely used: (1) directly check based on visual observations or using a measuring tape; (2) obtain coordinates of serval feature points by using total station. However, these methods are laborious, time-consuming, and insufficiently accurate. Recently, TLS has been widely used in quality assessment and control, such as surface-defect detection [45,46] or structure dimensional check in the Architecture, Engineering & Construction (AEC) industry [47–49]. Moreover, some studies have proposed using TLS to check the dimensions of structures. For example, Wang et al. [50] developed an automatic method to estimate the dimensions of a precast concrete panel using point clouds data. Bosch'e [51] proposed an approach to automatically detect and calculate the as-built dimensions of a steel structure based on laser scan data. Lee & Park [52] proposed a method using both TLS and an AI model to perform a dimensional check. And Nuttens et al. [53] provided a clear work flow to illustrate how to use TLS to measure the dimensional changes in two tunnels in Belgium and to generate results with submillimeter accuracy.

Although these studies provided good examples about the adoption of TLS technology in the AEC industry, they also admitted some issues about TLS measurement. First, TLS can quickly provide as-built data, but the scan strategy should be properly designed in order to provide good-quality data. Moreover, some techniques should be used to rapidly extract important data from dense scan data. In addition, the accuracy level of the TLS measurement should be improved to satisfy the requirements.

The contribution of the study lies in three aspects. First, the study proposes using TLS technology in the inspection of high formwork installation quality. Due to the unique complex structure of the high formwork and the labor and cost concerns of the inspection process, this study provides a reasonable and cost-effective way to conduct the inspection process. Secondly, the study provides an effective to quickly extract the data related to the important components of the high formwork such as vertical poles and horizontal tubes. Thirdly, the study used genetic algorithm optimized ANN model to improve the accuracy of the TLS measurements, which can provide coordinates of the cross points of the poles and tubes at millimeter level.

3. The Proposed Method

The proposed method used both TLS data and ANN model for evaluating installation quality for high formwork according to China building codes [8–10]. The major clauses of the building codes and regulations for the installation quality of high formwork are illustrated in Table 1. Therefore, the main tasks of the study are to evaluate the positions of the vertical poles, horizontal tubes, and diagonal bracing according to clauses i–iii, assess the longitudinal spacing of vertical pole according to clause iv, inspect the transverse spacing of vertical pole according to clause v, check the space between the two neighbour tubes (the lift height) according to clause vi, and check the position and the height of the bottom reinforcing tube according to clause vii.

Table 1. Classification of inspection terms.

Clause	Inspection Term	Requirements by China Building Codes	Acceptable Error
i	Standing pole	Every pole should be vertical to the ground; $\Delta \leq 1/500$ H and horizontal deviation $\Delta \leq \pm 50$ mm	5 mm
		Description diagram	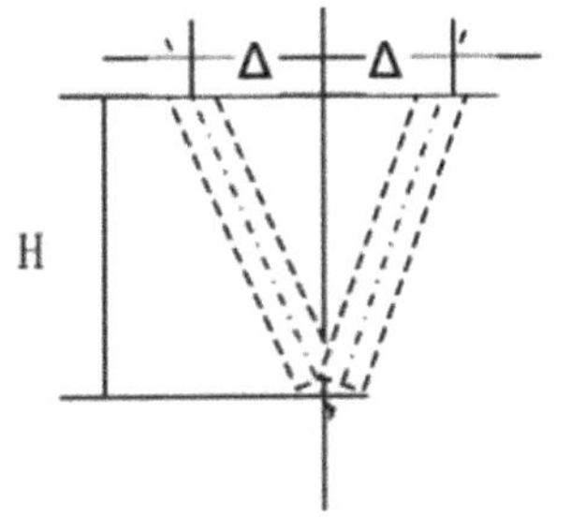
ii	Horizontal tube	Every tube should parallel to the ground	5 mm
		Description diagram	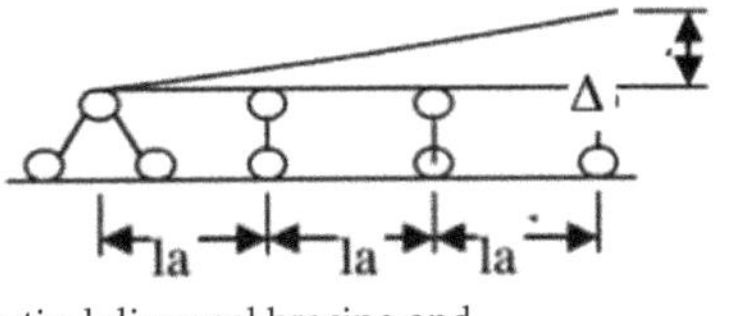
iii	Diagonal bracing	The angle of the vertical diagonal bracing and the ground should set between 45°–60° The angle of the horizontal diagonal bracing and the horizontal tube should set between 45°–60°	
iv	Longitudinal spacing of standing pole	The distance should less than 1.2 m	≤30 mm
v	Transverse spacing of standing pole	The distance should less than 1.2 m	≤30 mm
vi	Distance between two neighbouring tube (Lift height)	The space should less than 1.5 m	≤20 mm
vii	The bottom reinforcing tube	The height of the bottom tubes should equal to 350 mm; the transverse tubes should locate on the top of the longitudinal tubes.	

The proposed method for installation quality assessment of high formwork is carried out in four steps. First, a small-scale high formwork model set was built in the lab and scanned to collect multi-scan data, which were then registered in a common reference system. Secondly, RGB value, symmetry of the structure and shape fitting algorithm were used to extract the poles and tubes of the model set and all the cross-points of the poles and tubes were generated. After that, the model set was positioned on a moving equipment and moved along the X, Y and Z axes, from 1 mm to 10 mm at interval 1 mm, respectively. The model set was scanned at every movement in order to collect sufficient scan data for developing an ANN model that can help to generate accurate coordinates of the cross points at the millimetre level. Finally, another small-scale high formwork model set was established in the lab to validate the efficiency of the proposed method. The flowchart of the proposed method is described in Figure 1. The details of the four steps are illustrated as follows.

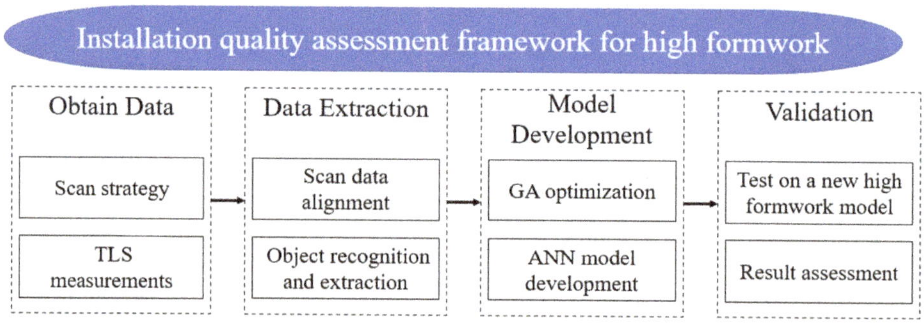

Figure 1. Flowchart of the installation quality assessment framework for a high formwork.

3.1. Scan Data Obtainment

The model set is a length 1.3 m × width 1.0 m × height 1.6 m formwork, consisting of steel vertical poles and horizontal tubes. The model set contains 77 standard poles and 270 tubes. The steel bars are connected via couplers, the longitudinal spacing of vertical pole is set as 12 cm, the transverse spacing of pole is 15 cm, and the lift height (the distance between two neighbouring tubes) is 10 cm. The simulated defects were purposefully set including: the leftmost longitudinal pole is not vertical to the ground but inclined 1 degree; the first top longitudinal tube is not parallel to the ground and is inclined 1 degree; the distance between the 5th and 6th longitudinal poles on the front side is 14 cm; the distance between the 5th and 6th transverse poles on the leftmost side is 17 cm; and the space between the 10th and 11th transverse tubes on the leftmost side is 12 cm. The details of the formwork dimensions are shown in Figure 2.

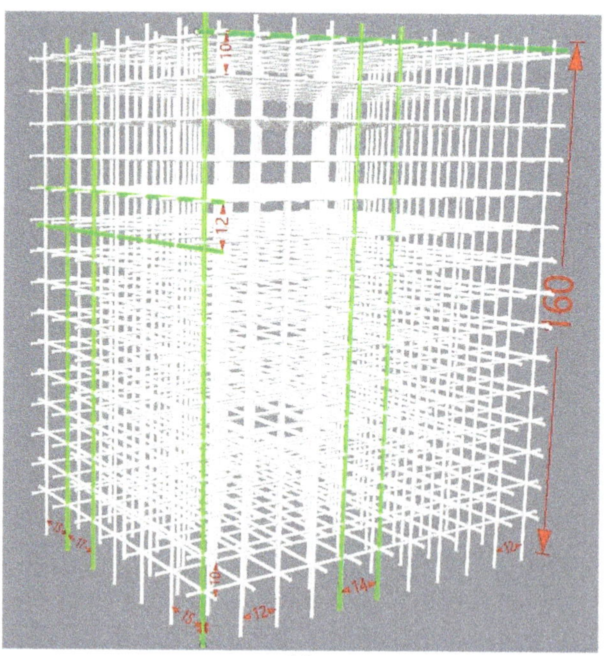

(**a**)

Figure 2. *Cont.*

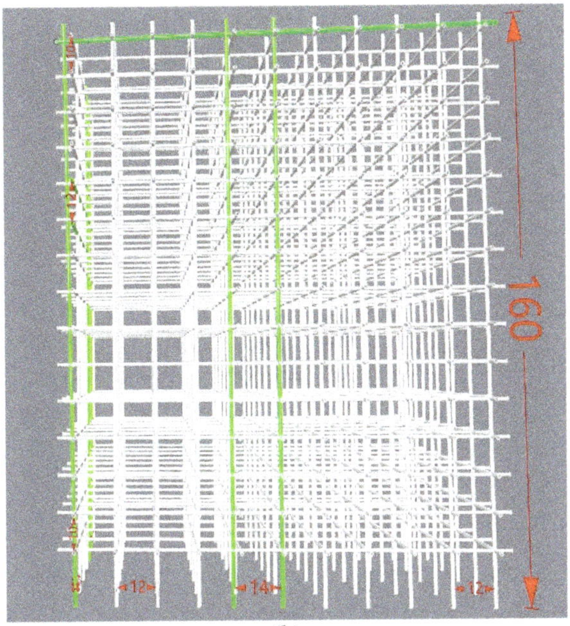

(**b**)

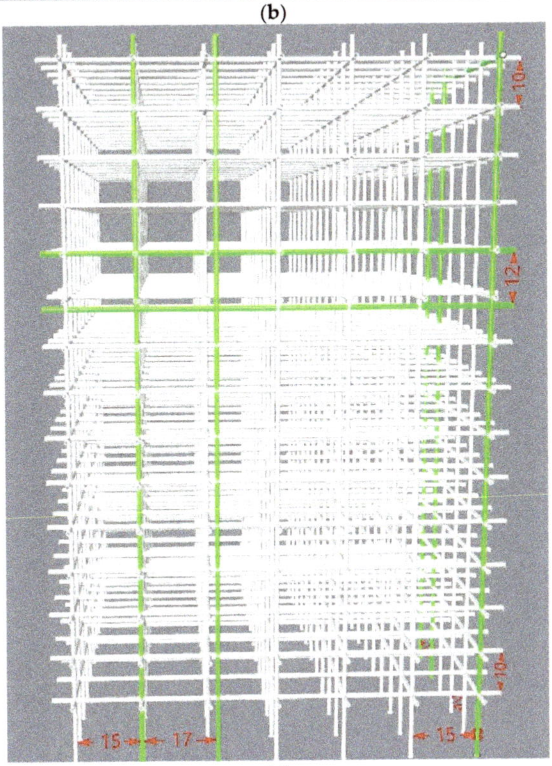

(**c**)

Figure 2. *Cont.*

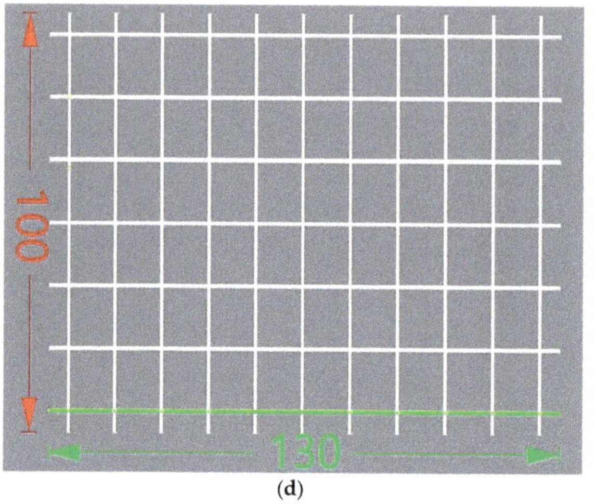

(**d**)

Figure 2. The high formwork model with simulated defects (**a**) The high formwork model, (**b**) The front side of the high formwork model, (**c**) The leftmost side of the high formwork model, (**d**) The plan view of the high formwork model.

3.1.1. Scan Strategy

The small-scale high formwork model set located on a moving equipment, the scanning position, and the control points for the laboratory trial are shown in Figure 3. The moving equipment can move freely along X, Y, and Z direction at submillimetre level. Targets were used for post-processing data registration in a common reference frame. A Trimble total station was used to survey control points and targets in order to build a control network. The observations were adjusted to minimize the errors. The resulting coordinates were corrected to within 1 mm. The Trimble TX5 3D laser scanner was used to carry out the scans and collect point cloud data. The small-scale model located on the moving equipment move from 1 mm to 10 mm at interval 1 mm along X, Y, and Z directions, respectively. In every 1 mm movement, the model was scanned once, and thus, the model was scanned 30 times in total.

3.1.2. Data Registration and Noise Removal

After the completion of the scan of the small-scale high formwork model set, it was necessary to conduct post processing of the raw point cloud data. Co-registration of multiple point clouds in a unit reference frame is important for further data processing. Two methods are usually adopted: homologous points identification and surface matching [54]. Homologous points identification needs several points indicated a same object that can be identified without spatial ambiguity in subsequent point clouds. Hence, targets-based registration was used in this study. A control network with more stable points should be established in order to periodically observe the targets and verify their stability. In this study, four stable locations were identified as TLS targets in such a way that they were geometrically well distributed.

The scans were registered using the targets and Iterative Closet Point (ICP) adjustment. The details of the ICP algorithm can be found in the study of Besl and McKay [55] and in the study of Sgrenzaroli and Wolfart [56]. The data processing was conducted using algorithms implemented in the scanner combined software JRC 3D Reconstructor. The combined software not only registers the multiple scans data but can also remove the noise data and reduce the data density to facilitate further data processing. Alternatively, point clouds data can be exported in many formats like ASCII for post processing in MATLAB.

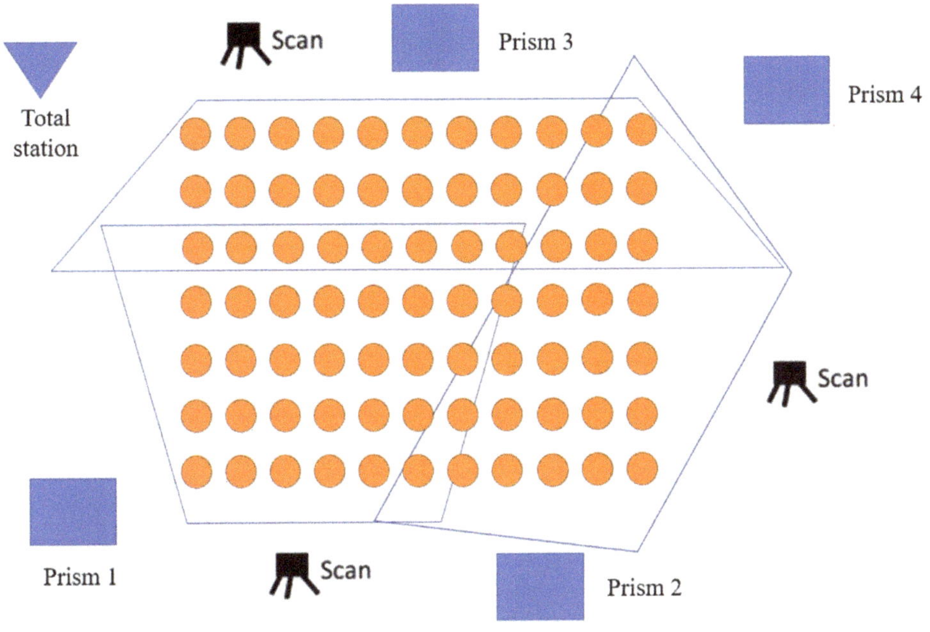

Figure 3. The layout of the TLS measurement.

3.2. Data Extraction

3.2.1. Removal of Mixed Pixels

The scan data of a high formwork usually include multiple objects such as vertical poles, horizontal tubes, couplers, braces, bolts, ground, timber formwork, and so on. Construction sites include other unexpected noise, including construction equipment, workers, and etc. In order to retain useful data relating to poles and tubes, the others must be removed. Unfortunately, mixed pixels that are a type of false measurement are always included in the laser scan data [49]. It occurs when a laser beam is split into two parts and falls on two different objects. Hence, the laser scanner obtains two reflective signals that are from two different objects, and then generates the mixed pixel measurements that cannot represent either of the two objects [57–60].

As mixed pixels adversely influence the recognition of the poles and tubes, it is necessary to remove the mixed pixels before further processing of the point data. Since mixed pixels are located at greater distances from their neighbours [57], an algorithm based on the average distance from one point to its k-nearest neighbours was employed in this study. If the average distance is bigger than a threshold value, the point is considered to be a mixed pixel. Moreover, in this study, colour information of the point data was used to filter out the mixed pixels. Steel poles and ledgers usually have substantially different colours so that pixels will experience large colour changes. The proposed method regards both the distance value and the average colour difference d_{RGB} between one point and its surrounding k neighbours. The colour difference can be calculated in Equation (1).

$$d_{RGB} = \frac{\sum_{i=1}^{k} \sqrt{(R_i - R_0)^2 + (G_i - G_0)^2 + (B_i - B_0)^2}}{k}, \quad (1)$$

where (x_0, y_0, z_0) and (R_0, G_0, B_0) are the coordinates and RGB values of this point, respectively, and (x_i, y_i, z_i) and (R_i, G_i, B_i) are the coordinates and RGB values of the ith neighbor, respectively.

Based on distance value and colour difference, a discriminant analysis can be conducted to classify the valid data and mixed pixels clearly, as shown in Figure 4. The irrelevant data can be removed from the scan data. This step aims to retain data associated with standard poles and ledgers and remove other scan data as outliers. The proposed method is expected to generate more reliable results than conventional methods.

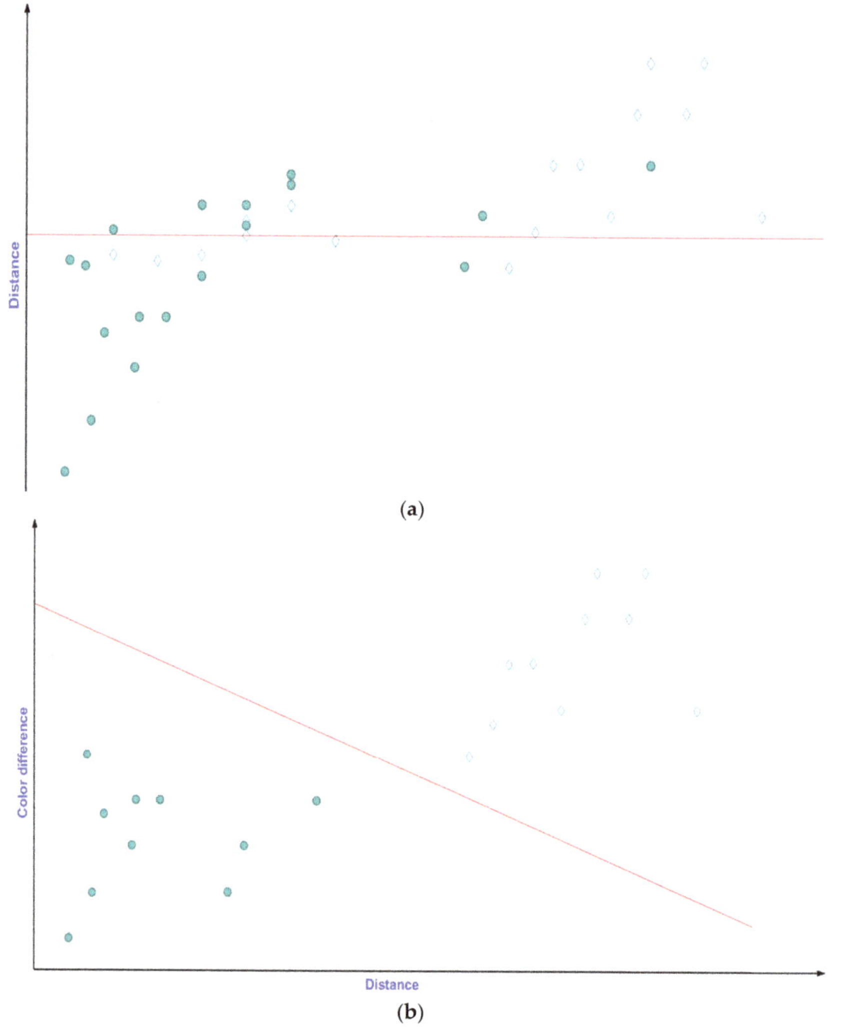

Figure 4. Remove mixed pixels (**a**) using distance value, (**b**) using both distance and color difference value.

3.2.2. Extraction of Scan Data Associated with Poles and Tubes

As the high formwork mainly composed of vertical cylindrical poles and horizontal cylindrical tubes, upright poles and horizontal tubes have distinctive geometries and positions that differ from other objects. Hence, both geometries and positions of the scan points were used for classification. The proposed data extract method consists of six steps, as shown in Figure 5.

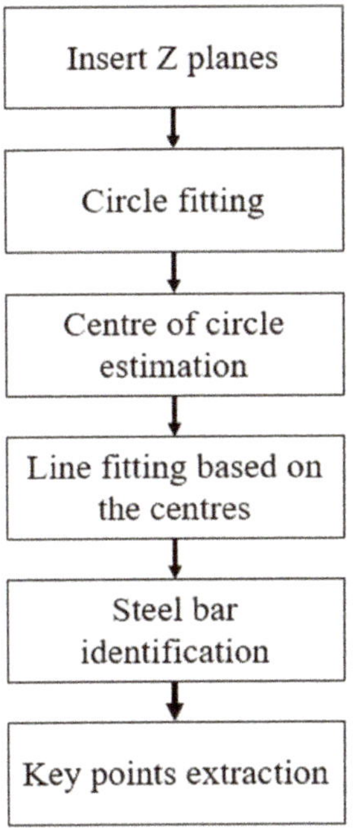

Figure 5. Data extraction method.

An array of Z-planes (Parallel to the XY plane) with a certain interval was inserted to the refined point cloud data and the data was projected onto the Z planes, as shown in Figure 6. For each slice, the point cloud data were projected on to the Z plane, as shown in Figure 7.

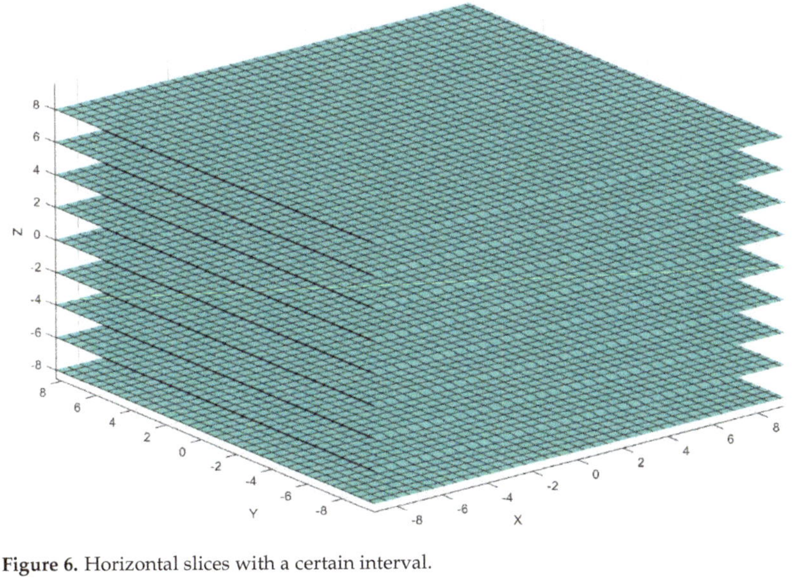

Figure 6. Horizontal slices with a certain interval.

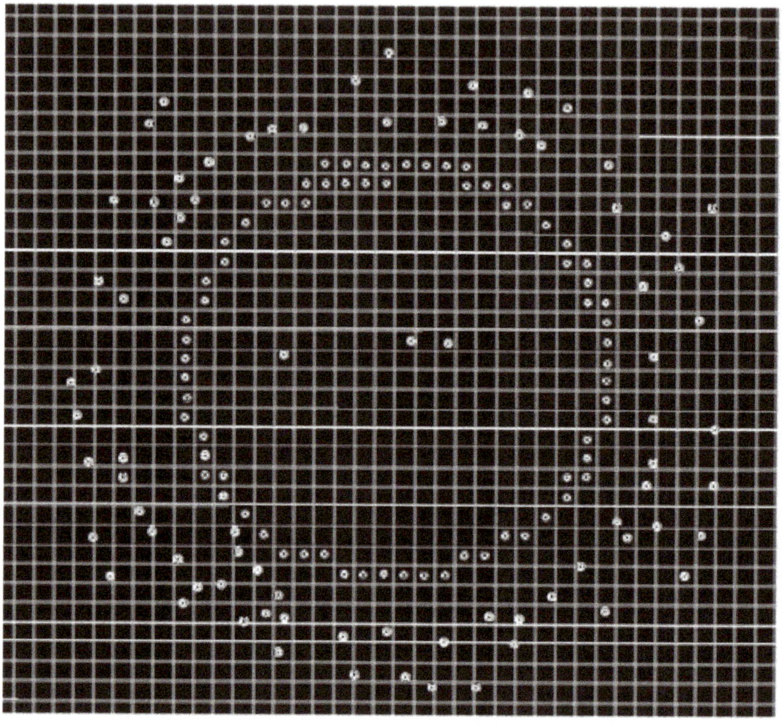

Figure 7. X-Y projection of point cloud data of a pole on a slice.

Since a vertical pole is circular in section, the data in every Z plane were clustered in a circle. The region growing algorithm [61] was used to add more points to the points clusters within a certain radius (the radius of a pole). The algorithm iteratively explores the point clusters and adds more neighbor points to the clusters in order to form a circular shape with a certain radius until all the points clusters in the array of Z planes have been checked. If a point does not belong to any of the clusters, it can be seen as an outlier to be removed. The remaining data in every Z plane were corrected in MATLAB by fitting a simple circle based on the section formation of a standard pole. Based on the points the boundary of the poles and tubes can be detected. The centre and diameter of a pole are then fitted according to the detected boundary, as shown in Figure 8. As the same pole has the similar XY coordinates, Equation (2) was used to evaluate whether the point belong to the pole or not.

$$(x - x_i)^2 + (y - y_i)^2 \leq p, \tag{2}$$

where x, y is the coordinates of the vertical pole; x_i, y_i is the coordinate of the i point; p is the threshold value.

The centre points of the circles on each Z planes belonging to the same pole can be determined, as shown in Figure 9. A line fitting algorithm in MATLAB was used to fit the centre points in each Z plane to obtain the vertical pole, as shown in Figure 10. Every vertical pole can be extracted. If the model set is subjected to deformation, data can be fitted with higher order polynomials for accurate representation. The same methods were used to extract every horizontal tube, although in that case an array of X-planes was inserted for the extraction of the longitudinal tubes and an array of Y-planes were inserted for the extraction of the transverse tubes. The thickness and the interval of the inserted planes are highly related to the size and density of the point data.

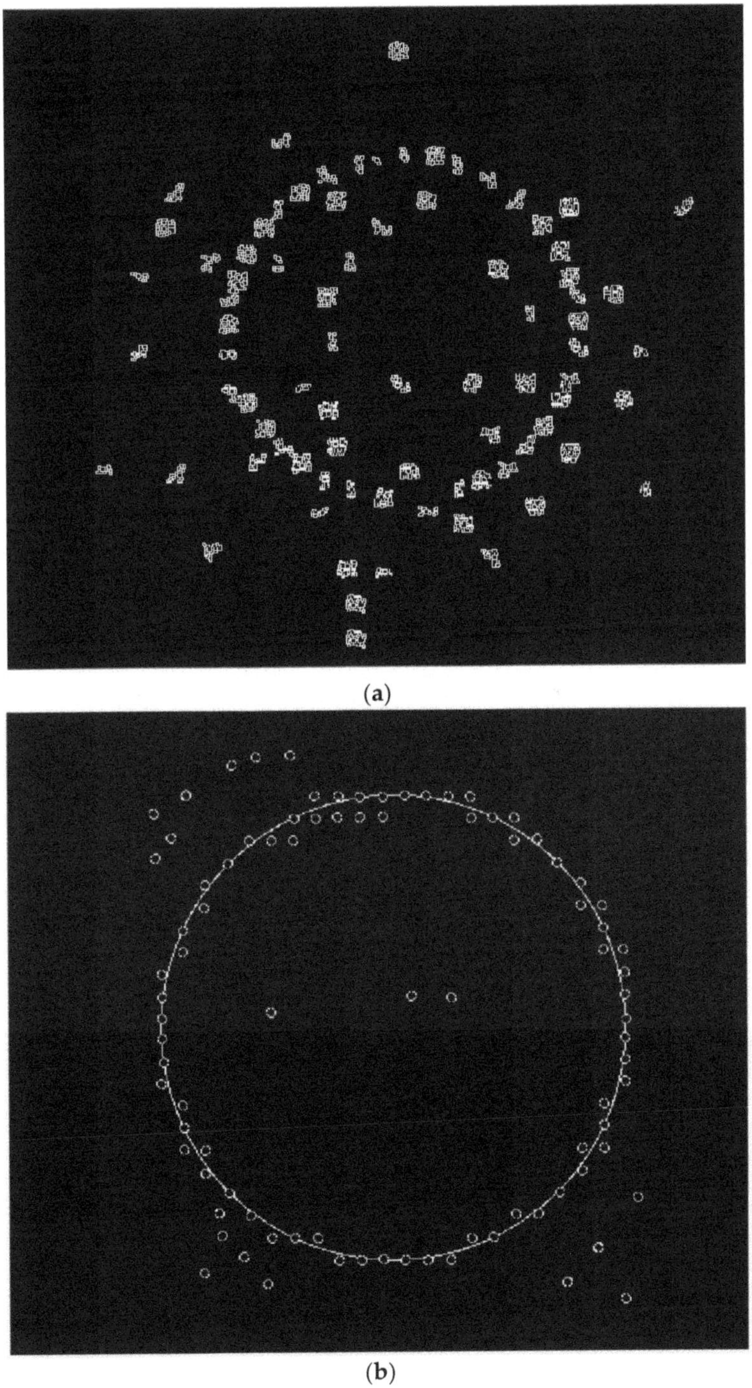

(a)

(b)

Figure 8. Boundary detection (a) rough boundary, (b) fine boundary.

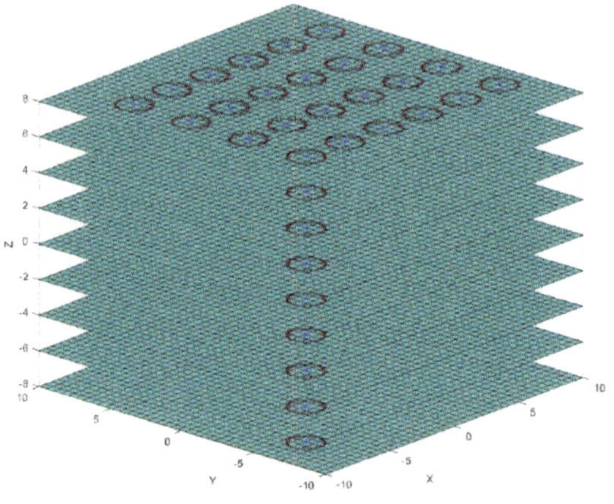

Figure 9. The centroid of fitted circles.

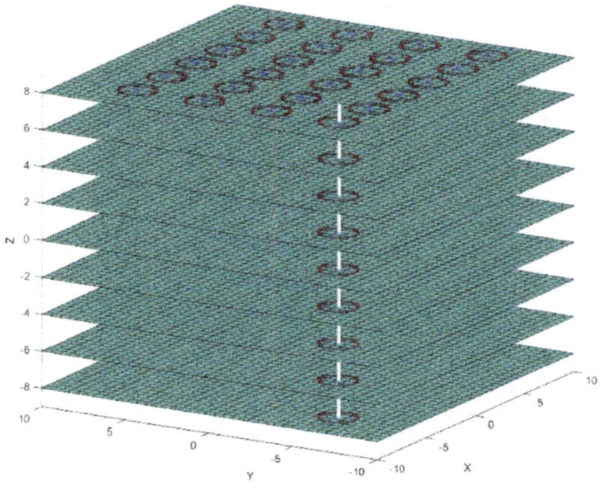

(**a**)

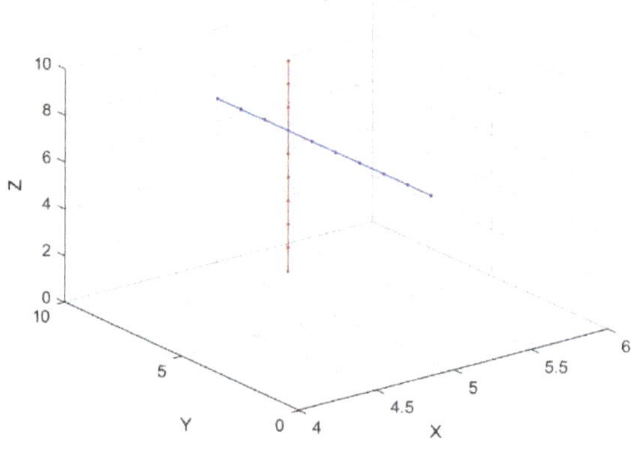

(**b**)

Figure 10. Identification of a vertical stand pole (**a**) potential line representing a vertical pole, (**b**) 3D line fitting.

This method has several potential advantages. First, it can collect sufficient data for individual poles to facilitate the recognition process. Second, the outliers can be removed effectively. Third, the incomplete scanning of standard poles due to the occlusion of line-of-sight by other objects can be corrected using the region growing technique.

The attained lines representing the poles were checked to determine whether they were vertical to the plane XOY. And the lines indicated tubes were checked to determine whether they were perpendicular to the plane YOZ. The projection of an inclined pole on the XOY plane is shown in Figure 11. If the distance between the leftmost and rightmost centroids are larger than a threshold value, it indicates the pole is not vertical to the XOY plane. Then the pole that is not vertical to the planes should be filtered out to check its inclination value. The pole should be considered a defect that is not follow the building code rule. The rule states that all poles should be perpendicular to the ground and all tubes should be parallel to the ground. Hence, the pole or the tube should be re-installed to correct the defect.

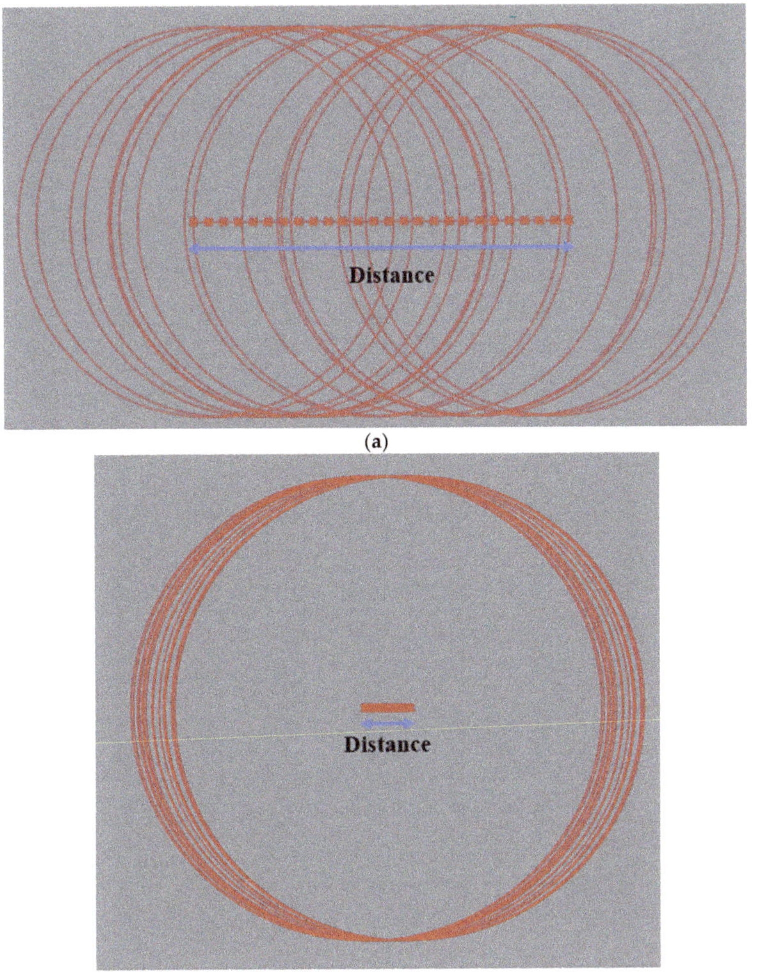

(b)

Figure 11. Intersections of a pole projected on XOY plane (**a**) inclined pole, (**b**) upright pole.

In this study, a pole and a tube of the small-scale high formwork model were inclined on purpose to validate the proposed method. The proposed method can identify the pole and the tube that are not properly installed. Next, the cross points of lines that represent poles and tubes were generated. The cross points are disk lock nodes, which are considered important points in this study. The further ANN model was developed to generate accurate coordinates of the key points in order to accurately calculate the spacing of the pole and the lift height. Such data are important for building code requirements.

The longitudinal spacing of pole can be calculated in Equation (3).

$$d_{li} = \sqrt{(x_{i+1} - x_i)^2}, \quad (3)$$

The transverse spacing of pole can be calculated in Equation (4).

$$d_{ti} = \sqrt{(y_{i+1} - y_i)^2}, \quad (4)$$

The lift height (space between two neighboring tubes) can be calculated in Equation (5).

$$s_i = \sqrt{(z_{i+1} - z_i)^2}, \quad (5)$$

where x_i, y_i, z_i indicate the coordinates of the i point.

3.3. ANN Structure and Training

The TLS technology was used as a basis for obtaining data for neutral network training. This can generate better predictions with acceptable accuracy. The flowchart of the proposed scheme is displayed in Figure 12.

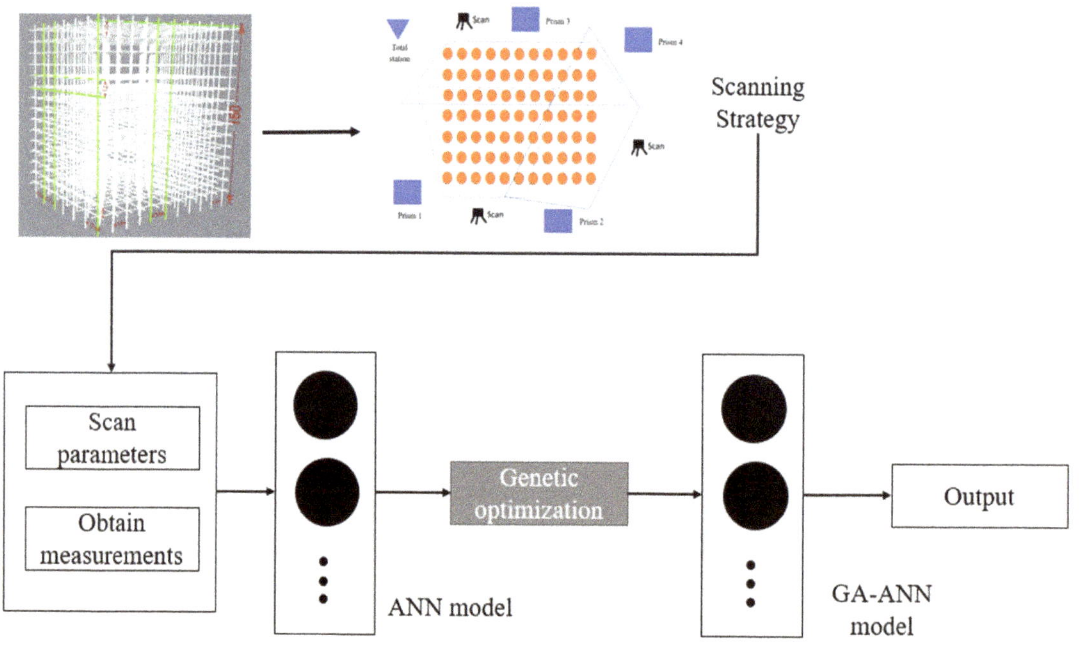

Figure 12. The flowchart of the TLS measurements optimized by using GA-ANN model.

3.3.1. ANN Training Data Collection

In order to improve the accuracy of the coordinates of the key points, the study needs to collect sufficient data to develop an ANN model. The spacing of the pole and the lift height of the small-scale high formwork model set were specified. The model set first was positioned on a movement equipment and moved along the X direction at interval 1 mm, and then the model set was moved along the Y direction at interval 1 mm from 1 mm-10 mm. Finally, the model set was moved along the Z direction from 1 mm to 10 mm at interval 1 mm. Every movement was scanned. Thus, the model set was scanned 30 times, and the collected 30 datasets were used to develop an ANN model to predict the coordinates of the cross points of the poles and tubes. The moving distance is specified as 1 mm, so that the coordinates of the key points at t moment can be calculated based on the coordinates at $t-1$ moment, as shown in Equation (6). The initial coordinates of the key points were generated from total station measurement.

$$P_i^t\begin{pmatrix} x_i^t \\ y_i^t \\ z_i^t \end{pmatrix} = P_i^{t-1}\begin{pmatrix} x_i^{t-1}+1\,\mathrm{mm}\left(\overrightarrow{X}\right) \\ y_i^{t-1}+1\,\mathrm{mm}\left(\overrightarrow{Y}\right) \\ z_i^{t-1}+1\,\mathrm{mm}\left(\overrightarrow{Z}\right) \end{pmatrix} \tag{6}$$

where $P_i^t\begin{pmatrix} x_i^t \\ y_i^t \\ z_i^t \end{pmatrix}$ indicates the coordinate of the i point at t moment; $x_i^{t-1}+1\,\mathrm{mm}(\overrightarrow{X})$ indicates the model move along the X axis at specified value 1 mm, so the x coordinate of the i point at t moment equal to the x coordinate of the point at $t-1$ moment add 1 mm, but the y and z coordinates of the i point at t moment are the same as that at $t-1$ moment.

3.3.2. ANN

The ANN model was used to solve the non-linear problem. The strong self-learning ability of the ANN can generate accurate predictions with high computational speed. Many kinds of neutral networks have been established, and multilayer feed forward is one of the most popular. Back propagation (BP) is one of commonly used algorithm, which can minimize ANN error properly [62]. Hence, the back propagation (BP) neutral network was adopted in this study.

Although ANN has been widely and successfully used in many studies, it has several disadvantages. For example, it is tedious to choose the number of hidden layers and the number of nodes at the hidden layer; the learning rate of ANN is usually decided randomly; and it is more likely to achieve local minima rather than global minima [63–65]. Moreover, the proper selection of parameters such as learning rate and momentum coefficient are important for model convergence progress. Hence, in this study, genetic algorithm (GA) was utilized to designate the parameters of the ANN model including the number of hidden layers, the number of nodes at the hidden layer, learning rate, and the momentum coefficient.

3.3.3. ANN Optimization by Genetic Algorithm

GA is a meta-heuristic, population-based searching algorithm based on natural selection [66,67]. In this method, each individual in the population can generate a solution by reproducing the individuals, and the solution should converge on the best one. In this study, an individual includes four genes such as the number of the hidden layer, the number of nodes at the hidden layer, learning rate and momentum coefficient, formed as shown in Equation (7).

$$i = (c_1,\ c_2,\ c_3,\ c_4), \tag{7}$$

where i represents an individual chromosome and c_1 to c_4 are binary codes for the four genes. A population is composed of a number of individual chromosomes, which can be expressed as $I = (i_1, i_2, i_3, \ldots\ i_n)$, n denotes the number of individual chromosomes.

GA uses three operators like selection, crossover and mutation to produce next generation chromosomes with better fit. The selection operator chooses the chromosomes with better fit based on the selection criteria defined in the algorithm. In the crossover operator the two selected parent chromosomes were merged from a new solution, which makes the probability of the best fit higher in the next generation. Different crossover rules can be used. In this study, a random crossover from two parents was employed. The mutation operator can make random changes to the chromosome features to avoid the local optimum. The details of the GA can be found in other studies [68,69]. The main parameters were used in Genetic Algorithm are shown in Table 2.

Table 2. Main parameters of GA algorithm used for optimizing ANN.

Parameter	Designation	Value
Population size	N_{pop}	50
Generation number	N_{gen}	100
Number of binary digits		10
Number of iterations		1000
Cross-over probability	P_c	0.95
Mutation probability	P_m	0.01
Fitness value		1

3.3.4. GA-ANN

The number of nodes at the input and output layers is equal to the input and output variables, respectively. The number of the hidden layer and the number of nodes at the hidden layer were determined by GA. The important parameters for an ANN model consist of Nh, Nnh, Lr, and Mc, and the four genes in the binary form were considered for each chromosome. The number of the hidden layer (Nh) was limited 1 to 2, the number of the node at each hidden layer (Nnh) is selected from 1 to 10, and the learning rate (Lr) and momentum coefficient (Mc) ranged from 0 and 1.

The TLS measuring coordinates were seen as inputs, the real value of the coordinates was regarded as output, and then the transfer model was estimated. The ANN models with different momentum coefficients are shown in Figure 13. The ANN3 with the momentum coefficient Mc = 0.876 perform the best compared with other ANN models. As shown in Figure 13, the momentum coefficient cand strongly influence the model performance.

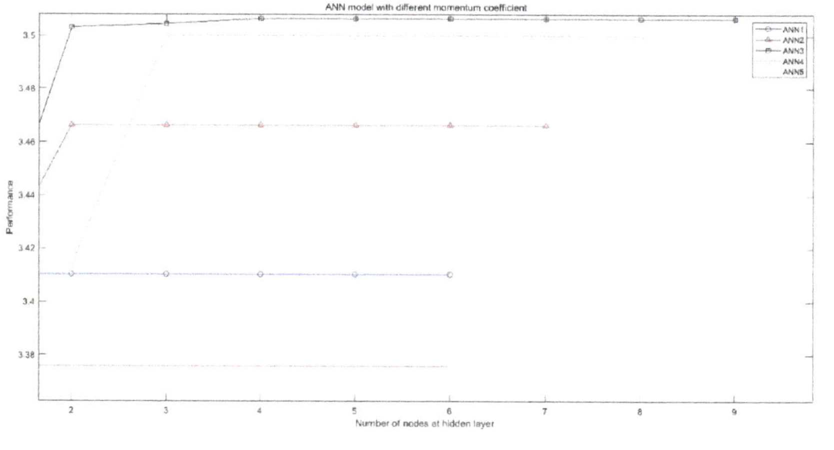

Figure 13. The model performance for ANN models with different momentum coefficients.

According to the results of the study [70], the position of the laser scanners has significant influence on data scanning accuracy. Hence, the study used the positions of the TLS as the ANN model inputs. In this study, the inputs of the ANN model were the coordinates of the key points, R distance, laser coefficients, and scan conditions. The output of the ANN model was the specified value (specified value of the set distance and space $d_{li} = 12$ cm; $d_{ti} = 15$ cm; $s_i = 10$ cm). The value of the cross points can be calculated by using Equations (3)–(5).

Sigmoid function was employed as a transfer function for output and hidden layers of the GA-ANN. The model training process was to adjust the model parameters in order to minimize the error between the output and the target. Mean Squared Error (MSE) was adopted to evaluate the model performance, it can be calculated in Equation (8).

$$\mathrm{MSE} = \frac{1}{n}\sum_{i=1}^{n}(p_i - x_i)^2, \tag{8}$$

where n is the number of data samples, p_i is the model estimate of the i sample, and x_i represents the actual value of the i sample.

During the training process, the number of nodes and layers and functions of the network were changed. MATLAB software was used to train and test the ANN model. The important parameters of the proposed ANN model were first encoded as the genes of the genetic algorithm (GA). Then, the fitness values of each individual chromosome were calculated based on the fitness function. After a series of GA selection, such as the crossover, the mutation and the duplication, the GA can discover the optimal individual chromosome corresponding to the optimal fitness values. The evolution process is shown in Figure 14. It should be noticed that GA can properly coverage the fitness value to the global optimum rather than any other local optimal by random initialization.

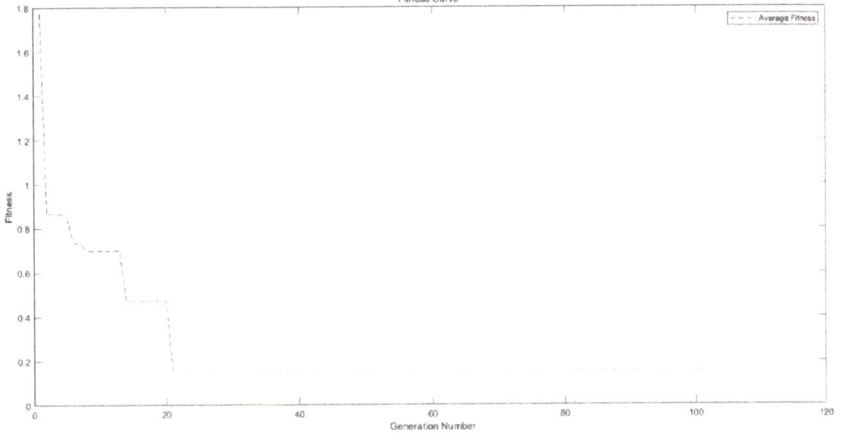

Figure 14. The evaluation process of GA.

According to the GA results, the best number for the hidden layer was found to be one; the best number for the nodes at the hidden layer is six. The structure of the proposed GA-ANN model is shown in Figure 15.

The optimum network results in the lowest MSE. The optimum configuration of the ANN model is shown in Table 3. Moreover, several network configurations with different nodes at hidden layer were developed to evaluate and compare the accuracy of the proposed GA-ANN model in estimating the variables. The MSE function was used to assess the performance of the proposed GA-ANN model. The prediction performance of the models for the data of simulated defects and the coordinates of key points are shown in

Table 4. The corresponding rating was provided by averaging each MSE for the models. As shown in Table 4, the proposed GA-ANN model achieves the best results among the other models. To assess the model performance of the prosed GA-ANN model, the study provided model performance value for ANN model and the ANN model optimised by using GA respectively. The model performance of the proposed GA-ANN model and the second-best ANN model compared with the actual value of the measurements are shown in Figure 16.

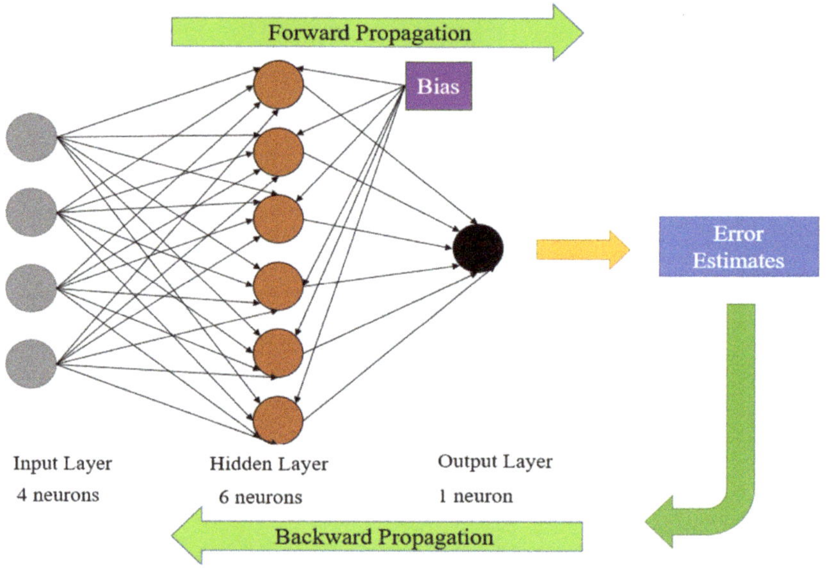

Figure 15. The structure of the proposed GA-ANN model.

Table 3. Optimum configuration of the ANN obtained from GA.

Gene	Designation	Value
Number of hidden layer (c1)	Nh	1
Number of hidden layer neurons (c2)	Nnh	6
Learning rate (c3)	Lr	0.125
Momentum coefficient (c4)	Mc	0.876

Table 4. Prediction performance of the proposed GA-ANN model versus ANN model.

Model	Nnh	Predicted Simulated Defects (MSE)					Predicted Coordinates (MSE)			Rank
		1	2	3	4	5	X	Y	Z	
ANN	2	0.713	0.529	0.871	0.405	0.679	0.862	0.837	0.893	7
	3	0.064	0.057	0.070	0.061	0.066	0.073	0.049	0.069	4
	4	0.021	0.011	0.018	0.023	0.016	0.019	0.012	0.014	2
	5	0.028	0.030	0.015	0.022	0.014	0.018	0.017	0.021	3
	7	0.143	0.135	0.164	0.145	0.105	0.179	0.101	0.141	5
	9	0.437	0.219	0.204	0.591	0.140	0.759	0.217	0.611	6
GA-ANN	6	0.005	0.004	0.004	0.003	0.004	0.002	0.003	0.002	1

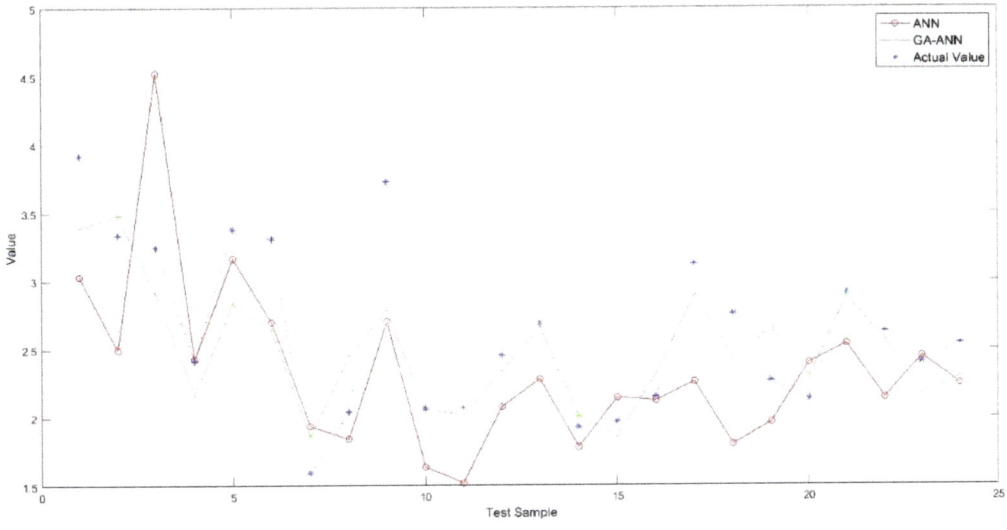

Figure 16. Model performance of the GA-ANN versus the ANN model and the actual value.

The obtained estimates from GA-ANN model versus the actual value are described in Figure 17. Obviously, the proposed model can generate reliable predictions as error is nearly zero. The R^2 results for the proposed GA-ANN model which are very close to 1 for training and testing dataset, as shown in Figure 18, explored that the proposed GA-ANN can generate high accuracy predictions.

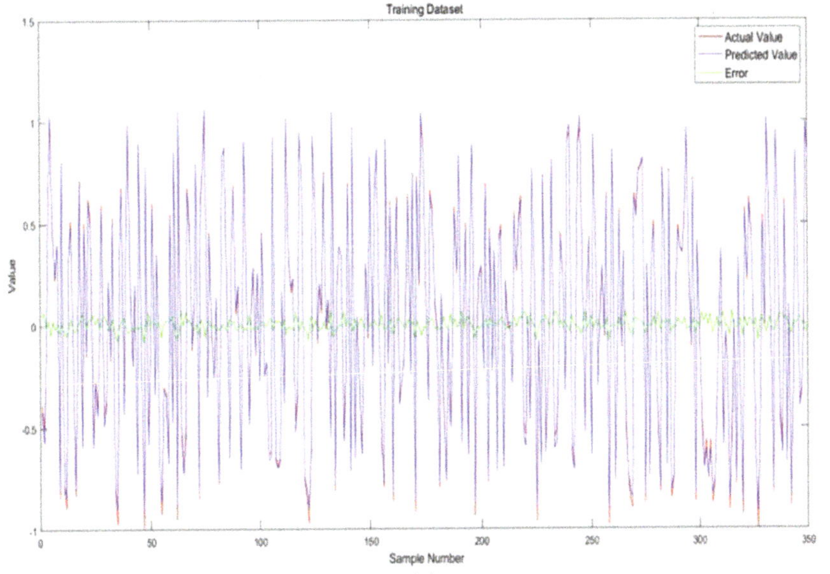

(a)

Figure 17. *Cont.*

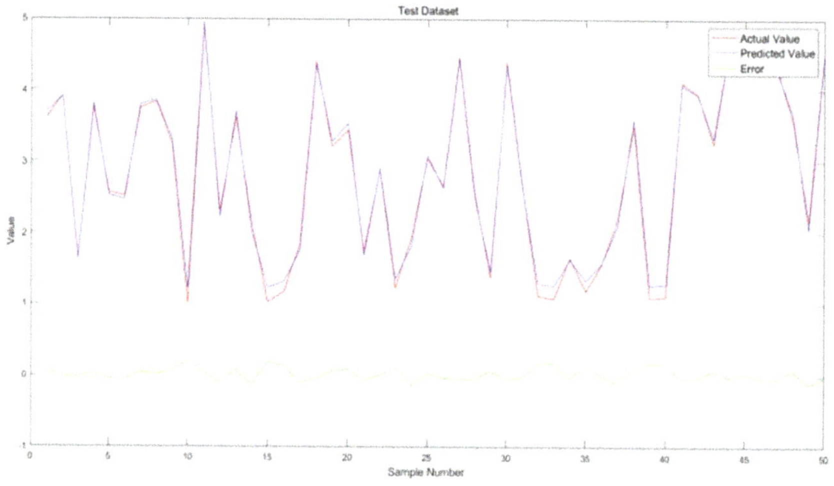

(b)

Figure 17. Actual versus predicted values of the measurements from GA-ANN model (**a**) training dataset (**b**) test dataset.

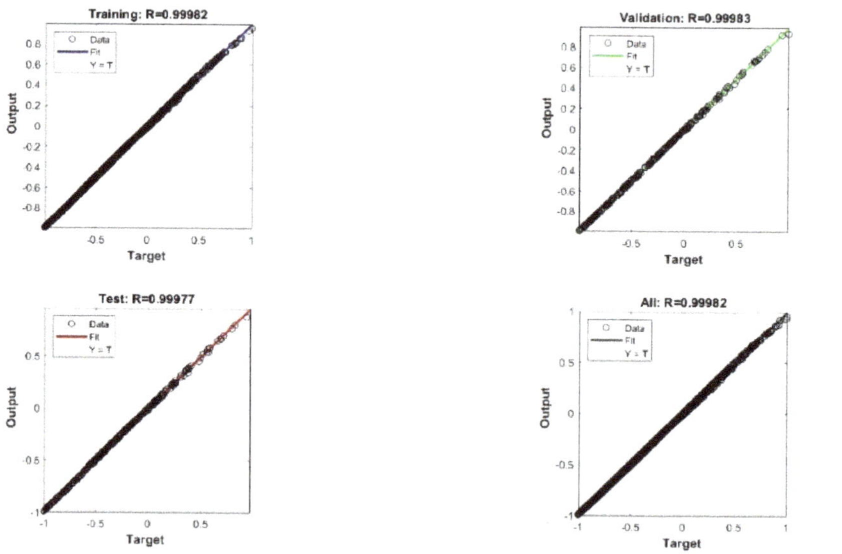

Figure 18. The R^2 of the proposed GA-ANN model.

According to [71], the R^2 of the ANN model indicates that the inputs provide sufficient information for the model predictions. Based on the results of [72], the Coefficient of Variation-Root Mean Square Error (CV-RMSE) should be <30%, CV-RMSEs of the ANN model is 12.5%, indicating that the developed ANN model meets the standard. Other model performance evaluations, such as Root Mean Square Error (RMSE) and Mean Absolute Error (MAE), also indicate that the model can generate predication with high accuracy.

3.3.5. Comparison with Other ML Algorithms

To better demonstrate the GA-ANN learning approach, the common methods (ANN, SVM and RF) that are often used are also applied in this study. Support vector machines (SVMs) are well-known supervised machine learning techniques that were proposed by Cortes and Vapnik [73] to solve classification problems, and then were extended to regression domain by Vapnik et al. [74]. For nonlinear problems, a nonlinear kernel function is utilized. Random forest (RF) is an effective machine learning method proposed in 2001 [75], which can be applied to classification, regression, and feature selection problems. RF is an ensemble learning model with a decision tree as the base classifier, combining bagging and random subspace theory.

To evaluate the performance of the suggested models, three different metrics, Mean square error (MSE), Correlation coefficient (R), and root mean square error equation (RMSE) are introduced. These statistical indicators assess the efficiency, linear relationship, and deviation experienced from the average values. Statistical indices including MSE, R, and RMSE gave an overall view of the precision and error of the model. The performance measurements of the models are shown in Table 5. GA-ANN model was superior, followed by ANN model, RF model and the least was SVM model. This might be indicated that, ANN, SVM, and RF are individual learning algorithms while GA-ANN is an optimized learning algorithm.

Table 5. The performance evaluation of the suggest models.

Model	R		RMSE		MSE	
	PSD	**PC**	**PSD**	**PC**	**PSD**	**PC**
GA-ANN	0.987	0.985	0.008	0.005	0.004	0.0035
ANN	0.918	0.873	0.023	0.045	0.018	0.015
SVM	0.880	0.798	0.187	0.215	0.157	0.173
RF	0.881	0.853	0.034	0.048	0.011	0.035

PSD = predicted simulated defects; PC = predicted coordinates.

3.4. *Validation*

A different small-scale high formwork model was installed with different values of distance between the poles and tubes and simulated defects in different area, as shown in Table 6. The different value set in the installation was to verify that the model can detect differences to improve the robustness of the training model.

Table 6. The detail of the second high formwork model set.

Term	Description
Length	60 cm
Width	70 cm
Height	100 cm
The distance between two neighboring horizontal ledgers (Lift height)	15 cm
The longitudinal spacing of vertical pole	8 cm
The transverse spacing of vertical pole	10 cm
The height of the bottom reinforcing ledger	5 cm
The angle of the diagonal brace	45°
The number of vertical poles	49
The number of horizontal ledgers	98
Simulated defects	
The distance between the 3rd and 4th horizontal ledgers on the longitudinal middle frame	18 cm
The longitudinal spacing of the 4th and 5th vertical pole on the longitudinal middle frame	10 cm
The transverse spacing of the 3rd and 4th vertical pole on the transvers middle frame	12 cm

The proposed GA-ANN model was used for predicting the simulated defects and the coordinates of the key points. The convergence process of the proposed GA-ANN model is displayed in Figure 19. The optimal result of the proposed GA-ANN model was obtained after the 100th iteration. In this study, the model was able to detect all the simulated defects and generate accurate coordinates of the key points.

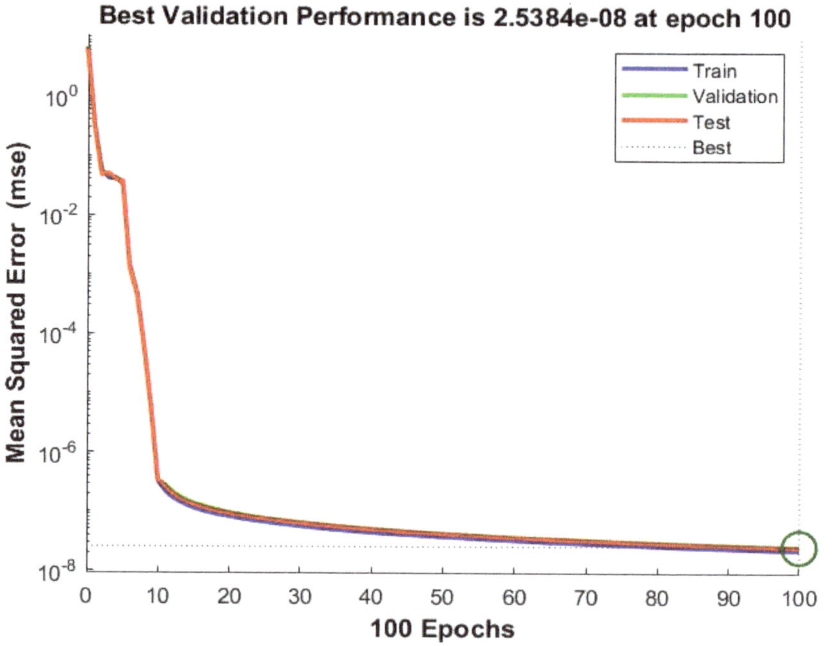

Figure 19. The converge process of the proposed GA-ANN model.

4. Case Study

4.1. Validation on a Real High Formwork

While the prior case studies show the effectiveness of the proposed method, a full in-site test on a real high formwork was carried out to evaluate the effectiveness of the proposed method. The dimensions of the high formwork were 5.5 m × 5.5 m × 6 m, consisting of 49 standard poles and 70 tubes. Both longitudinal and transverse spacing of pole are 0.8 m, while the lift height is 1.5 m. The formation of the high formwork is shown in Figure 20. The object provided an opportunity to validate the approach on a real structure.

As the results may have been influenced by the instrument position error between different point clouds, a control network was adopted to ensure that the horizontal position locates correctly within a millimetre. The error can be absorbed into +2 mm registration errors. Several scans from different viewpoints were collected so that the workflow efficiency could be evaluated in a real situation. Typically, a total of 10–15 min was spent at each position, including setup and scan time. The georeferencing process only took 5 min to complete. The point clouds of the high formwork are shown in Figure 21.

Figure 20. The photography of the high formwork.

Figure 21. Real point clouds of the high formwork.

4.2. Discussion of Results

Based on the proposed method, the Z planes were inserted at 5 cm interval from 1 cm to the height of the high formwork. Next, the lines representing the individual poles or tubes were generated. Then, the coordinates of the cross points were input the ANN model. Both longitudinal and transverse spacing of pole and lift height (the distance between two neighbouring ledgers) were checked. Moreover, the perpendicularity of pole and levelness of ledger were inspected. In addition, the angle of the diagonal bracing of the high formwork were checked. The standard error was calculated between real value and estimates. The standard error can be calculated in Equation (9).

$$\mathrm{SE} = \sqrt{\frac{\sum (Y - Y')^2}{N}}, \tag{9}$$

where Y is the real value, Y' is the estimated value, and N is the number of points to estimate. The overall accuracy is acceptable and the errors fall within the allowable tolerance range. The results are summarized in Table 7. The results indicate that the accuracy of the developed GA-ANN model is satisfied at the millimetre level. It can generate reliable results in real applications.

Table 7. Summary of the predicted error by using the proposed method.

Inspection Term	Specified Value	Error from the Proposed Method				
		Max	Min	Average	Standard Error	Percent of Pass
d_{li}	800	15	3	5	4.7	95%
d_{ti}	800	13	6	5.5	5	94.3%
s_i	1500	16	5	7	6.8	92%
A_m	60°	30′	10′	18′	15′	98%

Note: A_m indicates the angle of diagonal bracing.

The purpose of the validation is to examine the performance of the proposed approach dealing with real problems. Moreover, short range (about 10 m), simple scanned area, and scanning from one position can reduce errors. Results of the validation indicate the proposed method can detect and locate defects. Optimistically, it is capable of providing accurate measurements at the magnitude of the millimetre level. Based on the outcomes, the proposed method provides an effective tool for installation quality inspection for high formwork.

Based on the cost and time evaluation, the proposed method can provide an effective tool for quality inspection of a high formwork. The scanning time used for the inspection at every station was only twenty minutes, which suggests the superiority of the adoption of TLS in terms of both cost and time when compared to common methods. For example, a GNSS-based method, which recently was suggested for use in structure monitoring and inspection, has an error at the magnitude of the centimetre level, and requires extensive calculation and processing [76,77]. With such an error budget, it is unacceptable for installation quality inspection for high formwork. Moreover, to improve accuracy, it usually requires the addition of more stations which will significantly increase operation costs. Other methods, like unmanned aerial vehicle (UAV) imagery, are also recommended as an effective tool for installation quality inspection for high formwork. However, based on previous studies [78,79], it has average error of about 15mm. Moreover, it requires sophisticated processing to extract the required information from the raw imagery. In addition, and most importantly, considering factors such as registration errors, missing data or mixed pixels might cause the misidentification of important elements or increase errors in the measuring of the dimensions of high formwork.

Without the use of TLS, to safely conduct quality inspection in a cost- and time-effective manner, and with acceptable accuracy, would prove extremely difficult and labour-intensive.

Moreover, the data accuracy of TLS can be improved with the following suggestions: (1) conduct the scan without impacting the line-of-sight; (2) increase scanning resolution; (3) increase the number of scans; (4) scan the high formwork from all faces; (5) widen the range of scan stations.

5. Conclusions

The current study highlighted the advantages of the proposed method using TLS and ANN algorithms for installation quality inspection for a high formwork. When conducting the inspection operation, the non-invasive method, along with full area coverage measurements, are a benefit.

Adoption of the TLS uses less labour when compared to common inspection methods, avoids any labour-related risks, and can access difficult and dangerous areas, which will save significant money and time and greatly improve the performance of the inspection process. The collected point cloud data can be effectively processed using the proposed data extraction approach in this study and can rapidly extract the important elements of the high formwork.

The conventional ANN network was optimized via GA to achieve the optimum parameters of the ANN model including: the number of hidden layers, the number of nodes at the hidden layers, the learning rate and the momentum coefficients. The model performance evaluation parameters and generated result indicate that the proposed GA-ANN model can produce accurate predictions. The generated accurate coordinates of the important points that are the cross points of the poles and ledgers facilitate the spacing calculation between two neighbouring of pole or lift height (the distance between two neighbouring ledgers).

The proposed method can provide accurate results that facilitate quality inspection, safety management and decision making on the construction site. Moreover, the collected measurements can be kept properly, which facilitates further application in safety monitoring for high formwork during concrete cast-in. In addition, the study also provides an indication that measurements with tight tolerances can be achieved not only by using contact methods, but that the proposed methods using TLS outputs can also generate results with the required level of accuracy.

The benefits of the proposed method indicate that TLS can be successfully used in quality inspection and can generate accurate measurements in terms of the quantity and quality of the data points and timing. The measurements based on the proposed method are similar to the set values with acceptable error, and hence confirm the suitability of the proposed method. A successful approach for point cloud data extraction and measurements calculation can be used for other similar structures.

Furthermore, the laser scanning system can be utilized in combination with other techniques like photogrammetry to offer a hybrid method that can provide accurate measurements of the full covering area and the points of interest. TLS measurements can also be combined with BIM to vividly display the structure in real dimensions and state.

Author Contributions: Conceptualization, L.Z. and J.M.; methodology, L.Z.; software, L.Z.; validation, L.Z., J.M. and B.W.; resources, L.Z. and Z.L.; writing—original draft preparation, L.Z.; writing—review and editing, L.Z. and B.W.; visualization, L.Z. and H.Z. All authors have read and agreed to the published version of the manuscript.

Funding: This research was funded by Beijing University of Technology, grant number 314000514121010 and 047000513201. The APC was funded by the two grants.

Institutional Review Board Statement: Not applicable.

Informed Consent Statement: Not applicable.

Data Availability Statement: The data can be provided upon request.

Acknowledgments: The authors would like to thank Beijing University of Technology for its support through the research project. The authors would like to thank China Industry Associations for provide research data. In addition, I would like to thank all practitioners who contributed to this project. The authors really appreciate the reviewers of the paper for their kind advice and encouragement.

Conflicts of Interest: The authors declare no conflict of interest.

Abbreviations

Acronym	Meaning
AEC	Architecture, Engineering & Construction
ANN	Artificial Neutral Network
BP	Back Propagation
BIM	Building Information Modelling
CV-RMSEs	Coefficient of Variation-Root Mean Square Error
GA	Genetic Algorithm
GA-ANN	Genetic Algorithm optimized Artificial Neutral Network
GNSS	Global Navigation Satellite System
ICP	Iterative Closet Point
LiDAR	Light Detection and Ranging
Lr	Learning rate
MAE	Mean Absolute Error
MSE	Mean Squared Error
Mc	Momentum coefficient
Nh	Number of hidden layer
Nnh	Number of nodes at hidden layer
PC	Predicted Coordinates
PSD	Predicted Simulated Defects
RF	Random Forest
RMSE	Root Mean Square Error
RGB	Red, Green & Blue
SE	Standard Error
SVM	Support Vector Machine
TLS	Terrestrial Laser Scanner
UAV	Unmanned Aerial Vehicle

References

1. Sona, M.; Janakaraj, M. The Impact of Buildability Factors on Formwork in Residential Building Construction. *Int. J. Innov. Res. Sci. Eng. Technol.* **2019**, *8*, 1–10. [CrossRef]
2. Sagadevan, R.; Rao, B.N. Experimental and analytical investigation of structural performance of vertical concrete formworks. *Asian J. Civ. Eng.* **2019**, 1–12. [CrossRef]
3. Ramesh Kannan, M.; Helen Santhi, M. Automated constructability rating framework for concrete formwork systems using building information modelling. *Asian J. Civ. Eng.* **2018**, *19*, 387–413. [CrossRef]
4. Lee, D.; Lim, H.; Lee, D.; Cho, H.; Kang, K.-I. Assessment of Delay Factors for Structural Frameworks in Free-form Tall Buildings Using the FMEA. *Int. J. Concr. Struct. Mater.* **2019**, *13*, 1–11. [CrossRef]
5. Zhao, Z.G. *The Support System of the High Formwork Construction Techniques and Quality Management*, 1st ed.; China Architecture & Building Press: Beijing, China, 2016.
6. Xie, N. *Safety Control for the Support System of High Formwork*, 1st ed.; China Architecture & Building Press: Beijing, China, 2012.
7. Cai, X.F.; Zhuang, J.P.; Zhou, J.Z.; Zheng, Y.Q. *The Support System of Super High Formwork Investigation and Application*; China Architecture & Building Press: Beijing, China, 2012.
8. MoHURD. *Cuplok Scaffolding Safety Technical Standard*; Ministry of Housing and Urban-Rural Development: Beijing, China, 2009; Volume JGJ166-2008.
9. MoHURD. *Steel Tubular Scaffold with Couplers Safety Technical Standard*; Ministry of Housing and Urban-Rural Development: Beijing, China, 2011; Volume JGJ 130-2011.
10. MoHURD. *Disk Lock Steel Tubular Scaffold Safety Technical Standard*; Ministry of Housing and Urban-Rural Development: Beijing, China, 2021; Volume JGJ/T231-2021.
11. Jaafar, H.A.; Meng, X.; Sowter, A.; Bryan, P. New approach for monitoring historic and heritage buildings: Using terrestrial laser scanning and generalised Procrustes analysis. *Struct. Control Health Monit.* **2017**, *24*, e1987. [CrossRef]

12. Yang, Y.; Zhang, Y.; Tan, X. Review on Vibration-Based Structural Health Monitoring Techniques and Technical Codes. *Symmetry* **2021**, *13*, 1998. [CrossRef]
13. Van Dijk, N.P.; Gamstedt, E.K.; Bjurhager, I. Monitoring archaeological wooden structures: Non-contact measurements systems and interpretation as average strain fields. *J. Cult. Herit.* **2015**, *17*, 102–133. [CrossRef]
14. Yang, Y.; Cheng, Q.; Zhu, Y.; Wang, L.; Jin, R. Feasibility Study of Tractor-Test Vehicle Technique for Practical Structural Condition Assessment of Beam-Like Bridge Deck. *Remote Sens.* **2020**, *12*, 114. [CrossRef]
15. Yang, Y.; Lu, H.; Tan, X.; Chai, H.K.; Wang, R.; Zhang, Y. Fundamental mode shape estimation and element stiffness evaluation of girder bridges by using passing tractor-trailers. *Mech. Syst. Signal Processing* **2022**, *169*, 108746. [CrossRef]
16. Law, D.W.; Silcock, D.; Holden, L. Terrestrial laser scanner assessment of deteriorating concrete structures. *Struct. Control Health Monit.* **2018**, *25*, e2156. [CrossRef]
17. Barbarella, M.; Fiani, M.; Lugli, A. Landslide monitoring using multitemporal terrestrial laser scanning for ground displacement analysis. *Geomat. Nat. Hazards Risk* **2015**, *6*, 398–418. [CrossRef]
18. Guarnieri, A.; Masiero, A.; Vettore, A.; Pirotti, F. Evaluation of the dynamic processes of a landslide with laser scanners and Bayesian methods. *Geomat. Nat. Hazards Risk* **2015**, *5*, 614–634. [CrossRef]
19. Gordon, C.; Akinci, B.; Garrett, J.H. Formalism for construction inspection planning: Requirements and process concept. *J. Comput. Civ. Eng.* **2007**, *21*, 29–38. [CrossRef]
20. Akinci, B.; Boukamp, F.; Gordon, C.; Huber, D.; Lyons, C.; Park, K. A formalism for utilization of sensor systems and integrated project models for active construction quality control. *Autom. Constr.* **2006**, *15*, 124–138. [CrossRef]
21. Park, H.S.; Lee, H.M.; Adeli, H.; Lee, I. A new approach for health monitoring of structures: Terrestrial laser scanning. *Comput. Civ. Infrastruct. Eng.* **2007**, *22*, 19–30. [CrossRef]
22. Nie, Y.; Chen, Q.; Chen, T.; Sun, Z.; Dai, B. Camera and lidar fusion for road intersection detection road intersection detection. In Proceedings of the IEEE Symposium on Electrical & Electronics Engineering (EEESYM), Kuala Lumpur, Malaysia, 24–27 June 2012; pp. 273–276.
23. Volk, R.; Stengel, J.; Schultmann, F. Building information modeling (BIM) for existing buildings—Literature review and future needs. *Autom. Constr.* **2014**, *38*, 109–127. [CrossRef]
24. Mosalam, K.M.; Takhirov, S.M.; Park, S. Applications of laser scanning to structures in laboratory tests and field surveys. *Struct. Control. Health Monit.* **2014**, *21*, 115–134. [CrossRef]
25. Truong-Hong, L.; Laefer, D.F.; Hinks, T.; Carr, H. Combining an angle criterion with voxelization and the flying voxel method in reconstructing building models from LiDAR data. *Comput.-Aided Civ. Infrastruct. Eng.* **2013**, *28*, 112–129. [CrossRef]
26. Al-Neshawy, F.; Piironen, J.; Peltola, S.; Erving, A.; Heiska, N.; Nuikka, M.; Jari, P. Measuring the bowing of marble panels in building facades using terrestrial laser scanning technology. *J. Inf. Technol. Constr. (ITcon)* **2010**, *15*, 64–74.
27. Armesto-González, J.; Riveiro-Rodríguez, B.; González-Aguilera, D.; Rivas-Brea, M.T. Terrestrial laser scanning intensity data applied to damage detection for historical buildings. *J. Archaeol. Sci.* **2010**, *37*, 3037–3047. [CrossRef]
28. Mukupa, W.; Roberts, G.W.; Hancock, C.M.; Al-Manasir, K. A non-destructive technique for health assessment of fire-damaged concrete elements using terrestrial laser scanning. *J. Civ. Struct. Health Monit.* **2016**, *6*, 665–679. [CrossRef]
29. Pătrăucean, V.; Armeni, I.; Nahangi, M.; Yeung, J.; Brilakis, I.; Haas, C. State of research in automatic as-built modelling. *Adv. Eng. Inform.* **2015**, *29*, 162–171. [CrossRef]
30. Kim, C.; Son, H.; Kim, C. Fully automated registration of 3D data to a 3D CAD model for project progress monitoring. *Autom. Constr.* **2013**, *35*, 587–594. [CrossRef]
31. Rebolj, D.; Pučko, Z.; Babič, N.Č.; Bizjak, M.; Mongus, D. Point cloud quality requirements for Scan-vs-BIM based automated construction progress monitoring. *Autom. Constr.* **2017**, *84*, 323–334. [CrossRef]
32. Son, H.; Kim, C.; Kim, C. 3D reconstruction of as-built industrial instrumentation models from laser-scan data and a 3D CAD database based on prior knowledge. *Autom. Constr.* **2015**, *49*, 193–200. [CrossRef]
33. Pu, S.; Vosselman, G. Knowledge based reconstruction of building models from terrestrial laser scanning data. *ISPRS J. Photogramm. Remote Sens.* **2009**, *64*, 575–584. [CrossRef]
34. Budroni, A.; Boehm, J. Automated 3D reconstruction of interiors from point clouds. *Int. J. Archit. Comput.* **2010**, *8*, 55–73. [CrossRef]
35. Ochmann, S.; Vock, R.; Wessel, R.; Klein, R. Automatic reconstruction of parametric building models from indoor point clouds. *Comput. Graph.* **2016**, *54*, 94–103. [CrossRef]
36. Anil, E.B.; Tang, P.; Akinci, B.; Huber, D. Deviation analysis method for the assessment of the quality of the as-is Building Information Models generated from point cloud data. *Autom. Constr.* **2013**, *35*, 507–516. [CrossRef]
37. Kim, M.K.; Cheng, J.C.; Sohn, H.; Chang, C.C. A framework for dimensional and surface quality assessment of precast concrete elements using BIM and 3D laser scanning. *Autom. Constr.* **2015**, *49*, 225–238. [CrossRef]
38. Kim, M.K.; Sohn, H.; Chang, C.C. Automated dimensional quality assessment of precast concrete panels using terrestrial laser scanning. *Autom. Constr.* **2014**, *45*, 163–177. [CrossRef]
39. Zeibak-Shini, R.; Sacks, R.; Ma, L.; Filin, S. Towards generation of as-damaged BIM models using laser-scanning and as-built BIM: First estimate of as-damaged locations of reinforced concrete frame members in masonry infill structures. *Adv. Eng. Inform.* **2016**, *30*, 312–326. [CrossRef]

40. Riveiro, B.; Lourenço, P.B.; Oliveira, D.V.; González-Jorge, H.; Arias, P. Automatic morphologic analysis of quasi-periodic masonry walls from LiDAR. *Comput.-Aided Civ. Infrastruct. Eng.* **2016**, *31*, 305–319. [CrossRef]
41. Brigham, C.A.P.; Crider, J.G. A new metric for morphologic variability using landform shape classification via supervised machine learning. *Geomorphology* **2021**, *399*, 108065. [CrossRef]
42. Lee, J.; Son, H.; Kim, C.; Kim, C. Skeleton-based 3D reconstruction of as-built pipelines from laser-scan data. *Autom. Constr.* **2013**, *35*, 199–207. [CrossRef]
43. Czerniawski, T.; Nahangi, M.; Haas, C.; Walbridge, S. Pipe spool recognition in cluttered point clouds using a curvature-based shape descriptor. *Autom. Constr.* **2016**, *71*, 346–358. [CrossRef]
44. Holgado-Barco, A.; Gonz ´ alez-Aguilera, D.; Arias-Sanchez, P.; Martinez-Sanchez, J. Semiautomatic extraction of road horizontal alignment from a mobile LiDAR system. *Comput.-Aided Civ. Infrastruct. Eng.* **2015**, *30*, 217–228. [CrossRef]
45. Wojtkowska, M.; Kedzierski, M.; Delis, P. Validation of terrestrial laser scanning and artificial intelligence for measuring deformations of cultural heritage structures. *Measurement* **2021**, *167*, 1–18. [CrossRef]
46. Fawzy, H.E.-D. 3D laser scanning and close-range photogrammetry for buildings documentation: A hybrid technique towards a better accuracy. *Alex. Eng. J.* **2019**, *58*, 1191–1204. [CrossRef]
47. Kim, M.-K.; Thedjaa, J.P.P.; Wang, Q. Automated dimensional quality assessment for formwork and rebar of reinforced concrete components using 3D point cloud data. *Autom. Constr.* **2020**, *112*, 1–14. [CrossRef]
48. Liu, J.; Zhang, Q.; Wu, J.; Zhao, Y. Dimensional accuracy and structural performance assessment of spatial structure components using 3D laser scanning. *Autom. Constr.* **2018**, *96*, 324–336. [CrossRef]
49. Wang, Q.; Cheng, J.C.P.; Sohn, H. Automated Estimation of Reinforced Precast Concrete Rebar Positions Using Colored Laser Scan Data. *Comput.-Aided Civ. Infrastruct. Eng.* **2017**, *32*, 787–802. [CrossRef]
50. Wang, Q.; Kim, M.K.; Cheng, J.C.; Sohn, H. Automated quality assessment of precast concrete elements with geometry irregularities using terrestrial laser scanning. *Autom. Constr.* **2016**, *68*, 170–182. [CrossRef]
51. Bosché, F. Automated recognition of 3D CAD model objects in laser scans and calculation of as-built dimensions for dimensional compliance control in construction. *Adv. Eng. Inform.* **2010**, *24*, 107–118. [CrossRef]
52. Lee, K.H.; Park, H.P. Automated inspection planning of free-form shape parts by laser scanning. *Robot. Comput.-Integr. Manuf.* **2000**, *16*, 201–210. [CrossRef]
53. Nuttens, T.; Stal, C.; De Backer, H.; Schotte, K.; Van Bogaert, P.; De Wulf, A. Methodology for the ovalization monitoring of newly built circular train tunnels based on laser scanning: Liefkenshoek Rail Link (Belgium). *Autom. Constr.* **2014**, *43*, 1–9. [CrossRef]
54. Alba, M.; Scaioni, M. Comparison of techniques for terrestrial laser scanning data georeferencing applied to 3-D modelling of cultural heritage. *Int. Arch. Photogramm. Remote Sens. Spat. Inf. Sci.* **2007**, *7*, 1–8.
55. Besl, P.J.; McKay, N.D. A method for registration of 3-D shapes. *IEEE Trans. Pattern Anal. Mach. Intell.* **1992**, *14*, 239–256. [CrossRef]
56. Sgrenzaroli, M.; Wolfart, E. Accurate texture-mapped 3D models for documentation, surveying and presentation purposes. In Proceedings of the Scanning For Cultural Heritage Recording, Corfu, Greece, 1–2 September 2002; The ICOMOS/ISPRS Committee for Documentation of Cultural Heritage: Corfu, Greece, 2002.
57. Hebert, M.; Krotkov, E. 3D measurements from imaging laser radars: How good are they? In Proceedings of the IEEE/RSJ International Workshop on Intelligent Robots and Systems 1991, Osaka, Japan, 3–5 November 1991; pp. 359–364.
58. Tuley, J.; Vandapel, N.; Hebert, M. Analysis and removal of artifacts in 3-D LADAR data. In Proceedings of the 2005 IEEE International Conference on Robotics and Automation, Barcelona, Spain, 18–22 April 2005; pp. 2203–2210.
59. Tang, P.; Huber, D.; Akinci, B. A comparative analysis of depth-discontinuity and mixed-pixel detection algorithms. In Proceedings of the Sixth International Conference on 3-D Digital Imaging and Modeling (3DIM 2007), Montreal, QC, Canada, 21–23 August 2007; pp. 29–38.
60. Wang, Q.; Sohn, H.; Cheng, J.C. Development of a mixed pixel filter for improved dimension estimation using AMCW laser scanner. *ISPRS J. Photogramm. Remote Sens.* **2016**, *119*, 246–258. [CrossRef]
61. Adams, R.; Bischof, L. Seeded Region Growing. *IEEE Trans. Pattern Anal. Mach. Intell.* **1994**, *16*, 641–647. [CrossRef]
62. Simpson, P.K. *Artificial Neural Systems: Foundations, Paradigms, Applications, and Implementations*, 1st ed.; Pergamon Pr: Oxford, UK, 1990.
63. Joshi, S.; Pande, S. Intelligent process modeling and optimization of die-sinking electric discharge machining. *Appl. Soft Comput.* **2011**, *11*, 2743–2755. [CrossRef]
64. Momeni, E.; Nazir, R.; Jahed Armaghani, D.; Maizir, H. Prediction of pile bearing capacity using a hybrid genetic algorithm-based ANN. *Measurement* **2014**, *57*, 122–131. [CrossRef]
65. Khandelwal, M.; Marto, A.; Fatemi, S.A.; Ghoroqi, M.; Armaghani, D.J.; Singh, T.N.; Tabrizi, O. Implementing an ANN model optimized by genetic algorithm for estimating cohesion of limestone samples. *Eng. Comput.* **2017**, *34*, 307–317. [CrossRef]
66. Adeli, H.; Hung, S.L. *Machine Learning: Neural Networks, Genetic Algorithms, and Fuzzy Systems*, 1st ed.; John Wiley & Sons, Inc.: Hoboken, NJ, USA, 1994.
67. Sardinas, R.Q.; Santana, M.R.; Brindis, E.A. Genetic algorithm-based multi-objective optimization of cutting parameters in turning processes. *Eng. Appl. Artif. Intell.* **2006**, *19*, 127–133. [CrossRef]
68. Stone, R.; Krishnamurthy, K. A neural network thrust force controller to minimize delamination during drilling of graphite-epoxy laminates. *Int. J. Mach. Tools Manuf.* **1996**, *36*, 985–1003. [CrossRef]

69. Dhupal, D.; Doloi, B.; Bhattacharyya, B. Optimization of process parameters of Nd: YAG laser microgrooving of Al_2TiO_5 ceramic material by response surface methodology and artificial neural network algorithm. *J. Eng. Manuf.* **2007**, *221*, 1341–1350. [CrossRef]
70. Shen, Z.; Tang, P.; Kannan, O.; Cho, Y.K. As-built error modeling for effective 3d laser scanning on construction sites. In *ASCE International Workshop on Computing in Civil Engineering*; ASCE: Los Angeles, CA, USA, 2013; pp. 533–540.
71. Chakraborty, D.; Elzarka, H. Performance testing of energy models: Are we using the right statistical metrics? *J. Build. Perform. Simul.* **2018**, *11*, 433–448. [CrossRef]
72. Sun, Y.; Wang, S.; Xiao, F.; Gao, D. Peak load shifting control using different cold thermal energy storage facilities in commercial buildings: A review. *Energy Convers. Manag.* **2013**, *71*, 101–114. [CrossRef]
73. Cortes, C.; Vapnik, V. Support-vector networks. *Mach. Learn.* **1995**, *20*, 273–297. [CrossRef]
74. Vapnik, V.; Golowich, S.E.; Smola, A.J. Support vector method for function approximation, regression estimation and signal processing. *Adv. Neural Inf. Processing Syst.* **1997**, *9*, 281–287.
75. Breiman, L. Random forests. *Mach. Learn.* **2001**, *45*, 5–32. [CrossRef]
76. Oku Topal, G.; Akpinar, B. High rate GNSS kinematic PPP method performance for monitoring the engineering structures: Shake table tests under different satellite configurations. *Measurement* **2021**, *189*, 110451. [CrossRef]
77. Shen, N.; Chen, L.; Chen, R. Displacement detection based on Bayesian inference from GNSS kinematic positioning for deformation monitoring. *Mech. Syst. Signal Processing* **2022**, *167*, 108570. [CrossRef]
78. Palmer, L.; Franke, K.; Abraham Martin, R.; Sines, B.; Rollins, K.; Hedengren, J. Application and Accuracy of Structure from Motion Computer Vision Models with Full-Scale Geotechnical Field Tests. In Proceedings of the International Foundations Congress and Equipment Expo 2015 (IFCEE 2015), San Antonio, TX, USA, 17–21 March 2015; pp. 2432–2441.
79. Laefer, D.F.; Truong-Hong, L. Toward automatic generation of 3D steel structures for building information modelling. *Autom. Constr.* **2017**, *74*, 66–77. [CrossRef]

symmetry

MDPI

Article

Establishment of an Eleven-Freedom-Degree Coupling Dynamic Model of Heavy Vehicle-Pavement

Bo Liang [1,2], Jinghang Xiao [1,*] and Shirong Shi [3]

1 School of Civil Engineering, Chongqing Jiaotong University, Chongqing 400074, China; liang_laoshi@126.com
2 State Key Laboratory of Mountain Bridge and Tunnel Engineering, Chongqing Jiaotong University, Chongqing 400074, China
3 CISDI Engineering Co., Ltd., Chongqing 400013, China; shirong.shi@cisdi.com.cn
* Correspondence: xiaojinghangswust@126.com or 611190080004@mails.cqjtu.edu.cn; Tel.: +86-186-0809-8751

Abstract: Considering the actual situation of moving vehicles acting on road surfaces, a dynamic model of a heavy vehicle–road coupling system was established based on the traditional vehicle–road vibration model. Firstly, a seven-degree-of-freedom vehicle model was established, and the vibration characteristics of the road subsystem were considered part of the whole system. The excitation effect of road roughness on the vehicle model was considered, and the dynamic model of the coupling system was finally obtained by combining the displacement compatibility principle of the contact relationship between the wheels and the road surface. The results show that the maximum bending value of the surface course reaches 2.37 mm. The maximum shear stress in the middle part of the surface course is 43,858 Pa. The vertical dynamic stress in the middle part of the surface course is larger, reaching 119,373 Pa, while the value of the vertical dynamic stress in the subgrade is much smaller, reaching 5824 Pa. The coupling dynamic model can reflect the relationship between the moving vehicle and the road and the dynamic performance, which not only provides theoretical support for the design parameters of heavy-duty vehicles but also provides a reference for the design of road durability.

Keywords: heavy-duty vehicle; road; coupling model; dynamic response

Citation: Liang, B.; Xiao, J.; Shi, S. Establishment of an Eleven-Freedom-Degree Coupling Dynamic Model of Heavy Vehicle-Pavement. *Symmetry* **2022**, *14*, 250. https://doi.org/10.3390/sym14020250

Academic Editors: Yang Yang, Ying Lei, Xiaolin Meng, Jun Li and Juan Luis García Guirao

Received: 31 October 2021
Accepted: 11 January 2022
Published: 27 January 2022

1. Introduction

In recent years, heavy-duty vehicle transportation has become important all around the world [1,2]. The concept of dynamic interactions between the vehicle and the road surface has been widely studied, and a series of problems caused by the coupling vibrations of the vehicle structure and road have become very prominent. Thanks to the in-depth study of this issue by scholars, research on the dynamic interaction between vehicles and road systems [3,4], and the vehicle–road dynamics model of coupled systems [5–11], our understanding in this area is continuously developing. The key problems that hinder the long-term development of heavy haul transportation have been partially solved, including the impact coefficient of vehicles, the vibration of vehicles under the uneven road surface, and the instantaneous response of road vibrations.

As mentioned above, the dynamic characteristics of the roadbed are regarded as one of the important components of the system. At present, a vehicle–road system can be divided into a vehicle vibration model and a road subsystem model. The vehicle is the moving part, which is the main excitation part that causes the vehicle and road vibration. The pavement and the following parts are the road subsystem. Due to the long-term bearing of the dynamic load of the upper vehicle, including the repeated load of heavy vehicle movement, subgrade diseases form easily. In particular, the deformation of flexible asphalt pavement is often affected by many factors such as traffic congestion, joints, load capacity, and repeated vehicle loads [12,13]. All these factors affect the normal service of the road and make the road surface irregular, seriously affecting the normal operation of the road and even damaging it. The above analysis shows that the damage of roads not only involves damage to the road surface but also the excessive deformation of the roadbed. Therefore, when analyzing the dynamic interaction between vehicle and road, the vehicle vibration excitation is the key problem, which is, therefore, one of the main concerns of this work.

Scholars have studied the dynamic problems of vehicle and highway subgrades from different perspectives [14,15]. However, few of the existing models consider the vehicle–road system as a whole, and most models analyze the road dynamic performance under the dynamic load of vehicles. The literature [16] describes the lateral dynamics model of vehicles under road environment excitation. The authors of [17,18] studied the dynamic response of the vehicle–pavement coupling system based on the nonlinear Timoshenko beam method and the vehicle–road dynamic response of the multi-degree-of-freedom vehicle with a double-layer, rectangular, thin plate. Zhang et al. [19] studied the road deformation and crack propagation path under the action of a quarter of vehicles using a modified two-parameter foundation plate. The above analysis fails to fully reflect the interaction of a vehicle–subgrade system in the process of vehicle moving, and this is another main objective of the present study.

In the present study, we establish a seven-degree-of-freedom vehicle dynamic model and use the harmonic superposition method to simulate the process of road random vibration. The dynamic interaction between vehicle and road is analyzed theoretically. Then, the validity of the model is verified by numerical analysis, and the dynamic response is analyzed. The organization of the paper is as follows. In Section 2, the dynamic model and equilibrium equation of heavy load vehicles are established. In Section 3, the coupling dynamic equation of heavy load vehicle–road is established. Then, the vehicle–road system is analyzed from multiple angles through numerical simulation, and the validity of the established equation is verified in Section 4.

2. Vehicle Dynamic Model and Equilibrium Equation

2.1. Simplified Vehicle Road Dynamic Model

According to different research needs, traditional vehicle models can come in three different styles. They are the quarter two-degree-of-freedom vehicle model, the half four-degree-of-freedom vehicle model, and the whole vehicle model, respectively. The first two models are widely used due to their simple calculation, but the parameters of vibration are not fully considered and the calculation accuracy is not high. Based on the traditional model, a 7-DOF vehicle model is established in this paper, as shown in Figure 1. In this model, the car body and wheels are treated as rigid bodies. A vehicle consisting of a single body and several wheels passes through the line at a uniform speed; for the body itself, the freedom of nod vibration, deflection vibration, and vertical vibration will be considered [20], along with the vibration of the four wheels, for a total of seven degrees of freedom. Furthermore, it is assumed that the wheels of the vehicle are in close contact with the road surface. In this paper, we mainly study the vibration of the vehicle–road system caused by the excitation of road roughness. In particular, this paper studies the effects on vehicle ride performance, including comfort, safety, and irregularity. To simplify the analysis process, it was simplified into the 7-DOF vehicle structure model shown in Figure 1.

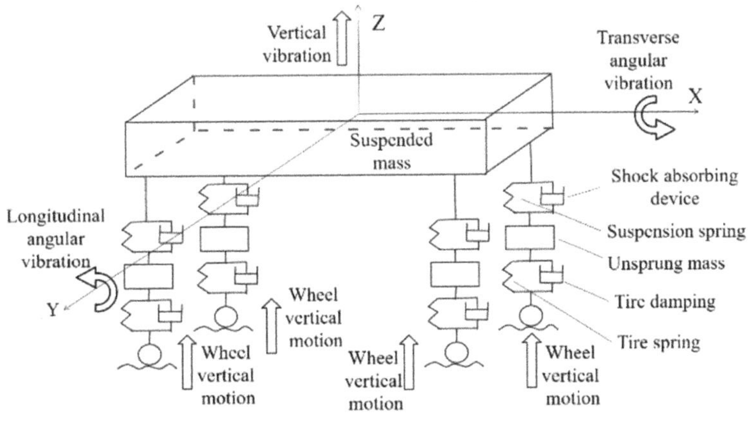

Figure 1. 7-DOF vehicle structure model.

2.2. Dynamic Balance of Vehicle–Road System

Based on the above description, combined with the vehicle vibration balance equation and the road vibration balance equation, the dynamic balance equation of the vehicle–road system is established. The stress analysis of the vehicle model and the body are shown in Figures 2 and 3, respectively.

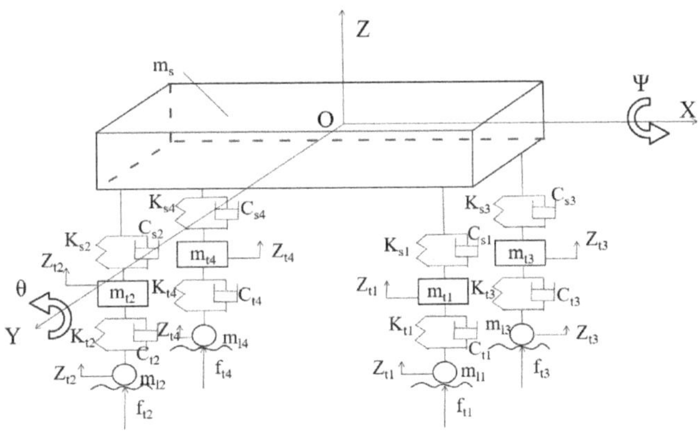

Figure 2. Stress analysis of vehicle model.

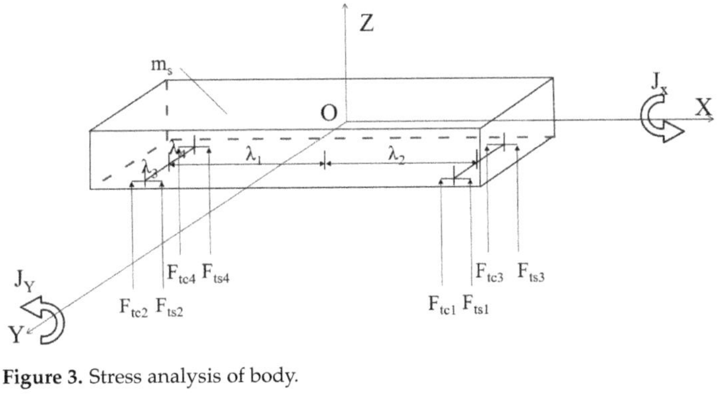

Figure 3. Stress analysis of body.

where

$$F_{tc1} = C_{s1}(\dot{Z}_s - \lambda_1\dot{\psi} - \lambda_3\dot{\theta} - \dot{Z}_{t1}),$$
$$F_{ts1} = K_{s1}(Z_s - \lambda_1\psi - \lambda_3\theta - Z_{t1}),$$
$$F_{tc2} = C_{s2}(\dot{Z}_s + \lambda_2\dot{\psi} - \lambda_3\dot{\theta} - \dot{Z}_{t2}),$$
$$F_{ts2} = K_{s2}(Z_s + \lambda_2\psi - \lambda_3\theta - Z_{t2}),$$
$$F_{tc3} = C_{s3}(\dot{Z}_s - \lambda_1\dot{\psi} + \lambda_4\dot{\theta} - \dot{Z}_{t2}),$$
$$F_{ts3} = K_{s3}(Z_s - \lambda_1\psi + \lambda_4\theta - Z_{t3}),$$
$$F_{tc4} = C_{s4}(\dot{Z}_s + \lambda_2\dot{\psi} + \lambda_4\dot{\theta} - \dot{Z}_{t4}),$$
$$F_{ts4} = K_{s4}(Z_s + \lambda_2\psi + \lambda_4\theta - Z_{t4}),$$

and F_{tc1}, F_{ts1}, F_{tc2}, F_{ts2}, F_{tc3}, F_{ts3}, F_{tc4}, and F_{ts4} represent the interaction force between the unsprung mass part of the vehicle and the body.

According to the force balance conditions in Figures 2 and 3, the balance equations of vertical vibration, pitch vibration, and roll vibration of the body are as follows [21–23]:

$$\begin{aligned}
&m_s\ddot{Z}_s + C_{s1}(\dot{Z}_s - \lambda_1\dot{\psi} - \lambda_3\dot{\theta} - \dot{Z}_{t1}) + K_{s1}(Z_s - \\
&\lambda_1\psi - \lambda_3\theta - Z_{t1}) + C_{s2}(\dot{Z}_s + \lambda_2\dot{\psi} - \lambda_3\dot{\theta} - \dot{Z}_{t2}) \\
&+K_{s2}(Z_s + \lambda_2\psi - \lambda_3\theta - Z_{t2})C_{s3}(\dot{Z}_s - \lambda_1\dot{\psi} + \\
&\lambda_4\dot{\theta} - \dot{Z}_{t2}) + K_{s3}(Z_s - \lambda_1\psi + \lambda_4\theta - Z_{t3}) + C_{s4} \\
&(\dot{Z}_s + \lambda_2\dot{\psi} + \lambda_4\dot{\theta} - \dot{Z}_{t4}) + K_{s4}(Z_s + \lambda_2\psi + \lambda_4\theta \\
&-Z_{t4}) = 0
\end{aligned} \tag{1}$$

$$\begin{aligned}
&J\ddot{\theta} + \lambda_1\left\{\begin{array}{c} C_{s1}(\dot{Z}_s - \lambda_1\dot{\psi} - \lambda_3\dot{\theta} - \dot{Z}_{t1}) + \\ K_{s1}(Z_s - \lambda_1\psi - \lambda_3\theta - Z_{t1}) + \\ C_{s3}(\dot{Z}_s - \lambda_1\dot{\psi} + \lambda_4\dot{\theta} - \dot{Z}_{t2}) + \\ K_{s3}(Z_s - \lambda_1\psi + \lambda_4\theta - Z_{t2}) \end{array}\right\} \\
&-\lambda_2\left\{\begin{array}{c} C_{s2}(\dot{Z}_s + \lambda_2\dot{\psi} - \lambda_3\dot{\theta} - \dot{Z}_{t2}) + \\ K_{s2}(Z_s + \lambda_2\psi - \lambda_3\theta - Z_{t2}) + \\ C_{s4}(\dot{Z}_s + \lambda_2\dot{\psi} + \lambda_4\dot{\theta} - \dot{Z}_{t4}) + \\ K_{s4}(Z_s + \lambda_2\psi + \lambda_4\theta - Z_{t4}) \end{array}\right\} = 0
\end{aligned} \tag{2}$$

$$\begin{aligned}
&J\ddot{\psi} + \lambda_3\left\{\begin{array}{c} C_{s1}(\dot{Z}_s - \lambda_1\dot{\psi} - \lambda_3\dot{\theta} - \dot{Z}_{t1}) + \\ K_{s1}(Z_s - \lambda_1\psi - \lambda_3\theta - Z_{t1}) + \\ C_{s2}(\dot{Z}_s + \lambda_2\dot{\psi} - \lambda_3\dot{\theta} - \dot{Z}_{t2}) + \\ K_{s2}(Z_s + \lambda_2\psi - \lambda_3\theta - Z_{t2}) \end{array}\right\} \\
&-\lambda_4\left\{\begin{array}{c} C_{s3}(\dot{Z}_s - \lambda_1\dot{\psi} + \lambda_4\dot{\theta} - \dot{Z}_{t2}) + \\ K_{s3}(Z_s - \lambda_1\psi + \lambda_4\theta - Z_{t3}) + \\ C_{s4}(\dot{Z}_s + \lambda_2\dot{\psi} + \lambda_4\dot{\theta} - \dot{Z}_{t4}) + \\ K_{s4}(Z_s + \lambda_2\psi + \lambda_4\theta - Z_{t4}) \end{array}\right\} = 0
\end{aligned} \tag{3}$$

The force analysis of the vehicle's unsprung mass model is shown in Figure 4.

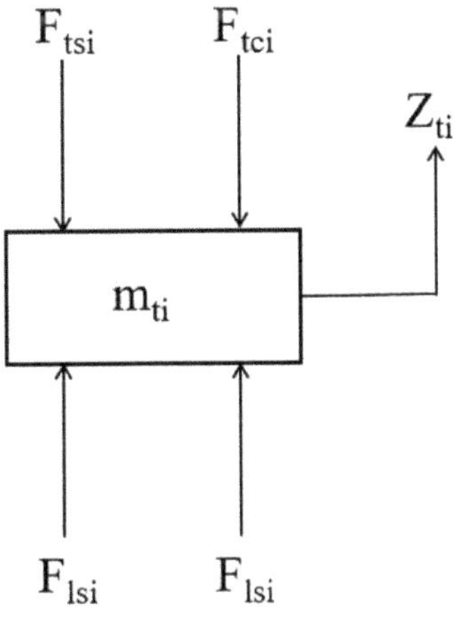

Figure 4. Force analysis of vehicle unsprung mass.

Where i = 1, 2, 3, and 4 represent the right front unsprung mass part, the right rear unsprung mass part, the left front unsprung mass part, and the left rear unsprung mass part, respectively. The force balance equations of the four unsprung mass parts are as follows:

$$\begin{aligned} &m_{t1}\ddot{Z}_{t1} + C_{t1}(\dot{Z}_{t1} - \dot{Z}_{l1}) + K_{t1}(Z_{t1} - Z_{l1}) - C_{s1}(\dot{Z}_s \\ &-\lambda_1\dot{\psi} - \lambda_3\dot{\theta} - \dot{Z}_{t1}) - K_{s1}(Z_s - \lambda_1\psi - \lambda_3\theta - Z_{t1}) = 0 \end{aligned} \tag{4}$$

$$\begin{aligned} &m_{t3}\ddot{Z}_{t3} + C_{t3}(\dot{Z}_{t3} - \dot{Z}_{l3}) + K_{t3}(Z_{t3} - Z_{l3}) - C_{s3}(\dot{Z}_s \\ &-\lambda_1\dot{\psi} + \lambda_4\dot{\theta} - \dot{Z}_{t2}) - K_{s3}(Z_s - \lambda_1\psi + \lambda_4\theta - Z_{t3}) = 0 \end{aligned} \tag{5}$$

$$\begin{aligned} &m_{t2}\ddot{Z}_{t2} + C_{t2}(\dot{Z}_{t2} - \dot{Z}_{l2}) + K_{t2}(Z_{t2} - Z_{l2}) - C_{s2}(\dot{Z}_s \\ &+\lambda_2\dot{\psi} - \lambda_3\dot{\theta} - \dot{Z}_{t2}) - K_{s2}(Z_s + \lambda_2\psi - \lambda_3\theta - Z_{t2}) = 0 \end{aligned} \tag{6}$$

$$\begin{aligned} &m_{t4}\ddot{Z}_{t4} + C_{t4}(\dot{Z}_{t4} - \dot{Z}_{l4}) + K_{t4}(Z_{t4} - Z_{l4}) - C_{s4}(\dot{Z}_s \\ &+\lambda_2\dot{\psi} + \lambda_4\dot{\theta} - \dot{Z}_{t4}) - K_{s4}(Z_s + \lambda_2\psi + \lambda_4\theta - Z_{t4}) = 0 \end{aligned} \tag{7}$$

The stress analysis of the wheel is drawn according to the stress of the vehicle wheel, as shown in Figure 5.

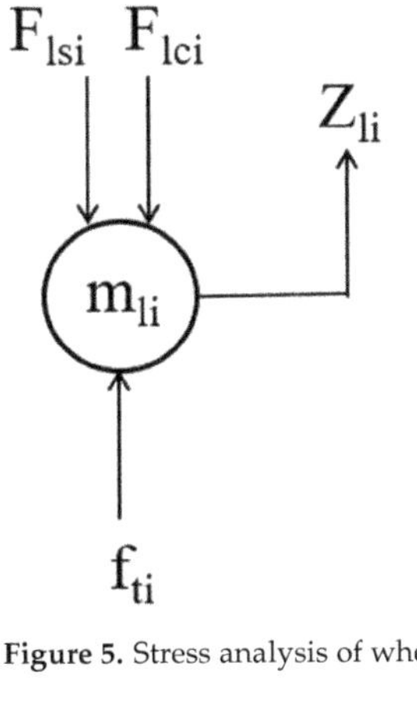

Figure 5. Stress analysis of wheel.

Where, i = 1, 2, 3, and 4 represent the four wheels, respectively. According to the force balance, the balance equations are obtained:

$$f_{t1} = C_{t1}(\dot{Z}_{t1} - \dot{Z}_{l1}) + K_{t1}(Z_{t1} - Z_{l1}) - m_{l1}\ddot{Z}_{l1} \tag{8}$$

$$f_{t2} = C_{t2}(\dot{Z}_{t2} - \dot{Z}_{l2}) + K_{t2}(Z_{t2} - Z_{l2}) - m_{l2}\ddot{Z}_{l2} \tag{9}$$

$$f_{t3} = C_{t3}(\dot{Z}_{t3} - \dot{Z}_{l3}) + K_{t3}(Z_{t3} - Z_{l3}) - m_{l3}\ddot{Z}_{l3} \tag{10}$$

$$f_{t4} = C_{t4}(\dot{Z}_{t4} - \dot{Z}_{l4}) + K_{t4}(Z_{t4} - Z_{l4}) - m_{l4}\ddot{Z}_{l4} \tag{11}$$

In the above equations and pictures (Equations (1)–(11) and Figures 1–5), m_s represents the body mass; θ is the body nod displacement angle, ψ represents the body turnover displacement angle; m_{ti} represents the unsprung mass of vehicle front and rear; m_{li} represents wheel mass; C_{ti} represents the damping coefficient of the corresponding wheel before and after the vehicle; C_{si} represents the damping coefficient of the corresponding suspension system before and after the vehicle; K_{ti} represents the stiffness coefficients of the corresponding wheels before and after the vehicle; K_{si} represent the stiffness coefficients of the front and rear suspension of the vehicle; Z_s represents the vertical vibration displacement of the body; Z_{ti} represent the vertical vibration displacements of the four unsprung mass parts corresponding to the front and rear of the vehicle; Z_{li} represents the vertical vibration displacement of the front and rear wheels of the vehicle; J_X represents the rotational inertia of the body around the x-axis; J_Y represents the rotational inertia of the body around the y-axis; and f_{ti} represents the dynamic loads of the four wheels.

Based on the balance equation of each part of the vehicle system mentioned above, the vector expression can be obtained as:

$$M\ddot{Z} + C\dot{Z} + KZ = F_t \tag{12}$$

where M represents the mass matrix of the vehicle; C represents the damping matrix of the vehicle; K represents the stiffness matrix of the vehicle; Z represents the displacement matrix of the vehicle; F_t represents the dynamic load vector acted on the road surface by the vehicle system. The above matrix expressions are as follows:

$$M = \begin{bmatrix} m & 0 & 0 & 0 & 0 & 0 & 0 & 0 & 0 & 0 & 0 \\ 0 & J_X & 0 & 0 & 0 & 0 & 0 & 0 & 0 & 0 & 0 \\ 0 & 0 & J_Y & 0 & 0 & 0 & 0 & 0 & 0 & 0 & 0 \\ 0 & 0 & 0 & m_{t1} & 0 & 0 & 0 & 0 & 0 & 0 & 0 \\ 0 & 0 & 0 & 0 & m_{t2} & 0 & 0 & 0 & 0 & 0 & 0 \\ 0 & 0 & 0 & 0 & 0 & m_{t3} & 0 & 0 & 0 & 0 & 0 \\ 0 & 0 & 0 & 0 & 0 & 0 & m_{t4} & 0 & 0 & 0 & 0 \\ 0 & 0 & 0 & 0 & 0 & 0 & 0 & m_{l1} & 0 & 0 & 0 \\ 0 & 0 & 0 & 0 & 0 & 0 & 0 & 0 & m_{l2} & 0 & 0 \\ 0 & 0 & 0 & 0 & 0 & 0 & 0 & 0 & 0 & m_{l3} & 0 \\ 0 & 0 & 0 & 0 & 0 & 0 & 0 & 0 & 0 & 0 & m_{l4} \end{bmatrix},$$

$$C = \begin{bmatrix} c_{11} & c_{12} & c_{13} & -c_{s1} & -c_{s2} & -c_{s3} & -c_{s4} & 0 & 0 & 0 & 0 \\ c_{21} & c_{22} & c_{23} & \lambda_1 c_{s1} & -\lambda_2 c_{s2} & \lambda_1 c_{s3} & -\lambda_2 c_4 & 0 & 0 & 0 & 0 \\ c_{31} & c_{32} & c_{33} & \lambda_3 c_{s1} & \lambda_3 c_{s2} & -\lambda_4 c_{s3} & -\lambda_4 c_4 & 0 & 0 & 0 & 0 \\ -c_{s1} & \lambda_1 c_{s1} & \lambda_3 c_{s1} & c_{s1}+c_{t1} & 0 & 0 & 0 & -c_{t1} & 0 & 0 & 0 \\ -c_{s2} & -\lambda_2 c_{s2} & \lambda_3 c_{s2} & 0 & c_2+c_{t2} & 0 & 0 & 0 & -c_{t2} & 0 & 0 \\ -c_{s3} & \lambda_1 c_{s3} & -\lambda_4 c_{s3} & 0 & 0 & c_3+c_{t3} & 0 & 0 & 0 & -c_{t3} & 0 \\ -c_{s4} & -\lambda_2 c_{s4} & -\lambda_4 c_{s4} & 0 & 0 & 0 & c_4+c_{t4} & 0 & 0 & 0 & -c_{t4} \\ 0 & 0 & 0 & -c_{t1} & 0 & 0 & 0 & c_{t1} & 0 & 0 & 0 \\ 0 & 0 & 0 & 0 & -c_{t2} & 0 & 0 & 0 & c_{t2} & 0 & 0 \\ 0 & 0 & 0 & 0 & 0 & -c_{t3} & 0 & 0 & 0 & c_{t3} & 0 \\ 0 & 0 & 0 & 0 & 0 & 0 & -c_{t4} & 0 & 0 & 0 & c_{t4} \end{bmatrix},$$

$$K = \begin{bmatrix} k_{11} & k_{12} & k_{13} & -k_{s1} & -k_{s2} & -k_{s3} & -k_{s4} & 0 & 0 & 0 & 0 \\ k_{21} & k_{22} & k_{23} & \lambda_1 k_{s1} & -\lambda_2 k_{s2} & \lambda_1 k_{s3} & -\lambda_2 k_{s4} & 0 & 0 & 0 & 0 \\ k_{31} & k_{32} & k_{33} & \lambda_3 k_{s1} & \lambda_3 k_{s2} & -\lambda_4 k_{s2} & -\lambda_4 k_{s4} & 0 & 0 & 0 & 0 \\ -k_{s1} & \lambda_1 k_{s1} & \lambda_3 k_{s1} & k_{s1}+k_{t1} & 0 & 0 & 0 & -k_{t1} & 0 & 0 & 0 \\ -k_{s2} & -\lambda_2 k_{s2} & \lambda_3 k_{s2} & 0 & k_{s2}+k_{t2} & 0 & 0 & 0 & -k_{t2} & 0 & 0 \\ -k_{s3} & \lambda_1 k_{s3} & -\lambda_4 k_{s3} & 0 & 0 & k_{s3}+k_{t3} & 0 & 0 & 0 & -k_{t3} & 0 \\ -k_{s4} & -\lambda_2 k_{s4} & -\lambda_4 k_{s4} & 0 & 0 & 0 & k_{s4}+k_{t4} & 0 & 0 & 0 & -k_{t4} \\ 0 & 0 & 0 & -k_{t1} & 0 & 0 & 0 & k_{t1} & 0 & 0 & 0 \\ 0 & 0 & 0 & 0 & -k_{t2} & 0 & 0 & 0 & k_{t2} & 0 & 0 \\ 0 & 0 & 0 & 0 & 0 & -k_{t3} & 0 & 0 & 0 & k_{t3} & 0 \\ 0 & 0 & 0 & 0 & 0 & 0 & -k_{t4} & 0 & 0 & 0 & k_{t4} \end{bmatrix},$$

$$Z = \begin{bmatrix} Z_s & \psi & \theta & Z_{t1} & Z_{t2} & Z_{t3} & Z_{t4} & Z_{l1} & Z_{l2} & Z_{l3} & Z_{l4} \end{bmatrix}^T,$$

$$C_{11} = c_{s1} + c_{s2} + c_{s3} + c_{s4},$$

$$C_{12} = C_{21} = -\lambda_1 c_{s1} + \lambda_2 c_{s2} - \lambda_1 c_{s3} + \lambda_2 c_{s4},$$

$$C_{13} = C_{31} = -\lambda_3 c_{s1} - \lambda_3 c_{s2} + \lambda_4 c_{s3} + \lambda_4 c_{s4},$$

$$C_{22} = {\lambda_1}^2 c_{s1} + {\lambda_2}^2 c_{s2} - {\lambda_1}^2 c_{s3} + {\lambda_2}^2 c_{s4},$$

$$C_{23} = C_{32} = \lambda_1 \lambda_3 c_{s1} - \lambda_2 \lambda_3 c_{s2} - \lambda_1 \lambda_4 c_{s3} + \lambda_2 \lambda_4 c_{s2},$$

$$C_{33} = {\lambda_3}^2 c_{s1} + {\lambda_3}^2 c_{s2} + {\lambda_4}^2 c_{s3} + {\lambda_4}^2 c_{s4},$$

$$K_{11} = K_{s1} + K_{s2} + K_{s3} + K_{s4},$$

$$K_{12} = K_{21} = -\lambda_1 K_{s1} + \lambda_2 K_{s2} - \lambda_1 K_{s3} + \lambda_2 K_{s4},$$

$$K_{13} = K_{31} = -\lambda_3 K_{s1} - \lambda_3 K_{s2} + \lambda_4 K_{s3} + \lambda_4 K_{s4},$$

$$K_{22} = {\lambda_1}^2 K_{s1} + {\lambda_2}^2 K_{s2} - {\lambda_1}^2 K_{s3} + {\lambda_2}^2 K_{s4},$$

$$K_{23} = K_{32} = \lambda_1 \lambda_3 K_{s1} - \lambda_2 \lambda_3 K_{s2} - \lambda_1 \lambda_4 K_{s3} + \lambda_2 \lambda_4 K_{s2},$$

$$K_{33} = {\lambda_3}^2 K_{s1} + {\lambda_3}^2 K_{s2} + {\lambda_4}^2 K_{s3} + {\lambda_4}^2 K_{s4},$$

$$F_t = \begin{bmatrix} 0 & 0 & 0 & 0 & 0 & 0 & 0 & f_{t1} & f_{t2} & f_{t3} & f_{t4} \end{bmatrix}^T$$

The boundary constraint conditions of the finite element model of the road multi-layer system need to be simplified according to the force and loading conditions of the road under the actual vehicle traveling conditions. When the vehicle is driving in the center of the road, the response at the far side of the road is negligible. Because of the thick soil foundation, the response at the bottom is too small and negligible. Therefore, it can be known from Saint-Venant's principle that longitudinal constraints are imposed on both sides of the principle and fixed displacement are imposed on the bottom to obtain the finite element model of the road [24–26], as shown in Figure 6.

Figure 6. Finite element model of road structure layer.

3. Establishment of Coupled Dynamic Equation between Heavy-Duty Vehicle and Road

3.1. Pavement Vibration Model

The roughness of the road surface is called unevenness in highway engineering. It originates from the accidental factors that cannot be controlled artificially in the process of construction and the irregular vertical undulation of the ground caused by repeated loads caused by vehicles. The unevenness of the road will cause the vibration of the vehicle. Based on the establishment of the road finite element model in the previous section, according to the forced vibration equation of the damped system with multiple degrees of freedom, the vibration balance equation of the road structure model can be obtained as follows:

$$[m]\{\ddot{\delta}\} + [c]\{\dot{\delta}\} + [k]\{\delta\} = \{f\} \tag{13}$$

where $[m]$ represents the overall mass matrix of the road finite element model, $[c]$ represents the overall damping matrix, $[k]$ represents the overall stiffness matrix, $\{\delta\}$ represents the displacement vector of the road finite element model, $\{\dot{\delta}\}$ represents the velocity vector, $\{\ddot{\delta}\}$ represents the acceleration vector, and $\{f\}$ represents the vertical load vector, including the dynamic load and static load.

To solve the common vibration problem caused by vehicles acting on the road, the vehicle vibration balance equation and road vibration balance equation in Equations (12) and (13) are needed to calculate the dynamic response of the whole system. To simplify the calculation without losing accuracy and reliability, the high-order term of vibration mode superposition is discarded, and only the contact between the vehicle and the road is considered; that is, the node between the wheel and the road. The vertical vibration displacement at any cross-section x position of the road pavement is obtained by superposition of the vibration mode functions of each order of the road:

$$z_r(x) = \sum_{n=1}^{N} A_n \phi_n(x_i) \tag{14}$$

where A_n represents the corresponding generalized coordinates and $\phi_n(x_i)$ is the vibration mode function at the corresponding road horizontal position x and the nth order.

3.2. Model of Road Roughness

The methods of road unevenness can be divided into static cross-section measurement, dynamic cross-section measurement, reaction level measurement system, and subjective evaluation method. The road smoothness is expressed as the superposition of some sine or cosine waves with random phases. For the superposition of N similar sine waves, the smoothness of a random road surface can be expressed as [24]:

$$Z(x) = \sum_{i=1}^{n} \sqrt{2} A_i \cdot \sin(2\pi \cdot x \cdot n_{mid\_i} + \theta_i) \tag{15}$$

where, $Z(x)$ represents the unevenness value of the random pavement, x represents the displacement of the pavement along the horizontal direction, θ represents the random number of $[0–2\pi]$, and n_{mid_i} represents the intermediate value of the spatial frequency of the pavement flatness between each cell.

Assuming the vehicle speed is v, let $x = vt$, and the first-order and second-order derivatives of the above formula can obtain the speed and acceleration of a random road, respectively.

$$\dot{Z}(x) = \sum_{i=1}^{n} 2\sqrt{2}\pi v n_{mid\_i} A_i \cdot cos(2\pi \cdot x \cdot n_{mid\_i} + \theta_i) \tag{16}$$

$$\ddot{Z}(x) = \sum_{i=1}^{n} -4\sqrt{2}\pi^2 v^2 n^2{}_{mid\_i} A_i \cdot \sin(2\pi \cdot x \cdot n_{mid\_i} + \theta_i) \tag{17}$$

3.3. Wheel–Road Displacement Coupled Vibration

It has been assumed that the vehicle is always in contact with the pavement during driving, and the vertical displacement of the pavement is divided into pavement flatness $z_p(x_i)$ and pavement vibration vertical displacement $z_r(x_i)$; then, the vertical displacement of the wheel can be expressed as:

$$z_{li} = z_r(x_i) + z_p(x_i) \tag{18}$$

where $z_p(x_i)$ is the road surface flatness value of tire i (i represents the wheel position) at the horizontal direction x of the road. Substituting Equation (14) into Equation (18), we get:

$$z_{li} = \sum_{n=1}^{N} A_n \phi_n(x_i) + z_p(x_i) \tag{19}$$

The contact between vehicle wheel and the road surface can be simplified according to Hertz's contact theory. Consider the coupling relationship between the wheel and the road surface to be a nonlinear contact force, as shown in Figure 7.

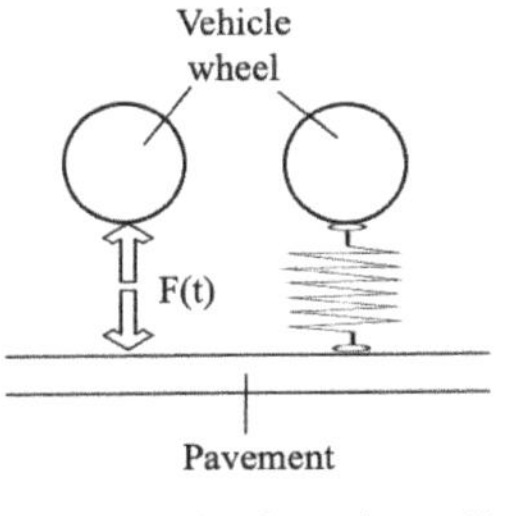

Figure 7. Wheel–road coupling vibration diagram.

According to the contact relation in Hertz's law, when further simplified the vertical contact force between the wheel and the road surface is expressed as:

$$F(t) = k_H(Z_w - Z_r) \tag{20}$$

3.4. Establishment of Vehicle–Road Coupling Dynamic Analysis Model

As mentioned above, the vehicle vibration balance equation and road vibration balance equation are combined to obtain the vibration coupling relationship model under the condition of displacement compatibility:

$$
\left\{
\begin{array}{l}
\begin{bmatrix} m_s & 0 & 0 & 0 & 0 & 0 & 0 \\ 0 & J_X & 0 & 0 & 0 & 0 & 0 \\ 0 & 0 & J_Y & 0 & 0 & 0 & 0 \\ 0 & 0 & 0 & m_{t1} & 0 & 0 & 0 \\ 0 & 0 & 0 & 0 & m_{t2} & 0 & 0 \\ 0 & 0 & 0 & 0 & 0 & m_{t3} & 0 \\ 0 & 0 & 0 & 0 & 0 & 0 & m_{t4} \end{bmatrix}
\begin{bmatrix} \ddot{Z}_s \\ \ddot{\psi} \\ \ddot{\theta} \\ \ddot{Z}_{t1} \\ \ddot{Z}_{t2} \\ \ddot{Z}_{t3} \\ \ddot{Z}_{t4} \end{bmatrix}
+ \left[\begin{bmatrix} c_{11} & c_{12} & c_{13} & -c_{s1} & -c_{s2} & -c_{s3} & -c_{s4} \\ c_{21} & c_{22} & c_{23} & \lambda_1 c_{s1} & -\lambda_2 c_{s2} & \lambda_1 c_{s3} & -\lambda_2 c_{s4} \\ c_{31} & c_{32} & c_{33} & \lambda_3 c_{s1} & \lambda_3 c_{s2} & -\lambda_4 c_{s3} & -\lambda_4 c_{s4} \\ -c_{s1} & \lambda_1 c_{s1} & \lambda_3 c_{s1} & c_{s1}+c_{t1} & 0 & 0 & 0 \\ -c_{s2} & -\lambda_2 c_{s2} & \lambda_3 c_{s2} & 0 & c_{s2}+c_{t2} & 0 & 0 \\ -c_{s3} & \lambda_1 c_{s3} & -\lambda_4 c_{s3} & 0 & 0 & c_{s3}+c_{t3} & 0 \\ -c_{s4} & -\lambda_2 c_{s4} & -\lambda_4 c_{s4} & 0 & 0 & 0 & c_{s4}+c_{t4} \end{bmatrix}
\left[\begin{bmatrix} \dot{Z}_s \\ \dot{\psi} \\ \dot{\theta} \\ \dot{Z}_{t1} \\ \dot{Z}_{t2} \\ \dot{Z}_{t3} \\ \dot{Z}_{t4} \end{bmatrix} \right] \right] \\
+ \begin{bmatrix} k_{11} & k_{12} & k_{13} & -k_{s1} & -k_{s2} & -k_{s3} & -k_{s4} \\ k_{21} & k_{22} & k_{23} & \lambda_1 k_{s1} & -\lambda_2 k_{s2} & \lambda_1 k_{s3} & -\lambda_2 k_{s4} \\ k_{31} & k_{32} & k_{33} & \lambda_3 k_{s1} & \lambda_3 k_{s2} & -\lambda_4 k_{s3} & -\lambda_4 k_{s4} \\ -k_{s1} & \lambda_1 k_{s1} & \lambda_3 k_{s1} & k_{s1}+k_{t1} & 0 & 0 & 0 \\ -k_{s2} & -\lambda_2 k_{s2} & \lambda_3 k_{s2} & 0 & k_{s2}+k_{t2} & 0 & 0 \\ -k_{s3} & \lambda_1 k_{s3} & -\lambda_4 k_{s3} & 0 & 0 & k_{s3}+k_{t3} & 0 \\ -k_{s4} & -\lambda_2 k_{s4} & -\lambda_4 k_{s4} & 0 & 0 & 0 & k_{s4}+k_{t4} \end{bmatrix}
\begin{bmatrix} Z_s \\ \psi \\ \theta \\ Z_{t1} \\ Z_{t2} \\ Z_{t3} \\ Z_{t4} \end{bmatrix} \\
= \begin{bmatrix} 0 \\ 0 \\ 0 \\ C_{t1}\left[\dot{z}_p(x_1) + \sum\limits_{n=1}^{N} \dot{A}_n \phi_n(x_1)\right] + K_{t1}\left[z_p(x_1) + \sum\limits_{n=1}^{N} A_n \phi_n(x_1)\right] \\ C_{t2}\left[\dot{z}_p(x_2) + \sum\limits_{n=1}^{N} \dot{A}_n \phi_n(x_2)\right] + K_{t1}\left[z_p(x_2) + \sum\limits_{n=1}^{N} A_n \phi_n(x_2)\right] \\ C_{t3}\left[\dot{z}_p(x_3) + \sum\limits_{n=1}^{N} \dot{A}_n \phi_n(x_3)\right] + K_{t3}\left[z_p(x_3) + \sum\limits_{n=1}^{N} A_n \phi_n(x_3)\right] \\ C_{t4}\left[\dot{z}_p(x_4) + \sum\limits_{n=1}^{N} \dot{A}_n \phi_n(x_4)\right] + K_{t4}\left[z_p(x_4) + \sum\limits_{n=1}^{N} A_n \phi_n(x_4)\right] \end{bmatrix} \\
\ddot{A}_n + 2\xi_n \omega_n \dot{A}_n + \omega_n^2 A_n = \sum\limits_{j=1}^{4} \Phi_n\left(x_j\right) \left\{ \begin{array}{c} \left(\frac{\lambda_j}{\lambda} m_s + m_{tj} + m_{lj}\right) g + K_{tj} Z_{tj} + C_{tj} \dot{Z}_{tj} \\ -\left\{ K_{tj}\left[z_p\left(x_j\right) + \sum\limits_{n=1}^{N} A_n \phi_n\left(x_j\right)\right] + C_{tj}\left[\dot{z}_p\left(x_j\right) + \sum\limits_{n=1}^{N} \dot{A}_n \phi_n\left(x_j\right)\right] \right\} \\ -m_{lj}\left[\ddot{Z}_p\left(x_j\right) + \sum\limits_{n=1}^{N} \ddot{A}_n \Phi_n\left(x_j\right)\right] \end{array} \right\}
\end{array}
\right.
\tag{21}
$$

In the above equation, the total number of equations is N + 7, where N represents the modal equations of the road subsystem and generally takes the low-order vibration mode. The dynamic equilibrium equation will be solved by the Newmark-Beta method. The vibration equation of the standard vehicle–road system can be obtained by rearranging Equation (21):

$$[M]\left\{\ddot{Y}\right\}+[C]\left\{\dot{Y}\right\}+[K]\{Y\}=\{F\} \tag{22}$$

where $[M]$ represents the total mass matrix, $[C]$ represents the total damping matrix, $[K]$ represents the total stiffness matrix, and $[F]$ represents the total load matrix. The matrix expressions are as follows:

$$C^1=\begin{bmatrix} 2\xi_1\omega_1+\sum\limits_{j=1}^{4}\Phi_1(x_j)C_{tj}\Phi_1(x_j) & \sum\limits_{j=1}^{4}\Phi_1(x_j)C_{tj}\Phi_2(x_j) & \cdots & \sum\limits_{j=1}^{4}\Phi_1(x_j)C_{tj}\Phi_n(x_j) \\ \sum\limits_{j=1}^{4}\Phi_2(x_j)C_{tj}\Phi_1(x_j) & 2\xi_2\omega_2+\sum\limits_{j=1}^{4}\Phi_2(x_j)C_{tj}\Phi_2(x_j) & \cdots & \sum\limits_{j=1}^{4}\Phi_2(x_j)C_{tj}\Phi_n(x_j) \\ \vdots & \vdots & \ddots & \vdots \\ \sum\limits_{j=1}^{4}\Phi_n(x_j)C_{tj}\Phi_1(x_j) & \sum\limits_{j=1}^{4}\Phi_n(x_j)C_{tj}\Phi_2(x_j) & \cdots & 2\xi_n\omega_n+\sum\limits_{j=1}^{4}\Phi_n(x_j)C_{tj}\Phi_n(x_j) \end{bmatrix},$$

$$C^2=\begin{bmatrix} 0 & 0 & 0 & -C_{t1}\Phi_1(x_1) & -C_{t2}\Phi_1(x_2) & -C_{t3}\Phi_1(x_3) & -C_{t4}\Phi_1(x_4) \\ 0 & 0 & 0 & -C_{t1}\Phi_2(x_1) & -C_{t2}\Phi_2(x_2) & -C_{t3}\Phi_2(x_3) & -C_{t4}\Phi_2(x_4) \\ \vdots & \vdots & \vdots & \vdots & \vdots & \vdots & \vdots \\ 0 & 0 & 0 & -C_{t1}\Phi_n(x_1) & -C_{t2}\Phi_n(x_2) & -C_{t3}\Phi_n(x_3) & -C_{t4}\Phi_n(x_4) \end{bmatrix},$$

$$C^3=\begin{bmatrix} 0 & 0 & 0 & 0 \\ 0 & 0 & 0 & 0 \\ 0 & 0 & 0 & 0 \\ -C_{t1}\Phi_1(x_1) & -C_{t1}\Phi_2(x_1) & \cdots & -C_{t1}\Phi_n(x_1) \\ -C_{t2}\Phi_1(x_2) & -C_{t2}\Phi_2(x_2) & \cdots & -C_{t2}\Phi_n(x_2) \\ -C_{t3}\Phi_1(x_3) & -C_{t3}\Phi_2(x_3) & \cdots & -C_{t3}\Phi_n(x_3) \\ -C_{t4}\Phi_1(x_4) & -C_{t4}\Phi_2(x_4) & \cdots & -C_{t4}\Phi_n(x_4) \end{bmatrix},$$

$$C^4=\begin{bmatrix} c_{11} & c_{12} & c_{13} & -c_{s1} & -c_{s2} & -c_{s3} & -c_{s4} \\ c_{21} & c_{22} & c_{23} & \lambda_1 c_{s1} & -\lambda_2 c_{s2} & \lambda_1 c_{s3} & -\lambda_2 c_{s4} \\ c_{31} & c_{32} & c_{33} & \lambda_3 c_{s1} & \lambda_3 c_{s2} & -\lambda_4 c_{s3} & -\lambda_4 c_{s4} \\ -c_{s1} & \lambda_1 c_{s1} & \lambda_3 c_{s1} & c_{s1}+c_{t1} & 0 & 0 & 0 \\ -c_{s2} & -\lambda_2 c_{s2} & \lambda_3 c_{s2} & 0 & c_{s2}+c_{t2} & 0 & 0 \\ -c_{s3} & \lambda_1 c_{s3} & -\lambda_4 c_{s3} & 0 & 0 & c_{s3}+c_{t3} & 0 \\ -c_{s4} & -\lambda_2 c_{s4} & -\lambda_4 c_{s4} & 0 & 0 & 0 & c_{s4}+c_{t4} \end{bmatrix},$$

$$K^1=\begin{bmatrix} \omega_1^2+\sum\limits_{j=1}^{4}\Phi_1(x_j)K_{tj}\Phi_1(x_j) & \sum\limits_{j=1}^{4}\Phi_1(x_j)K_{tj}\Phi_2(x_j) & \cdots & \sum\limits_{j=1}^{4}\Phi_1(x_j)K_{tj}\Phi_n(x_j) \\ \sum\limits_{j=1}^{4}\Phi_2(x_j)K_{tj}\Phi_1(x_j) & \omega_2^2+\sum\limits_{j=1}^{4}\Phi_2(x_j)K_{tj}\Phi_2(x_j) & \cdots & \sum\limits_{j=1}^{4}\Phi_2(x_j)K_{tj}\Phi_n(x_j) \\ \vdots & \vdots & \ddots & \vdots \\ \sum\limits_{j=1}^{4}\Phi_n(x_j)K_{tj}\Phi_1(x_j) & \sum\limits_{j=1}^{4}\Phi_n(x_j)K_{tj}\Phi_2(x_j) & \cdots & \omega_n^2+\sum\limits_{j=1}^{4}\Phi_n(x_j)K_{tj}\Phi_n(x_j) \end{bmatrix},$$

$$K^2=\begin{bmatrix} 0 & 0 & 0 & -K_{t1}\Phi_1(x_1) & -K_{t2}\Phi_1(x_2) & -K_{t3}\Phi_1(x_3) & -K_{t4}\Phi_1(x_4) \\ 0 & 0 & 0 & -K_{t1}\Phi_2(x_1) & -K_{t2}\Phi_2(x_2) & -K_{t3}\Phi_2(x_3) & -K_{t4}\Phi_2(x_4) \\ \vdots & \vdots & \vdots & \vdots & \vdots & \vdots & \vdots \\ 0 & 0 & 0 & -K_{t1}\Phi_n(x_1) & -K_{t2}\Phi_n(x_2) & -K_{t3}\Phi_n(x_3) & -K_{t4}\Phi_n(x_4) \end{bmatrix},$$

$$K^3 = \begin{bmatrix} 0 & 0 & 0 & 0 \\ 0 & 0 & 0 & 0 \\ 0 & 0 & 0 & 0 \\ -K_{t1}\Phi_1(x_1) & -K_{t1}\Phi_2(x_1) & \cdots & -K_{t1}\Phi_n(x_1) \\ -K_{t2}\Phi_1(x_2) & -K_{t2}\Phi_2(x_2) & \cdots & -K_{t2}\Phi_n(x_2) \\ -K_{t3}\Phi_1(x_3) & -K_{t3}\Phi_2(x_3) & \cdots & -K_{t3}\Phi_n(x_3) \\ -K_{t4}\Phi_1(x_4) & -K_{t4}\Phi_2(x_4) & \cdots & -K_{t4}\Phi_n(x_4) \end{bmatrix},$$

$$K^4 = \begin{bmatrix} k_{11} & k_{12} & k_{13} & -k_{s1} & -k_{s2} & -k_{s3} & -k_{s4} \\ k_{21} & k_{22} & k_{23} & \lambda_1 k_{s1} & -\lambda_2 k_{s2} & \lambda_1 k_{s3} & -\lambda_2 k_{s4} \\ k_{31} & k_{32} & k_{33} & \lambda_3 k_{s1} & \lambda_3 k_{s2} & -\lambda_4 k_{s3} & -\lambda_4 k_{s4} \\ -k_{s1} & \lambda_1 k_{s1} & \lambda_3 k_{s1} & k_{s1}+k_{t1} & 0 & 0 & 0 \\ -k_{s2} & -\lambda_2 k_{s2} & \lambda_3 k_{s2} & 0 & k_{s2}+k_{t2} & 0 & 0 \\ -k_{s3} & \lambda_1 k_{s3} & -\lambda_4 k_{s3} & 0 & 0 & k_{s3}+k_{t3} & 0 \\ -k_{s4} & -\lambda_2 k_{s4} & -\lambda_4 k_{s4} & 0 & 0 & 0 & k_{s4}+k_{t4} \end{bmatrix},$$

$$\{F\} = \begin{Bmatrix} \sum_{j=1}^{4} \Phi_1(x_{ij})\left[\left(\frac{\lambda_j}{\lambda}m_s + m_{tj} + m_{lj}\right)g - K_{tj}z_p(x_j) - C_{tj}\dot{z}_p(x_j) - m_{lj}\ddot{Z}_p(x_j)\right] \\ \sum_{j=1}^{4} \Phi_2(x_{ij})\left[\left(\frac{\lambda_j}{\lambda}m_s + m_{tj} + m_{lj}\right)g - K_{tj}z_p(x_j) - C_{tj}\dot{z}_p(x_j) - m_{lj}\ddot{Z}_p(x_j)\right] \\ \vdots \\ \sum_{j=1}^{4} \Phi_n(x_{ij})\left[\left(\frac{\lambda_j}{\lambda}m_s + m_{tj} + m_{lj}\right)g - K_{tj}z_p(x_j) - C_{tj}\dot{z}_p(x_j) - m_{lj}\ddot{Z}_p(x_j)\right] \\ 0 \\ 0 \\ 0 \\ K_{t1}z_p(x_1) + C_{t1}\dot{z}_p(x_1) \\ K_{t2}z_p(x_1) + C_{t2}\dot{z}_p(x_1) \\ K_{t3}z_p(x_1) + C_{t3}\dot{z}_p(x_1) \\ K_{t4}z_p(x_1) + C_{t4}\dot{z}_p(x_1) \end{Bmatrix},$$

$$\{Y\} = \begin{Bmatrix} A_1 \\ A_2 \\ \vdots \\ A_n \\ Z_s \\ \psi \\ \theta \\ Z_{t1} \\ Z_{t2} \\ Z_{t3} \\ Z_{t4} \end{Bmatrix}, \{\dot{Y}\} = \begin{Bmatrix} \dot{A}_1 \\ \dot{A}_2 \\ \vdots \\ \dot{A}_n \\ \dot{Z}_s \\ \dot{\psi} \\ \dot{\theta} \\ \dot{Z}_{t1} \\ \dot{Z}_{t2} \\ \dot{Z}_{t3} \\ \dot{Z}_{t4} \end{Bmatrix}, \{\ddot{Y}\} = \begin{Bmatrix} \ddot{A}_1 \\ \ddot{A}_2 \\ \vdots \\ \ddot{A}_n \\ \ddot{Z}_s \\ \ddot{\psi} \\ \ddot{\theta} \\ \ddot{Z}_{t1} \\ \ddot{Z}_{t2} \\ \ddot{Z}_{t3} \\ \ddot{Z}_{t4} \end{Bmatrix}$$

$$[C] = \begin{bmatrix} C^1 & C^2 \\ C^3 & C^4 \end{bmatrix}, [K] = \begin{bmatrix} K^1 & K^2 \\ K^3 & K^4 \end{bmatrix}$$

It is difficult to see from the above matrices that the total mass matrix $[M]$, the total damping matrix $[C]$, and the total stiffness matrix $[K]$ change with the road position x (x_1, x_2, x_3, and x_4 represent the positions of the four wheels). Therefore, the vehicle road system coupling dynamic balance equations can be regarded as a highly second-order nonlinear differential equation. By solving the equation, the vertical vibration displacement Z_s, vertical velocity $\dot{Z}_s$, and vertical acceleration $\ddot{Z}_s$ of the vehicle body; the vertical vibration displacement Z_{t1}, Z_{t2}, Z_{t3}, and Z_{t4} of each unsprung mass part of the vehicle body, the vertical velocity $\dot{Z}_{t1}$, $\dot{Z}_{t2}$, $\dot{Z}_{t3}$, and $\dot{Z}_{t4}$; and the vertical acceleration $\ddot{Z}_{t1}$, $\ddot{Z}_{t2}$, $\ddot{Z}_{t3}$ and $\ddot{Z}_{t4}$ can be obtained. At the same time, the nodal displacement θ, nodal velocity $\dot{\theta}$, nodal acceleration $\ddot{\theta}$, and deflection displacement ψ of the vehicle body can also be obtained using the deflection velocity $\dot{\psi}$ and deflection acceleration $\ddot{\psi}$.

The vehicle is regarded as a multi-stiffness system. According to D'Alembert's principle, considering the dynamic balance conditions of each rigid body in the system, the road vibration balance equation and the vehicle dynamic balance equation are listed according to the coupling model under the contact force condition in Figure 8.

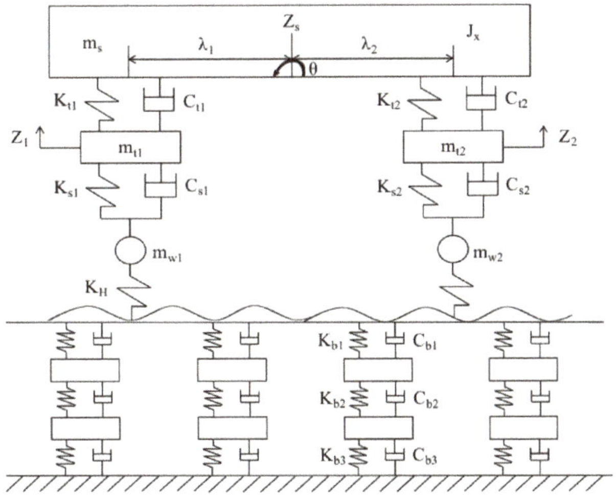

Figure 8. Coupling vibration relation model under contact force.

The dynamic balance equation of the vehicle–road system is obtained by combining Equation (20). The total number of equations is N + 7, where N represents the number of modal equations of the path sub-model.

$$[M]\{\ddot{Y}\}+[C]\{\dot{Y}\}+[K]\{Y\}=\{F\} \tag{23}$$

The Newmark-β and Park integral methods are used to solve the above equations, and the responses of the vehicle and road in the time domain, such as the force, displacement, and vibration velocity, can be obtained [27].

4. Theoretical Analysis of Vehicle–Road System

According to the vehicle–road coupling dynamic equation and the dynamic balance equation listed above, the following vehicle driving mass and road design parameters can be obtained using the Newmark-β method in a self-programmed program. It is assumed that the heavy vehicle has a mass of 20 t and runs at a speed of 80 km/h on a class B road. The road structure adopts planar 8-node PLANE82 unit, and the road model takes 100 m along the road longitudes. The calculated step time is 0.02. See Appendices A and B for detailed parameters.

4.1. Maximum Deflection of Pavement

According to the relevant parameters, an instantaneous dynamics calculation is carried out in this paper. Figure 9 shows the curve of deflection value change at a node of road surface under a moving load. Table 1 shows the deflection values of each layer of the road structure in the vertical direction. It can be seen that the bending value of the surface layer is the largest, and the bending value of the surface layer and surface layer changes relatively little. When the subgrade is used, the bending settlement value changes greatly, and the fundamental reason is that the material properties of subgrade soil differ greatly from the surface layer, base layer, and cushion layer. The smaller the elastic modulus of the material is, the larger the vertical deformation is. As the bottom layer of the soil foundation is set as a constraint in the calculation, its displacement is zero.

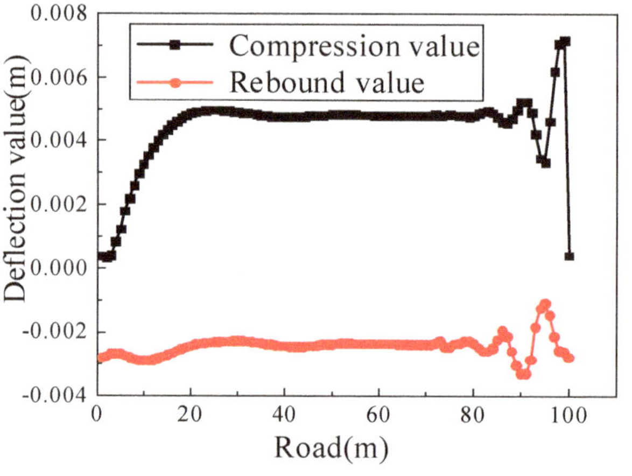

Figure 9. Variation curve of pavement deflection value.

Table 1. Deflection value of road structure in the vertical direction.

Location of Structural Layer	Deflection Value (mm)	Location of Structural Layer	Deflection Value (mm)
Surface course	2.37	Surface subgrade	2.34
Base course	2.35	Upper part of the subgrade	1.93
Subbase course	2.34	Middle part of the subgrade	1.06
Bed course	2.34	Bottom part of the subgrade	0

4.2. Stress State Analysis of Pavement

4.2.1. Tensile Stress Analysis

The surface of the asphalt pavement undergoes deflection changes under the coupled dynamic load of the vehicle. At the same time, the road structure layer also undergoes bending and tensile stress. Under repeated vehicle loads, this bending and tensile stress will cause bending and tensile fatigue damage to the asphalt road structure.

Figure 10 shows the time–history curve of the maximum tensile stress at each node of each structural layer on the road. It can be seen from the figure that the tensile stress on the road is sometimes positive and sometimes negative, but it is the maximum at the road surface course and the base course.

Figure 11 shows the horizontal stress nephogram at the position of the maximum tensile stress at each node of the road structure layer. From this figure, it can be seen that the maximum tensile stress mainly occurs in the road surface course at the vehicle load and the combination part of the road surface course, bed course, and subgrade at the load, while the combination part of the road surface layer and the cushion is the largest. This shows that each structural layer of the road is relatively easy to damage in these parts. For the tensile stress of the road structure, it is easy to cause the cracking of road pavement on the surface course. However, there is a large tensile stress at the joint between the structural layers, which will increase the wear between the structural layers and directly lead to the damage of the road.

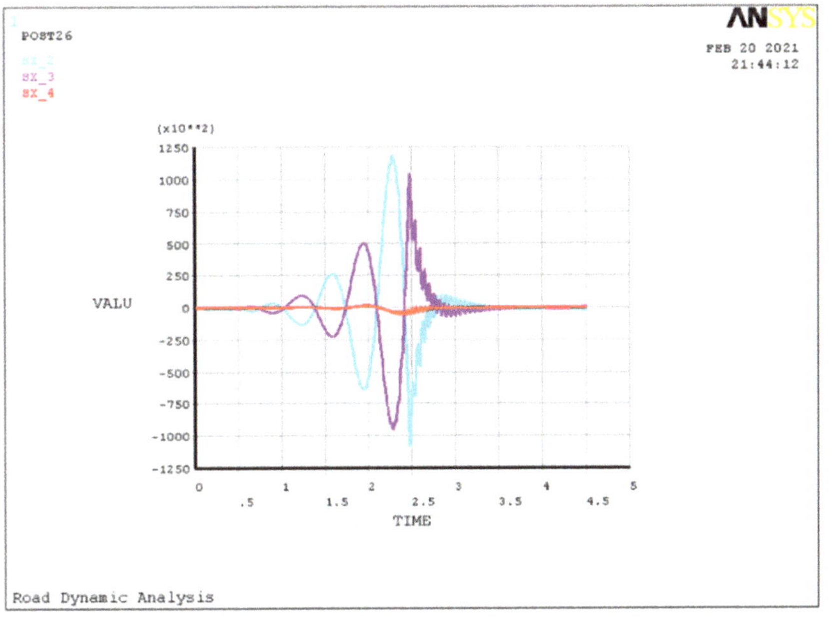

Figure 10. Time–history curve of maximum tensile stress at each node of road structure layer. (Note: $\times 10^{**}2$ means $\times 10^2$, .5 means 0.5).

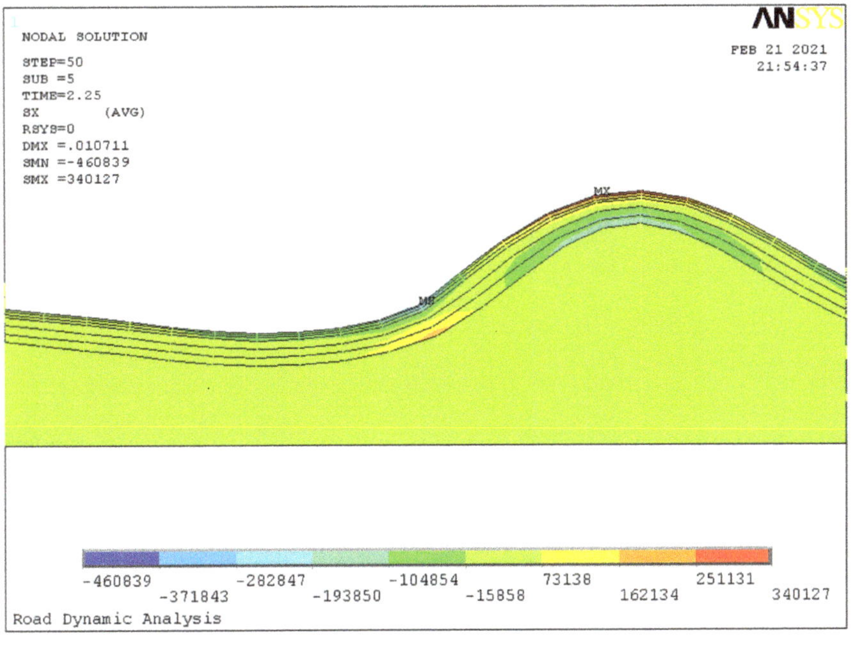

Figure 11. Horizontal stress nephogram at the position of maximum tensile stress. (Note: .010711 means 0.010711).

4.2.2. Shear Stress Analysis

According to the existing service conditions of the highway, the asphalt pavement is mainly damaged by shear stress. In asphalt pavement, it is easy to cause the shear failure of asphalt structural layers, especially under high temperatures. Shear failure is one of the most common problems with asphalt pavements.

Figure 12 shows the time–history variation diagram of the horizontal shear stress of each node in the pavement structure layer under the action of the coupled dynamic load. It can be seen from the figure that the horizontal shear stress of the road surface is the most obvious, while the horizontal shear stress of the base and subgrade surface is relatively small, indicating that the horizontal sheer force of the asphalt road structure mainly has an effect on the asphalt road surface, which reminds us to pay special attention to the shear condition of the road surface in the design and construction process. At the same time, it can be seen that the horizontal shear stress can be positive or negative within the range of influence, and its value is approximately symmetric.

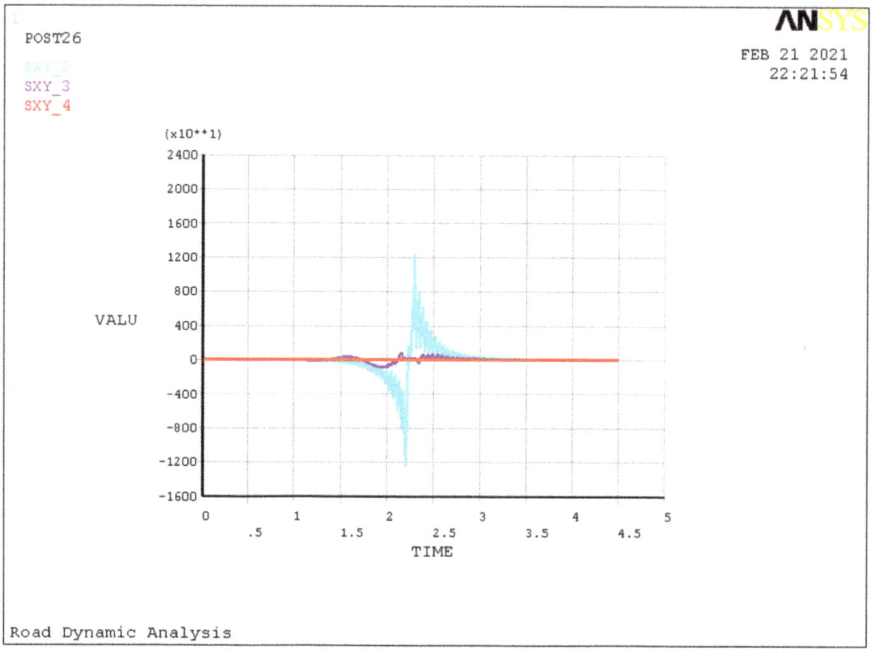

Figure 12. Time–history diagram of shear stress. (Note: $\times 10^{**}1$ means $\times 10^1$, .5 means 0.5).

Figure 13 shows the variation in the shear stress of nodes in the pavement structure layer at the midpoint of the road, along with the depth of the road section. According to the data in the figure, the proportion of cushion and roadbed with horizontal shear stress below 1 m is almost zero. The shear stress of asphalt pavement mainly occurs in the road surface and the base layer, the maximum shear stress is in the middle of the surface layer, and the value of shear stress is 43,858 Pa.

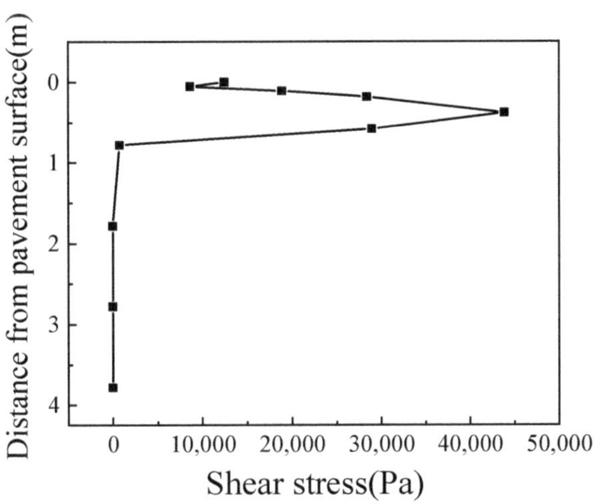

Figure 13. Diagram of shear stress variation with road depth.

4.3. Vertical Dynamic Stress Analysis

Figure 14 shows the variation of road vertical dynamic stress with depth when the vehicle travels to about 50 m. It can be seen from the figure that, like the horizontal shear stress, the value of the vertical dynamic stress in the number of road surface layers and the base layer is larger, and the dynamic stress in the middle of the surface layer is 119,373 Pa. The value of vertical dynamic stress in the road subgrade is much smaller, and the vertical dynamic stress in the middle of the subgrade is 5824 Pa. In the vertical direction of the road, the vertical dynamic stress of each layer of the road structure decreases rapidly. In contrast, the dynamic stress on the surface of the asphalt base course is 0.602 times the dynamic stress on the surface course, and when it reaches the bottom of the subbase, the dynamic stress decreases to 0.038 times the surface stress.

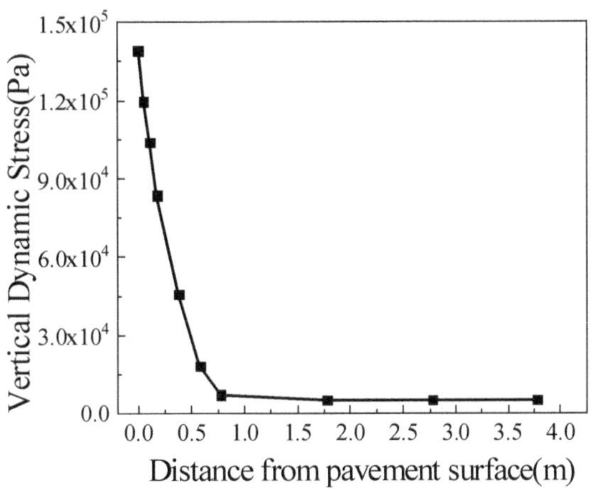

Figure 14. Diagram of vertical dynamic stress variation with road depth.

Figure 15 shows the time–history variation diagram of the vertical dynamic stress of each layer. It can be seen from the diagram that the vertical dynamic stress of road is sometimes positive and sometimes negative, and the positive value shows tension while the negative value shows compression, and each value shows the maximum value when the asphalt road surface course.

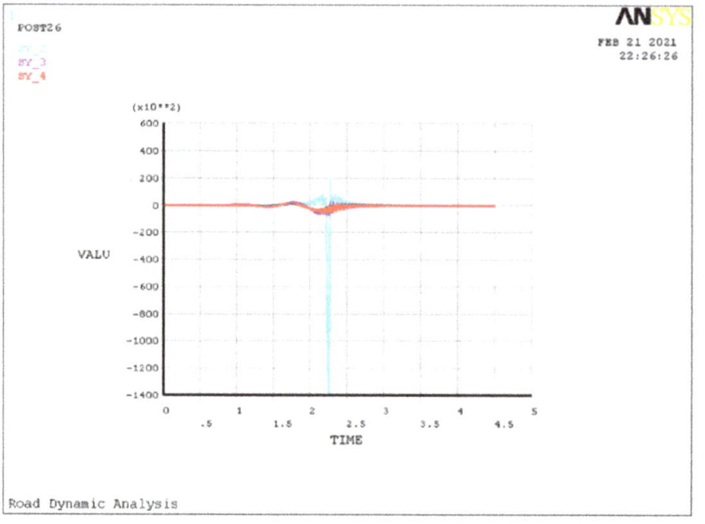

Figure 15. Diagram of time–history variation of vertical dynamic stress of each layer. (Note: ×10**2 means $\times 10^2$, .5 means 0.5).

Figure 16 shows the distribution of vertical dynamic stress on asphalt roads. It can be seen from the figure that vertical dynamic stress is mainly concentrated in asphalt road surface and base, within a relatively small range of the whole road length. Due to the alternating action of vertical dynamic stress tension and compression, the damage of road surface is accelerated.

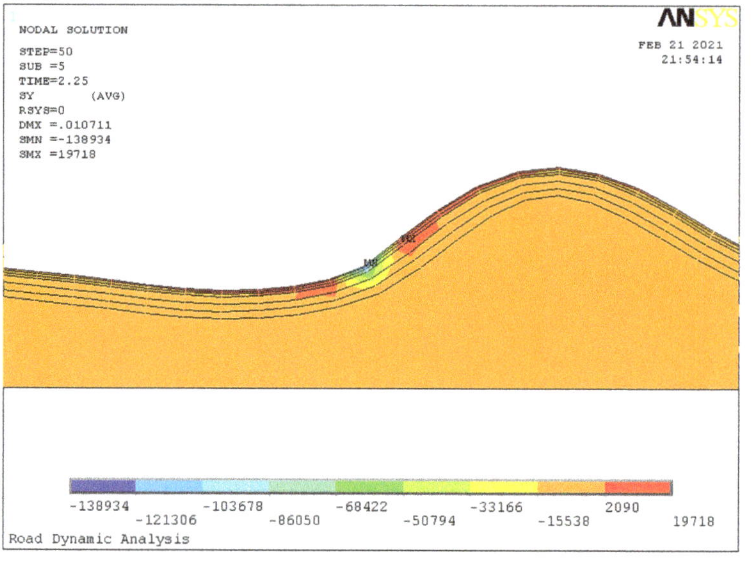

Figure 16. Vertical dynamic stress nephogram. (Note: .010711 means 0.010711).

4.4. Stress Analysis of Pavement Considering Tire Starting Force and Braking Force

When the vehicle is running on the road, in addition to the dynamic load in the vertical direction of the vehicle, it also receives the starting force and braking force in the horizontal direction when the vehicle starts and brakes. In fact, when the vehicle is driving on the road, the horizontal starting and braking force are the reasons for the large shear stress on the road surface. China's urban road design code also requires us to consider the horizontal force applied to the road when the vehicle's braking starts.

The horizontal shear stress change mileage curve of the road under the action of vehicle vertical coupling dynamic load, horizontal starting force, and braking force is shown in Figures 17 and 18. The horizontal shear stress of surface course and the base course is obvious during vehicle starting and braking. Compared with normal conditions, the horizontal shear stress of the road during starting and braking is larger. This is not only the main factor affecting vehicle starting and braking but also a factor affecting road service life.

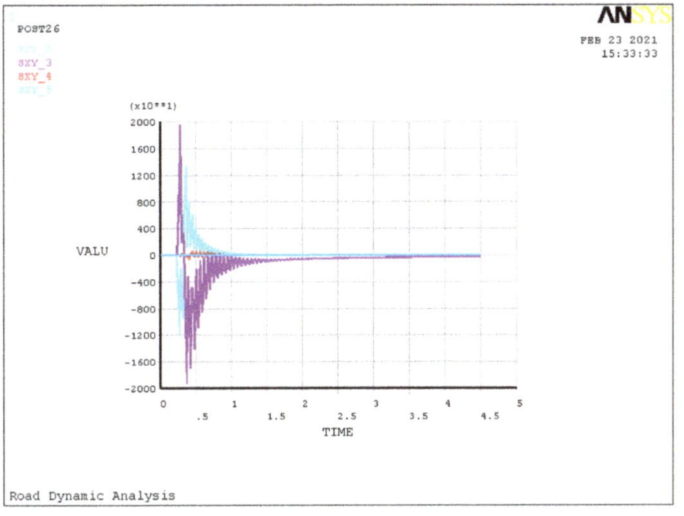

Figure 17. Diagram of shear stress time–history at startup. (Note: $\times 10^{**}1$ means $\times 10^1$, .5 means 0.5).

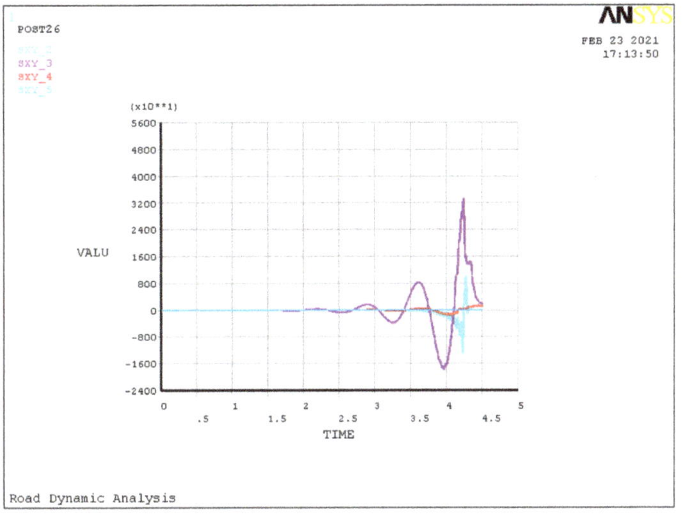

Figure 18. Diagram of shear stress time–history during braking. (Note: $\times 10^{**}1$ means $\times 10^1$, .5 means 0.5).

In the process of vehicle starting and braking, the road surface course and the base course are stressed the most, and the shear stress of the base course is larger than that of the surface course, which is more obvious when the vehicle is braking. From the perspective of the vertical road displacement, the vehicle can affect the front of the road when it is started, and the vehicle can still affect the back road when it is braked, as shown in Figures 19 and 20.

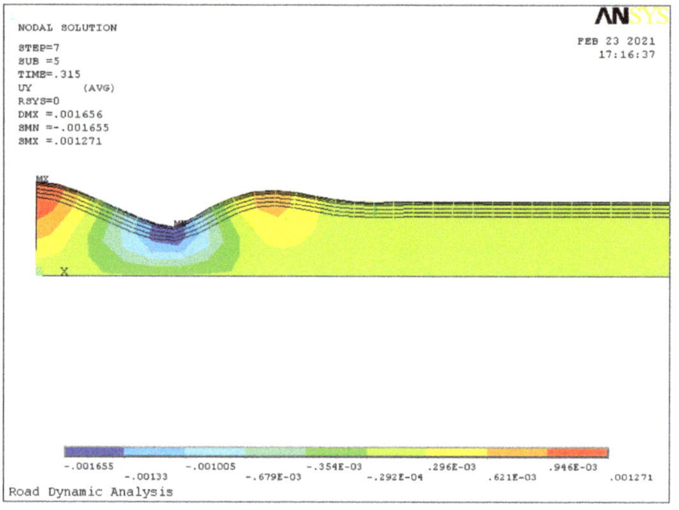

Figure 19. Vertical displacement nephogram at startup. (Note: .315 means 0.315, and so on with the other numbers).

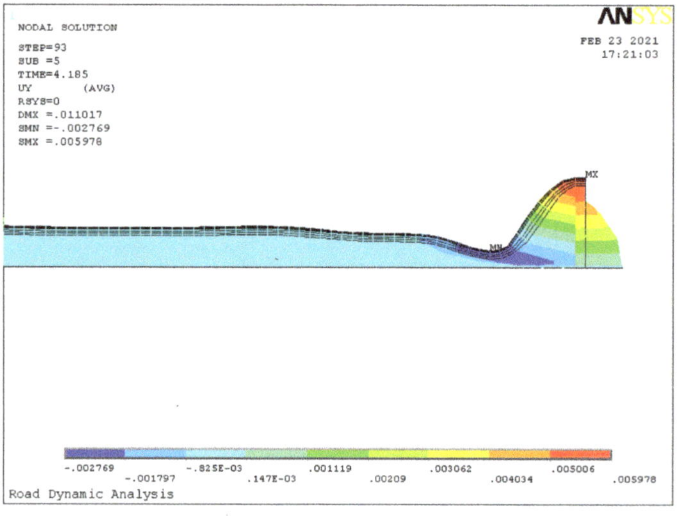

Figure 20. Vertical displacement nephogram during braking. (Note: .011017 means 0.011017, and so on with the other numbers).

5. Conclusions

The heavy-duty vehicle road coupling model not only adopts the vibration coupling relationship under the condition of displacement compatibility between vehicle and road but also tries to adopt the vibration coupling relationship under the condition of contact force between vehicle and road. Further research is conducted on the coupling relationship, which provides a new idea for the future development of a vehicle–road coupling vibration model. This paper studies the vehicle–road coupling dynamic analysis model and vehicle–road interaction analysis under heavy load conditions, which will play a significant role in the whole design method and theory of vehicles and roads and will also have very important application value to future engineering practice.

In this paper, the subgrade is regarded as part of the vehicle road dynamic system, and the vehicle road coupling dynamic model under heavy load is established. The results show that the equation can better reflect the dynamic performance of a vehicle–road system; Additionally, it better reflects the coupling vibration relationship and interaction between the vehicle and the road. The model will provide a theoretical basis for a more accurate analysis of the interaction between heavy-duty vehicles and roads and then provide a theoretical basis for important parameters in heavy-duty vehicle design and road durability design. The coupling dynamic equation established in this paper can analyze the instantaneous dynamic performance of the coupling system by the finite element method.

The two-dimensional finite element model is used in the analysis of road coupled dynamic load in this paper, and the three-dimensional finite element model can be used to simulate the spatial dynamic response of the road. This will be the focus of future research.

Author Contributions: J.X. wrote the manuscript, B.L. designed the proposed control strategy and conceived the structure of the paper, S.S. completed theoretical derivation and numerical simulation. All authors have read and agreed to the published version of the manuscript.

Funding: This research was funded by the Chongqing Science and technology research project under Grant No. CSTC2010AC6078.

Institutional Review Board Statement: Not applicable.

Informed Consent Statement: Not applicable.

Data Availability Statement: Not applicable.

Conflicts of Interest: The authors declare no conflict of interest.

Appendix A

Table A1. Heavy vehicle parameters.

Symbols of Vehicle Parameters	Value
Body Mass (m_s/kN)	180
Inertial mass (x-axis) (J_X/kN·m^2)	20,000
Inertial mass (y-axis) (J_Y/kN·m^2)	7200
Rear non-suspension mass (m_{t2}/kN)	15.55
Front non-suspension mass (m_{t1}/kN)	3.85
Wheel weight (m_{l1}/kN)	0.15
Vehicle net weight (G/kN)	200
Rear suspension stiffness coefficient (K_{S2}/kN·m^{-1})	4800
Front suspension stiffness coefficient (K_{S1}/kN·m^{-1})	1200
Rear suspension damping coefficient (C_{S2}/kN·m^{-1}·s^{-1})	20
Front suspension damping coefficient (C_{S1}/kN·m^{-1}·s^{-1})	5
Rear-wheel stiffness coefficient (K_{t2}/kN·m^{-1})	9600

Table A1. *Cont.*

Symbols of Vehicle Parameters	Value
Rear-wheel damping coefficient (C_{t2}/kN·m^{-1}·s^{-1})	24
Front-wheel damping coefficient (C_{t1}/kN·m^{-1}·s^{-1})	6
The distance from the center of gravity of the car to the rear axle (λ_2/m)	1.5
The distance from the center of gravity of the car to the front axle (λ_1/m)	2.5
The distance from the center of gravity of the car to the right wheel (λ_4/m)	0.9
The distance from the center of gravity of the car to the left wheel (λ_3/m)	0.9

Appendix B

Table A2. The Parameter of Road Layers.

Material Parameters		Thickness (cm)	Elastic Modulus (MPa)	Poisson's Ratio	Density (Kg·m^{-3})
Surface course	SAC-16	5	1400	0.25	2500
	AC-20	6	1200	0.25	2500
	AC-25	7	1000	0.25	2500
Base course	6% cement stabilized macadam	20	1000	0.25	2400
Subbase course	5% cement stabilized macadam	20	900	0.25	2400
Bed course	4% cement stabilized macadam	20	600	0.25	2300
Subgrade		300	60	0.4	1900

References

1. Jo, Y.; Oh, C.; Kim, S. Estimation of heavy vehicle-involved rear-end crash potential using WIM data. *Accid. Anal. Prev.* **2019**, *128*, 103–113. [CrossRef] [PubMed]
2. Li, Q.; Liu, J.Q. Asphalt pavement evenness deterioration analysis based on the vehicle–pavement interaction. *J. Vib. Shock* **2018**, *37*, 76–81, discussion 116.
3. Lyu, Z.; Qian, J.G.; Shi, Z.H.; Gao, Q. Dynamic responses of layered poroelastic ground under moving traffic loads considering effects of pavement roughness. *Soil Dyn. Earthq. Eng.* **2020**, *130*, 105996. [CrossRef]
4. Walter, V.W. New Resonances and Velocity Jumps in Nonlinear Road-vehicle Dynamics. *Procedia IUTAM* **2016**, *19*, 209–218.
5. Krishnanunni, C.G.; Rao, B.N. Decoupled technique for dynamic response of vehicle–pavement systems. *Eng. Struct.* **2019**, *191*, 264–279. [CrossRef]
6. Zhang, J.H.; Guo, P.; Lin, J.W.; Wang, K.N. A mathematical model for coupled vibration system of road vehicle and coupling effect analysis. *Appl. Math. Model.* **2016**, *40*, 1199–1217. [CrossRef]
7. Zhang, F.; Feng, D.C.; Ling, X.Z.; Li, Q.L. Vertical Coupling Dynamic Model of Heavy Truck-pavement-subgrade. *China J. Highw. Transp.* **2015**, *28*, 1–12.
8. Xu, H.L.; Yuan, Y.; Qu, T.J.; Li, Q.L. Dynamic model for a vehicle–pavement coupled system considering pavement roughness. *J. Vib. Shock* **2014**, *33*, 152–156.
9. Liang, B.; Luo, H.; Ma, X.N. Dynamic model of vertical vehicle-subgrade coupled system under secondary suspension. *Appl. Math. Mech.* **2007**, *28*, 769–778. [CrossRef]
10. Shi, S.R. Coupled Dynamic Analysis Model of Vehicle and Road under Condition of Heavy Vehicle and Interaction Analysis of Vehicle and Road. Master's Thesis, Chongqing Jiaotong University, Chongqing, China, 2012.
11. Yang, Y.; Lu, H.; Tan, X.; Chai, H.K.; Wang, R.; Zhang, Y. Fundamental mode shape estimation and element stiffness evaluation of girder bridges by using passing tractor-trailers. *Mech. Syst. Signal Process.* **2022**, *169*, 108746. [CrossRef]
12. Cong, L.; Yang, F.; Guo, G.H.; Ren, M.D.; Shi, J.C.; Tan, L. The use of polyurethane for asphalt pavement engineering applications: A state-of-the-art review. *Constr. Build. Mater.* **2019**, *225*, 1012–1025. [CrossRef]
13. Yang, Y.; Zhang, Y.; Tan, X. Review on Vibration-Based Structural Health Monitoring Techniques and Technical Codes. *Symmetry* **2021**, *13*, 1998. [CrossRef]
14. Gui, S.R.; Chen, S.S.; Wan, S. Sensitivity Analysis of Vehicle-Bridge Coupling Random Vibration Based on Road Roughness Spectral Function. *J. Vib. Meas. Diagn.* **2018**, *38*, 353–359, discussion 422–423.
15. Li, H.Y.; Yang, S.P.; Li, S.H. Dynamical analysis of an asphalt pavement due to vehicle–road interaction. *J. Vib. Shock* **2009**, *28*, 86–89; discussion 102–205.

16. He, Y.; Lu, X.Y.; Chu, D.F.; Wu, C.Z. Reliability Estimation of Vehicle Lateral Dynamic Under Vehicle-road-environment Coupling Actions. *Automot. Eng.* **2019**, *41*, 800–806.
17. Ding, H.; Yang, Y.; Chen, L.Q.; Yang, S.H. Vibration of vehicle–pavement coupled system based on a Timoshenko beam on a nonlinear foundation. *J. Sound Vib.* **2014**, *333*, 6623–6636. [CrossRef]
18. Yang, S.H.; Li, S.H.; Lu, Y.J. Investigation on dynamical interaction between a heavy vehicle and road pavement. *Veh. Syst. Dyn.* **2010**, *48*, 923–944. [CrossRef]
19. Zhang, J.N.; Yang, S.H.; Li, S.H.; Ding, H.; Lu, Y.J.; Si, C.D. Study on crack propagation path of asphalt pavement under vehicle–road coupled vibration. *Appl. Math. Model.* **2021**, *101*, 481–502. [CrossRef]
20. Yang, Y.; Yang, L.; Wu, B.; Yao, G.; Li, H.; Robert, S. Safety Prediction Using Vehicle Safety Evaluation Model Passing on Long-Span Bridge with Fully Connected Neural Network. *Adv. Civ. Eng.* **2019**, *2019*, 8130240. [CrossRef]
21. Wang, J.; Li, T.J.; Meng, L.Q. Study on Dynamic Characteristics of Vehicle Based on the Whole Vehicle Model with 7 Freedom. *J. Anhui Sci. Technol. Univ.* **2013**, *27*, 72–76.
22. Liu, J.B.; Du, X.L. *Structural Dynamics*, 1st ed.; China Machine Press: Beijing, China, 2004; pp. 264–280.
23. Deng, X.J. *Road Bed &Road Surface Project*, 1st ed.; China Communication Publishing: Beijing, China, 2001; pp. 27–55.
24. Lu, Y.J.; Yang, S.H.; Li, S.H.; Chen, L.Q. Numerical and experimental investigation on stochastic dynamic load of a heavy-duty vehicle. *Appl. Math. Model.* **2010**, *34*, 2698–2710. [CrossRef]
25. Othman, M.I.A.; Said, S.; Marin, M. A novel model of plane waves of two-temperature fiber-reinforced thermoelastic medium under the effect of gravity with three-phase-lag model. *Int. J. Numer. Methods Heat Fluid Flow* **2019**, *29*, 4788–4806. [CrossRef]
26. Marin, M.; Othman, M.I.A.; Seadawy, A.R.; Carstea, C. A domain of influence in the Moore–Gibson–Thompson theory of dipolar bodies. *J. Taibah Univ. Sci.* **2020**, *14*, 653–660. [CrossRef]
27. Wu, B.; Zhang, L.L.; Yang, Y.; Liu, L.J.; Ni, Z.J. Refined Time-domain Buffeting Analysis of a Long-span Suspension Bridge in Mountainous Urban Terrain. *Adv. Civ. Eng.* **2020**, *2020*, 4703169. [CrossRef]

symmetry

MDPI

Article

Seismic Mitigation of Curved Continuous Girder Bridge Considering Collision Effect

Zhengying Li *, **Shaobo Kang** and **Chuan You**

MOE Key Laboratory of New Technology for Construction of Cities in Mountain Area, School of Civil Engineering, Chongqing University, Chongqing 400030, China; kang0119@cqu.edu.cn (S.K.); 20076250@cqu.edu.cn (C.Y.)

* Correspondence: lizhengy@cqu.edu.cn

Abstract: Due to structural irregularity, curved bridgesaremore likely to cause non-uniform collisions and unseating between adjacent components when subjected to earthquakes. Based on the analysis of the collision response of curved bridges duringearthquakes, and according to the seismic characteristics of curved bridges, research was carried out on pounding mitigation and unseating prevention measures. A curved bridge with double column piers was taken as an engineering example, and a finite element model of curved bridges thatcould consider the non-uniform contact collision between adjacent components was built with ABAQUS software. Viscoelastic dampers, viscous dampers, and a lead rubber bearing were selected as the damping devices, and a steel wire rope-rubber mat was used as the pounding mitigation device to form the combinatorial seismic mitigation system. Based on the principle of energy dissipation combined with constraints, three kinds of combined seismic mitigation case were determined; a seismic response analysis was then performed. The results indicated that the three kinds of combined seismic case were effective atreducing the response topounding force, stress, damage, girder torsion and displacement, and achieved the goals of seismic mitigation and unseating prevention.

Keywords: curvedcontinuous girder bridge; collision response; seismic mitigation; pounding mitigation and unseating prevention

Citation: Li, Z.; Kang, S.; You, C. Seismic Mitigation of Curved Continuous Girder Bridge Considering Collision Effect. *Symmetry* **2022**, *14*, 129. https://doi.org/10.3390/sym14010129

Academic Editor: Jan Awrejcewicz

Received: 10 December 2021
Accepted: 6 January 2022
Published: 11 January 2022

1. Introduction

Over the past decades, curved continuous girder bridges have been widely used in transportation networks that featuregeometric restrictions and constraints on site space, such as complicated interchanges and river crossings. However, curved bridges are vulnerable to earthquakes. Research in previous studies showed that curved bridges could sustain severe damage due to the coupling between bending and torsional forces or the displacement caused by complex vibrations [1–3]. The combination of their horizontal curvature and the irregularity of their adjacent segments means that curved bridges feature complex dynamic characteristics, which cause non-uniform collisions along the contact surface [4,5]. In addition, large pounding forces during collisions can amplify the relative displacement, which may lead to girder unseating and the collapse of the bridge. Chiyu Jiao et al. [6] conducted a shake table experiment on a 1/25-scale curved bridge model to investigate the influence of collision between adjacent girders on the seismic response of bridges. It was found that the collision was non-uniform along the contact surface, and the girder-to-girder collision could induce significantly large in-plane rotation of the adjacent bridges, which could substantially increase the global displacement demands of the bridges.

Damage to curved bridges is mainly due to pounding between girders or between girders and the abutment, aswas commonly observed during past major earthquakes [7]. Damage reports oncurved bridges during the 1995 Hyogo-Ken Nanbu earthquake in Japan indicated that pounding can lead to local damage and the collapse of bridge decks [8].

The failure of girder ends and bearing damage due to the pounding of adjacent simply-supported spans were reported after the 2001 Bhuj earthquake in Gujarat, India [9].The collapse of curved bridges, such as Baihua bridge, wasreported as a consequence of the 2008 Wenchuan earthquake. It was noted that the collapse only occurred in one of the curved segments of the 18span bridge while the other straight segments remained in location [10]. The destruction of bridges was also observed in the 2010 Chile earthquake and in the 2011 Christchurch earthquake [11,12]; most of the bridge damage, including the phenomenon of unseating, was caused by collisions between adjacent structures, especially the damage to curved bridges.

Reports from past major earthquakes and research findings from analytical and experimental researches highlight the seismic vulnerability of curved bridges and the seismic pounding impact on bridges. Therefore, seismic mitigation and collision prevention measures on curved Bridges have been highly valued in academic and engineering circles.

A variety of seismicmitigationtechnologieshave been proposed for bridges, including seismic isolation bearing, passive energy dissipation devices, tuned damper, and semi-active damping devices. Seismic isolation devices, such as rubber or lead rubber bearings, have been used to reduce seismic forces [13–16]. However, the use of lead rubber bearings can lead to large displacements, which consequently increase the possibility of pounding between adjacent segments or even unseating damage. In order to mitigate possible pounding and unseating damage, some researchers recommended the installation of rubber shock absorbers and restrainers. A few researchers have investigated the effectiveness of using steel restrainers [17–21] or Shape Memory Alloy (SMA) [22] and rubber bumpers together to mitigate pounding and unseating damage between adjacent decks. Felix [17] performeda comparative analysis of curved viaducts with cablerestrainers and different isolation bearings, and pointed out that cablerestrainers could reduce the probability of girder unseating. Raheem [19] conducted numerical studies on an isolated bridge with cablerestrainers and a natural rubber shock absorber. It was found that the use of rubber shock absorbers at the expansion gaps can significantly reduce the pounding forces, since the absorbers can reduce the impact stiffness and the cablerestrainers can effectively prevent girder unseating. Chiyu Jiao [21] performed a shake table experiment on a curved bridge model with a pounding buffer zone made of a natural rubber pad or aluminum foam at the expansion joint location to mitigate the pounding effect. The results show that the buffer zones of rubber and aluminum can effectively reduce seismic impact forces, and hence alleviate the localized damage. Meanwhile, the application of viscous dampers or viscoelastic dampers to bridges has received significant attention in recent years [23–26]. Due to the large resistance force and energy dissipation capacity, viscous dampers offer wide application prospects [27]. The main advantage of viscous dampers is that they are only activated during earthquakes, and theyshow no resistance force under slow relative segment movement.

Although there are some studieson seismic mitigation and unseating prevention, relatively few studies of curved continuous girder bridges have comprehensively considered the measures of seismic energy dissipation and pounding or unseating prevention together. In order to improve the seismic performance of curved girder bridges, it is necessary to carry out further research on the reasonable combination of seismicmitigationtechnologies and pounding mitigation measures together. For this purpose, this paper presents acomparative numerical analysis of the impact of different combinations of unseating restrainer with different isolation bearings or dampers on the dynamic response of the curved continuous girder bridge to seismic shock. Based on the principle of energy dissipation combined with constraints, this paper optimizes the combination of unseating restrainers and damping devices to form three seismic mitigation strategies for the curved continuous girder bridge. Athree-dimensional non-linear model of anentire bridge structural system is established; the model includes adjacent bridge superstructures with different sizes and the presence of expansion joints, as well as considering the unbalanced distribution of pounding forces across the contact surface and the nonlinear characteristics of bearings and dampers. Furthermore, the seismic response of the bridge structural systems is studied by dynamic nonlinear analysis. This study sheds some light on the benefits and limitations of restraining devices and damping or isolation devices when acting in combination. The results presented could assist bridge engineers in selecting damping devices to effectively mitigate damage tothis type of bridge structure.

2. Bridge Analysis Model

The bridge model takes the middle bridge section of the Jinjiang Interchange of Lipan Highway as an example. The schematic plan is shown in Figure 1. The curved ramp bridge has two joints, four spans, and three expansion gap joints, and the width of expansion joint is 50 mm. In order to analyze the collision of the curved bridge during anearthquake, three expansion joints in the original bridge are retained and abutments are set at both ends. In the figure, the first span is between the right abutment and the pier ①, and the second, third, and fourth spans are from the pier ① to the left abutment in turn. The span length of the first, second, and third spans is 18.33 m, respectively, and the length of the fourth span is 26.18 m. The radius of the curvature of the curved bridge is 50 m.The constraint conditions of the bridge support are shown in Figure 1, and the original structure of the bridge uses ordinary rectangular plate rubber bearings. The superstructure of the bridge uses cast-in-placereinforced concrete box girders, and the substructure piers are double-columnreinforced concrete round piers with a height of 20 m. The specific sections are shown in Figure 2.

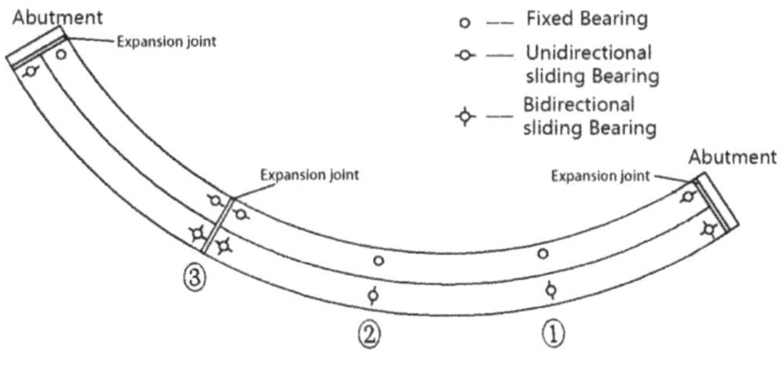

Figure 1. Plan of the curved bridge (unit: m).

(a) (b)

Figure 2. Cross-sectional dimensions of curved bridge (unit: mm). (**a**) Transverse profile. (**b**) Longitudinal elevation.

2.1. Bridge Main Structure Model

This paper establishes afinite element model with ABAQUS software to carry out a structural dynamic nonlinear time-history analysis, considering the non-uniform collision response between the expansion joints of the curved continuous girder bridge and the nonlinearity of the piers, damping devices and supports.

The superstructure of the bridge is a reinforced concrete box girder. Because of the asymmetrical collision response between the expansion joints, the use of beam elements is not applicable. Considering the solid element integral model, the three-dimensional solid reduction integral element C3D8R is used to simulate the bridge deck [28]. In order to better simulate the collision response without excessive calculation workload, the bridge deck main girder uses the multi-scale method to divide the grid [29]. Refining the grid at the expansion joint, the grid size near the expansion joint is 0.18 m, and the grid size at other parts is 0.50 m. The pier adopts a separated model; the three-dimensional solid reduction integral element C3D8R is used to simulate the concrete; the two-node bar element T3D2 is used to simulate the reinforcement [28].

The concrete intensive grade is C40, and the concrete damage plasticity model is used as the constitutive model [30], which describes the stiffness degradation of concrete during an earthquake through the tensile and compressive damage factors. For C40 concrete, the modulus of elasticity is 3.25×10^4 N/mm^2, the density is 2400 Kg/m^3, and the Poisson's coefficient recommended is 0.2. In the plastic phase of concrete material, the stress-strain relationship and damage factor of the constitutive model are shown in Table 1. Furthermore, the characteristics of the longitudinal and transverse reinforcement bars are defined based on the bilinear elastic-plastic model shown in Figure 3. The elastic modulus of the longitudinal reinforcement and transverse reinforcement bars is 2.0×10^5 N/mm^2 and 2.1×10^5 N/mm^2, respectively, and the yield strengths are 360 N/mm^2 and 270 N/mm^2, respectively. The limit strain is recommended to be 0.1; the density is 7850 Kg/m^3, and the Poisson's coefficientis 0.3. At the strengthening stage, the elastic modulus E_h is 1/100 of the initial elastic modulus E_s.

Table 1. Calculation parameters of concrete damage plasticity model.

Compressive Strength (N/mm^2)	Inelastic Strain	Compressive Damage Factor	Tensile Strength (N/mm^2)	Cracking Strain	Tensile Damage Factor
15.5934	0.000000	0.0000	3.0042	0.000000	0.0000
17.6459	0.000017	0.0151	1.2476	0.000196	0.5901
19.5436	0.000040	0.0305	0.7419	0.00033	0.7419
23.1160	0.000104	0.0643	0.5406	0.000454	0.8092
24.3117	0.000136	0.0778	0.4326	0.000575	0.8473
25.5944	0.000177	0.0941	0.3646	0.000695	0.8721
27.7023	0.000275	0.1274	0.3176	0.000814	0.8895
29.2035	0.000393	0.1620	0.2829	0.000933	0.9024
29.5360	0.000431	0.1726	0.2561	0.001051	0.9125
30.1015	0.000529	0.1978	0.2347	0.00117	0.9205
30.4000	0.000684	0.2352	0.2026	0.001406	0.9326
30.0218	0.000861	0.2754	0.1794	0.001643	0.9413
29.0579	0.001057	0.3175	0.1365	0.00235	0.9571
27.7390	0.001263	0.3593	0.0478	0.010006	0.9877
26.2440	0.001475	0.3995	0.0374	0.014126	0.9908
24.6966	0.001689	0.4372	0.0336	0.016481	0.9920
23.1748	0.001902	0.4721	0.0180	0.040028	0.9962
21.7238	0.002113	0.5042	0.0178	0.040616	0.9963

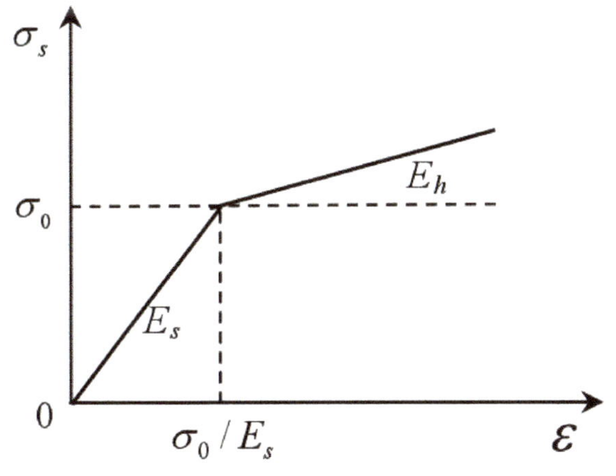

Figure 3. Constitutive model of reinforcement bar.

2.2. Contact Collision Simulation

During an earthquake, the actual collision between adjacent structures of a curved bridge is not a single problem ofnormal frontal contact, but a complex problem ofspatial contact. The collision between adjacent components that is caused by the in-plane rotation of the curved bridge is arbitrary and asymmetrical. In order to consider anon-uniform collision, the master–slave surface method [28] is used. This method is mainly applied to deal with the contact between two surfaces, and could simulate the non-uniform collision between adjacent contact surfaces at the expansion joints with relative accuracy. The master–slave surface method involvestwo surfaces in contact with each other as master and slave surfaces, respectively, and the nodes on the master and slave surfaces are defined as master and slave nodes, respectively. Duringthe analysis process, the following instruction s should be followed: first, search for the contact pairs on the master and slave surfaces; next, calculate the effect of the contact pairs. The algorithms involved mainly include search algorithm and contact algorithm. The surface-to-surface contact automatic search algorithm [28] is used in this paper, and the penalty function method [28] is used as the contact algorithm. The penalty function method is used to obtain the contact force by multiplying a penalty parameter related to the element stiffness by the contact surface intrusion value [31]. The calculation of the contact force mainly includes the normal contact force and tangential friction force. In ABAQUS software, the "surf-to-surf" command adopted to simulate the normal contact force is set as "hard contact", and the penalty friction is adopted to simulate tangential force [28]. According to [32], the tangential friction coefficient is set as 0.5.

There are three types of contact pairs in the bridge model. One is formed by the adjacent structural surfaces at the expansion joints of the curved bridge. The expansion joint at the abutment position takes the abutment contact surface as the master surface, the middle expansion joint takes the contact surface of long unit girder as the master surface, and the corresponding surface is the slave surface, including the free surface of the rubber pad. Another is formed by the top surface of the sliding support and the bottom surface of the girder at the corresponding position, with the bottom surface of the girder as the master surface and the top surface of the support as the slave surface. The other is formed by the contact surfaces of the other components, including the upper and lower surfaces of the fixed support and its adjacent contact surfaces, the contact surfaces between cylindrical piers and cap beams, and the contact surfaces between rubber pads and the girder attached to it.The three-dimensional finite element model of the whole structure and its components is shown in Figure 4. In order to express the collision response at different expansion joints more clearly, they were named as shown in Figures 4 and 5. The three expansion joints are respectively named as short unit expansion joints, middle expansion joints, and long unit expansion joints; the positions of the collision surface are respectively named as the short-unit abutment position, the short-unit middle position, the long-unit middle position, and the long-unit abutment position; the nodes at both ends of the expansion joint are numbered.

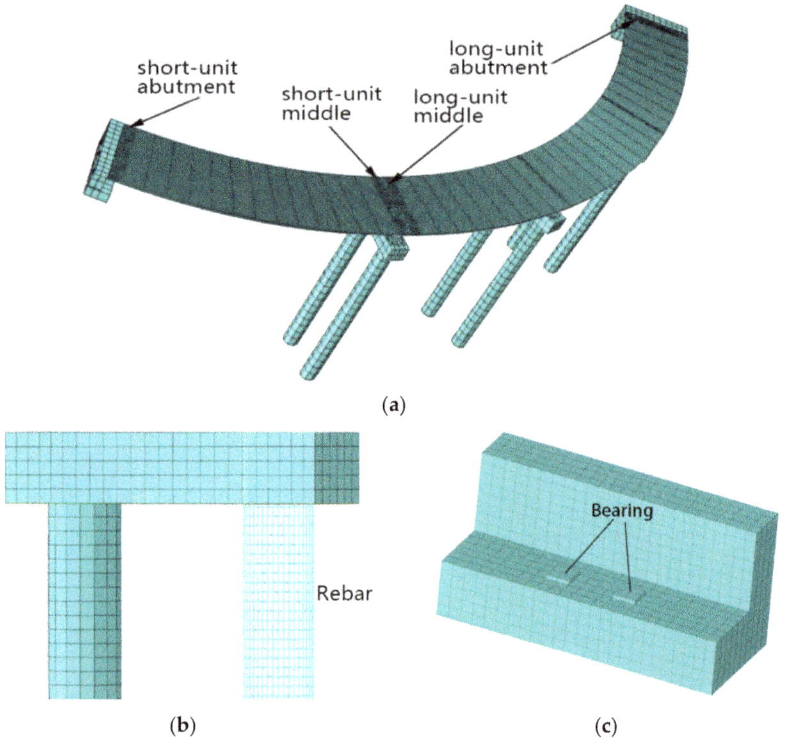

Figure 4. Finite element model of curved bridge. (**a**) Overall structure model. (**b**) Double-column pier. (**c**) Abutment and bearing.

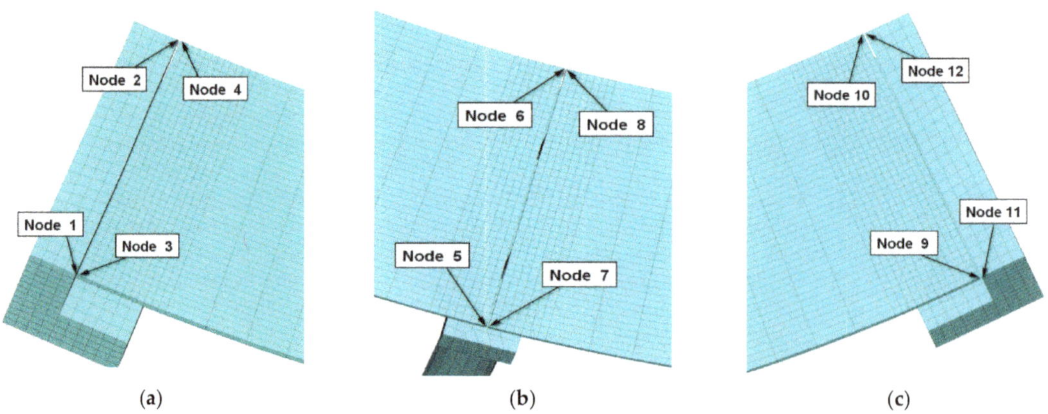

Figure 5. The name and node number of expansion joints. (**a**) Short unit expansion joints. (**b**) Middle expansion joints. (**c**) Long unit expansion joints.

2.3. Seismic Mitigation and Unseating Prevention Devices

In order to reduce the collision response of the curved girder bridge and improve the seismic performance during seismic shock, a relatively effective method is to install unseating restrainers or damping devices inside and outside the expansion joints. The restraint effect of the unseating restrainer and the energy consumption effect of the damping device can effectively reduce the torsion and collision reaction of the beam to achieve the goal of pounding mitigation and unseating prevention. The seismic mitigation and unseating prevention devices used in this paperinclude viscoelastic dampers, viscous dampers, steel strand cables-rubber pads, and lead-core rubber bearings.

Steel strand cable restrainers are usually used to prevent the girder from falling, but traditional steel strand cables can only bear axial tension. Therefore, they prevent unseating but offer no pounding mitigation function. In order to achieve the goal of seismic mitigation and unseating prevention function simultaneously, the two ends of the steel stranded cable are anchored at the webs of the beams on both sides of the expansion joint, and rubber pads are set between the expansion joints, as shown in Figure 6. The setting of rubber pads on the contact surfaces of each expansion joint isshown in Figure 7. The restrainer can prevent excessive relative displacement between the beams to perform the unseating prevention function. The rubber pad can buffer collisions with adjacent beams, and can also absorb the energy generated by the collision between the beams and avoid the direct collision of adjacent beams, thereby mitigating pounding.

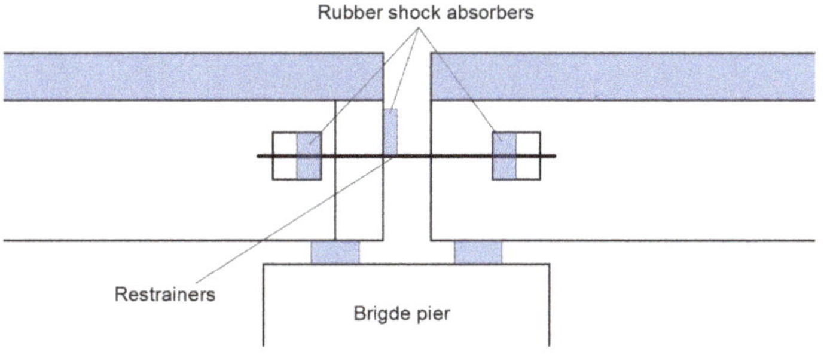

Figure 6. Steel wire rope-rubber mat device.

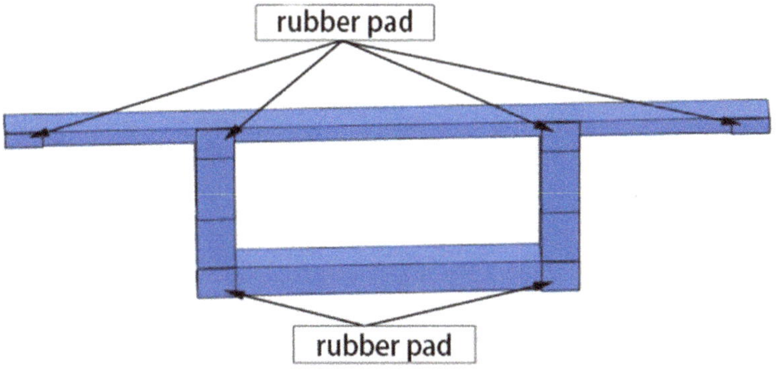

Figure 7. The location of rubber pads.

The pounding mitigation and unseating prevention device is composed of the Strand Cable-Rubber Blanket (SCRB), which is mainly composed of steel strand cables and rubber pads. Among these, the rubber pad is made of general rubber, and its elastic modulus is taken as the elastic modulus of natural rubber. The cable restrainer used on the bridge usually is the steel stranded cable. The mechanical model of the cable restrainer can be the bilinear model. Its pre-yield stiffness is K_1, its post-yield stiffness is K_2, and it generally takes K_2 = 0.05 K_1 [12]. The elongation of the steel cable, whose length is L, is ΔL, and the pre-yield stiffness can be calculated using the following formula:

$$K_1 = \frac{E' A_0}{L} \tag{1}$$

$$E' = \frac{TL}{A_0 \Delta L} = E \frac{(6L^3 + 1.0671S^3)L}{(6S + 1.0671L)S^3} \tag{2}$$

where E' is the elastic modulus of steel strand,

T is the total tension of steel strand,

E is the elastic modulus of steel wire, generally taken as 2×10^5 N/mm^2,

S is the length of inner and outer wire of a twist pitch, $S = \sqrt{4\pi^2 R^2 + L^2}$, $R = \frac{d_1 + d_2}{2} = 1.0165d_1$,

R is the distance between the centerline of the outer wire and the centroid of the steel strand,

d_1 is the cross-section diameter of outer wire,

d_2 is the cross-section diameter of inner wire.

Consider the use of lead rubber bearings (LRB) and laminated rubber bearings (GJZ) for seismic isolators. The shear force-displacement hysteretic curve of laminated rubber bearings, which can be approximately linearized, is narrow and long, [33]. The stiffness of the lead rubber bearing includes vertical stiffness, horizontal stiffness, and horizontal equivalent stiffness. The horizontal stiffness includes two important parameters, namely the pre-yield stiffness and the post-yield stiffness. The horizontal restoring force model can be simulated by the bilinear model.The pre-yield stiffness of the support is used in the elastic stage, and the post-yield stiffness is used in the plastic stage. According to industry standards [34], the ratio of the pre-yield and post-yield stiffness is between 0.15 and 0.16; in this paper, it is taken as 0.154.

Viscoelastic Dampers (VEDs) arecomposed of viscoelastic material, which is usually polymer, constrained steel plate, and related parts. Viscoelastic dampers depend on the damping of some viscoelastic material, and they feature initial stiffness. Theircommon mechanical calculation models include the Kelvin model, the standard linear and equivalent standard solid model, etc. The most commonly used model to describe the properties of Viscoelastic Dampers is the Kelvin model. The Kelvin model [35] is composed of a spring and a damper in parallel. This model is used in this article, and its mechanical equation is as follows:

$$f = ku + c\dot{u} \tag{3}$$

where u is the displacement acting on the model,

f is the output force of the damper,

k is the spring equivalent stiffness,

c is the equivalent damping coefficient.

Viscous Fluid Damper (VFDs) arecomposed of pistons, cylinders, orifices, and other related components. Viscous dampersare velocity-dependent and can dissipate seismic energy without adding any initial stiffness to the bridge structure. The mechanical model of VFDs adopt the Maxwellmodel, which is created by connecting the spring and dampers in series. Due to the series of viscous elements, deformation will increase infinitely under any small external force, so the Maxwell model is essentially viscous. According to [36], if the actual frequency is less than the cut-off frequency of 4 Hz, the effect of the spring stiffness on the displacement is negligible. Therefore, the relationship between the damping force and the relative speed can be expressed by the following formula:

$$f = -c\dot{u}^{\alpha} \tag{4}$$

where c is the damping coefficient,

α is the speed index.

In order to achieve a better effect of pounding mitigation and unseating prevention, the aforementioned damping devices and unseating restrainers were optimized in combination. The four cases obtained are shown in Table 2. In the table, Case A represents the original structure, in which all supports are GJZ, and there is no unseating restrainer at the expansion joints. All the supports in Case F are GJZ, and the VED and SCRB are set inside and outside the expansion joints. All the supports in case G are GJZ, and the VFD and SCRB are set inside and outside the expansion joints. All the supports in case H are LRB, and SCRB is set inside and outside all the expansion joints.

Table 2. The arrangement of various cases.

Case	Support Type		VED	VFD	SCRB
	GJZ	LRB			
A	•	○	○	○	○
F	•	○	•	○	•
G	•	○	○	•	•
H	○	•	○	○	•

Note: • means setting, ○ means not setting.

The design parameters of the damping devices and unseating restrainers selected in the calculation conditions are shown in Table 3.

Table 3. The parameters of the devices.

Device	Parameter Value
GJZ	Support length and width : $l_1 \times l_2 \times t = 600 \times 480 \times 100$ mm Elastic modulus : $E = 38.4$ Mpa Shear modulus : $G = 1$ Mpa
VED	Equivalent stiffness : $K = 4000$ KN/m Equivalent damping coefficient : $C = 600$ KN/m/s
VFD	Damping coefficient : $C = 600$ KN/(m/s)$^{0.3}$ Speed index : $\alpha = 0.3$
SCRB	Pre–yield stiffness : K'_1= 254238 KN/m Post–yield stiffness : K'_2= 12711.9 KN/m Yield force: F_y = 2605.7 KN Rubber pad length and width : $l_1 \times l_2 \times t$= 360 × 180(360) × 30 mm Elastic modulus : $E = 7.84$ Mpa
LRB	Piers and : Pre–yield stiffness : K_u= 21.97 KN/mm Post–yield stiffness : K_d= 3.38 KN/mm Vertical stiffness : K_l= 7350 KN/mm Yield force: F_y = 209 KN Pier : Pre–yield stiffness : K_u= 17.75 KN/mm Post–yield stiffness : K_d= 2.73 KN/mm Vertical stiffness : K_l= 6000 KN/mm

3. Dynamic Analysis Method and Ground Motion Input

Implicit and explicit algorithms can be used to solve dynamic equations, sincethe low frequency component is usually the main component in the dynamic response of the bridge structure. Considering the calculation accuracy, a larger time step is allowed; therefore, the unconditionally stable implicit algorithm, the Newmark method, is used to solve the dynamic equation. The contact collision process is the nonlinear problem. Nonlinear dynamic equations require iterative solutions that adopt the modified Newton-Raphson iterative method [37]. The structure in this study considers Rayleigh damping.

To determine the α and β factors, the radial frequency of the first mode and the highest mode to achieve a 90% modal mass participation wereconsidered, respectively. We tookthe damping coefficient as 0.05, the calculated massdamping coefficient $\alpha = 0.4158$, and the stiffness damping coefficient $\beta = 0.006$.

The bridge is located in Renhe District, Panzhihua City, Sichuan Province, China, with site category I_1 (i.e., with a shear wave velocity, VS, in the range of 250 to 500 m/s, and with a thickness of covering soil of less than 5 m). The seismic fortification intensity is determined to be 7 degrees (the peak acceleration of rare earthquakes is 3.1 m/s^2). Through a modal analysis, it was determined that the basic period of the bridge structure is 0.7966 s. The wave selection software, compiled by the Structural Engineering and Disaster Prevention Center of Chongqing University, was used to input the basic information aboutthe bridge structure and site for wave selection, and MEX00003 was finally selected. The Fourier spectra and the acceleration waveform of the scaled ground motion are shown in Figure 8. Due to the far-field type of the selected motions, the vertical component of the motion was neglected; only the horizontal seismic ground motion was considered.

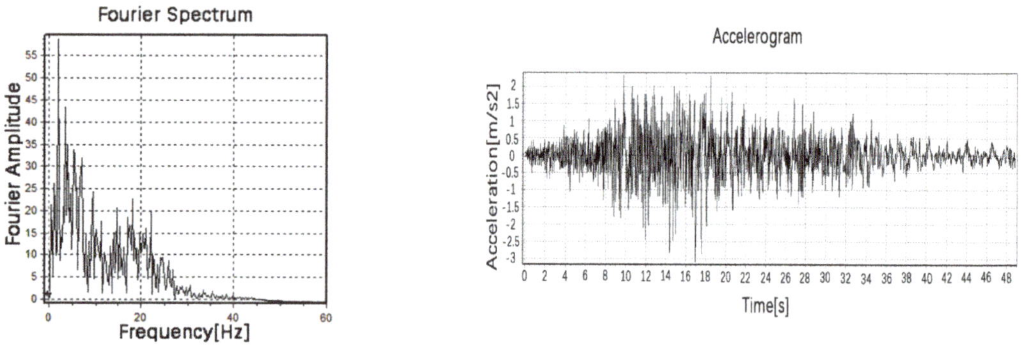

(**a**) (**b**)

Figure 8. MEX00003 ground motion. (**a**) Fourier spectrum of ground motion. (**b**) Acceleration of the ground motion.

When analyzing the seismic response of curved bridges, the seismic input should be carried out separately along the connecting direction of two adjacent bridge piers to determine the most unfavorable seismic horizontal input direction. The curved bridge model established in this paper features three piers. According to the requirements of the specification, the ground motion should be input along the connection direction of piers ① and ② and the connection direction of piers ② and ③. Comparing the effect of collision response of curved bridges at different input angles of ground motion, it was found that the connecting direction of piers ② and ③ is the most unfavorable direction.Therefore, the seismic response of the structure was mainly analyzed when the seismic wave was input along the connecting direction of piers ② and ③.

4. Seismic Response Analysis

4.1. Contact Force between Adjacent Structures at the Expansion Joint

A comparative analysis of several cases was carried out on the displacement, stress, contact force, and damage of the bridge structure obtained through the dynamic analysis. Compare the peak value of the normal contact force at each expansion joint of the three combined shock absorption cases with the case A. The peak value of the normal contact force of each case is shown in Figure 9. The normal contact force time history of cases A and G are listed in Figures 10 and 11.

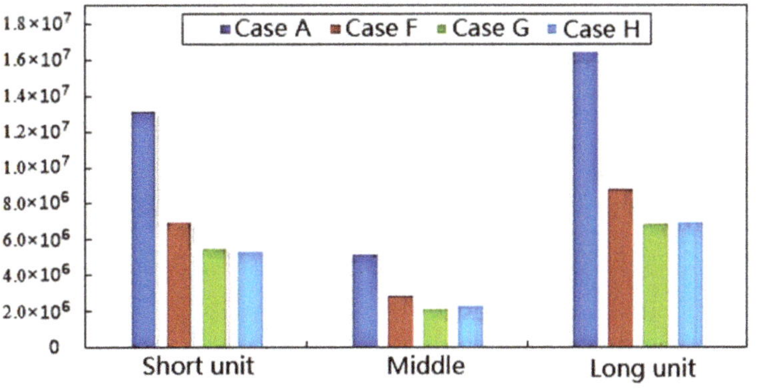

Figure 9. The maximumnormal contact force of expansion joints adjacent structures of every case (unit:N).

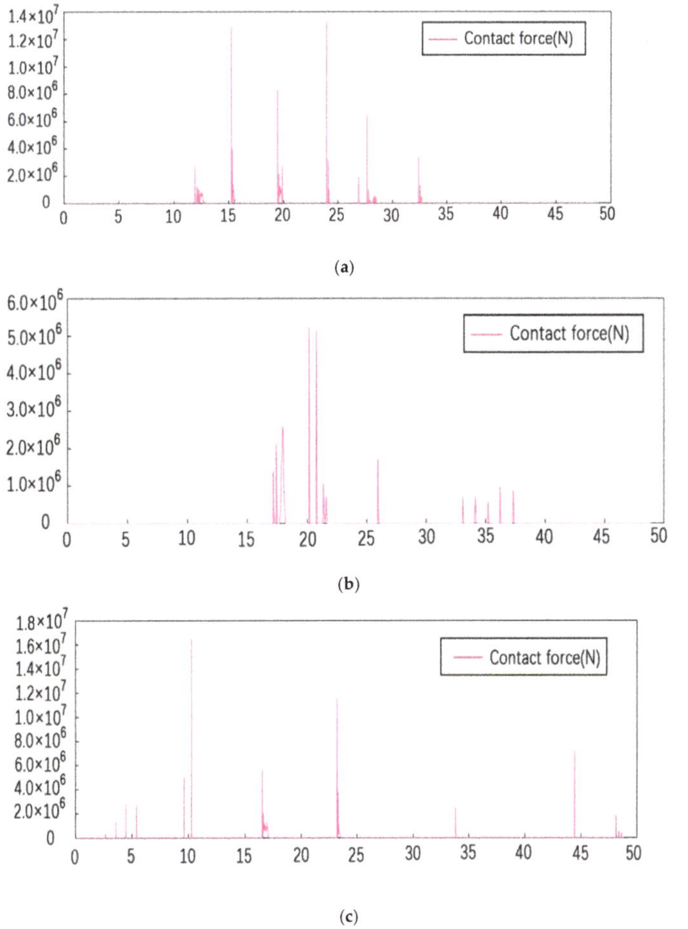

Figure 10. The normal contact force time-histories of expansion joints adjacent structures of case A (time:s). (**a**) Short unit expansion joints (**b**) Middle expansion joints (**c**) Long unit expansion joints.

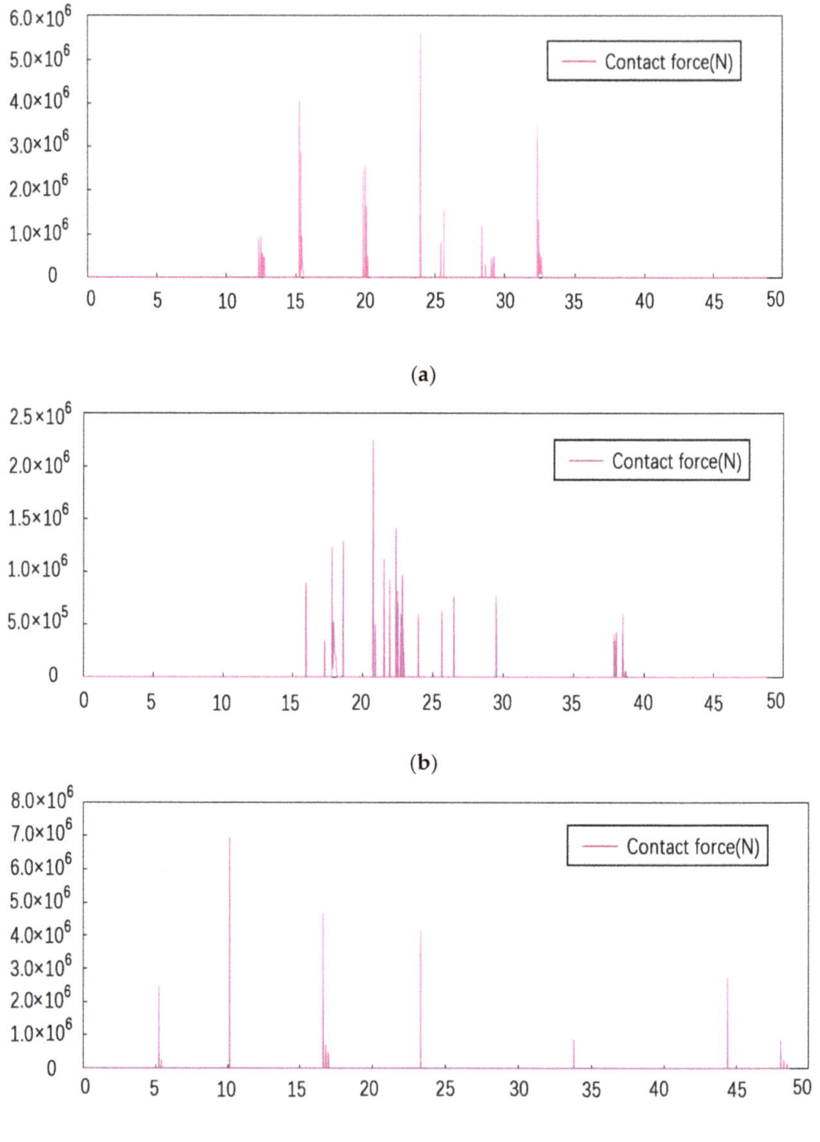

(**a**)

(**b**)

(**c**)

Figure 11. The normal contact force timehistories of expansion joints adjacent structures of case G (time:s). (**a**) Short unit expansion joints. (**b**) Middle expansion joints. (**c**) Long unit expansion joints.

From the analysis in Figure 9, the combined shock absorption working case provides more effective pounding mitigation. The maximum contact force of case F (the combination of VED and SCRB) is reduced by 47.6% at the maximum; case G (the combination of VFD and SCRB) is reduced by 57.9% at the maximum; and case H (the combination of LRB and SCRB) is reduced by 57.4% at the maximum. Thisindicates that the energy dissipation effect of dampers or seismic isolator plus the cushioning effect of rubber pads can greatly reduce the contact force between adjacent beams. Analyzing the anti-collision effect of each combination device from the perspective of contact force, the energy dissipation and seismic mitigation effect of the VFD and LRB combined with SCRB are more effective.

From the comparison of Figures 10 and 11, it can be seen that in cases G and A, the contact force time history curves are similar and the moment of the maximum contact force is almost the same, but the maximum contact force is quite different. For case A, the times corresponding to the peak contact force at the three expansion joints (the peak contact force) are 23.95 s (13.22 MN), 20.16 s (5.23 MN), and 10.26 s (16.45 MN), and for case G, they are 23.98 s (5.56 MN), 20.72 s (2.25 MN), and 10.24 (6.92 MN). This shows that the maximum contact force was greatly reduced after installing the combined VFD and SCRB devices.

4.2. Contact Stress of the Contact Surface of the Expansion Joint

Extracting the normal contact stress of the contact surface at the expansion joint, the anti-collision effect of the combined case was analyzed from the perspectiveof stress. The position of each collision surface is shown in Figure 4, and the maximum stress of each combined case is shown in Figure 12. The peak stress of cases A and G, where the effect of the seismic mitigation is the strongest, is shown in Figures 13 and 14.

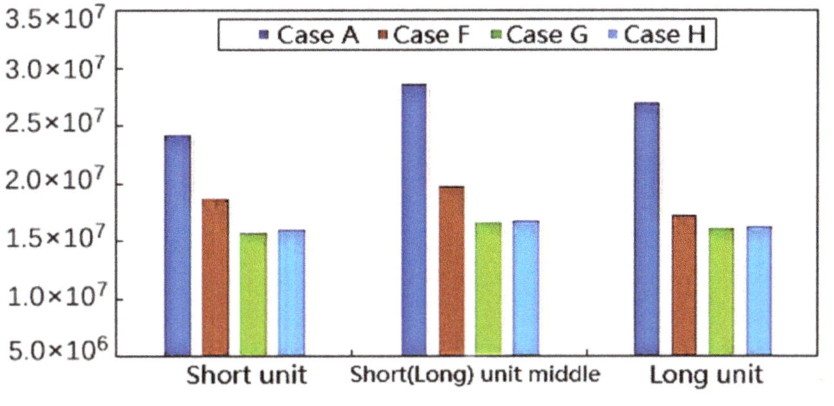

Figure 12. The maximum normal contact stress of adjacent surface at expansion joints in every case (unit:N/m^2).

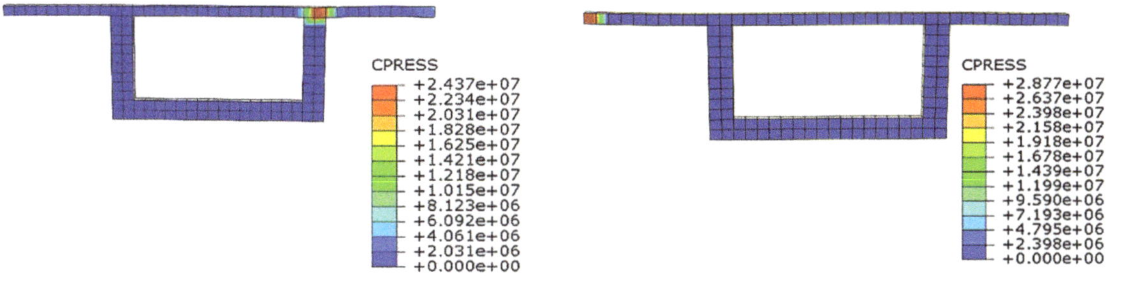

(**a**) (**b**)

Figure 13. *Cont.*

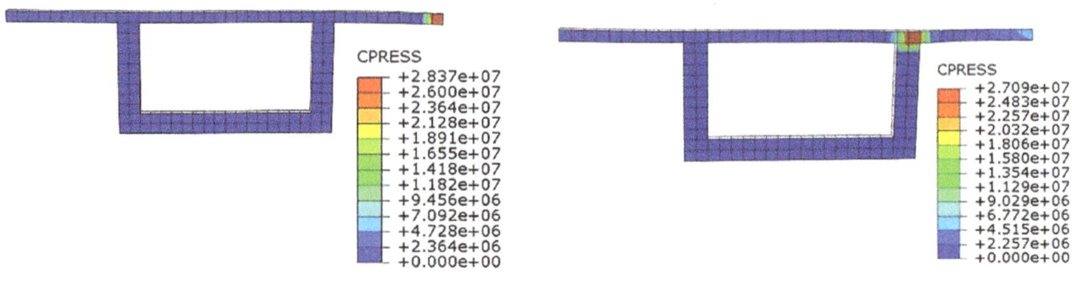

(**c**) (**d**)

Figure 13. The normal contact stress of adjacent surface at expansion joints in case A(unit: N/m^2). (**a**) The contact surface of the short unit expansion joint at the abutment position (t = 23.95 s). (**b**) The contact surface of the short unit expansion joint in the middle position (t = 20.16 s). (**c**) The contact surface of the long unit expansion joint at the abutment position in the middle position (t = 20.16 s). (**d**) The contact surface of the long unit expansion joint at the abutment position (t = 10.26 s).

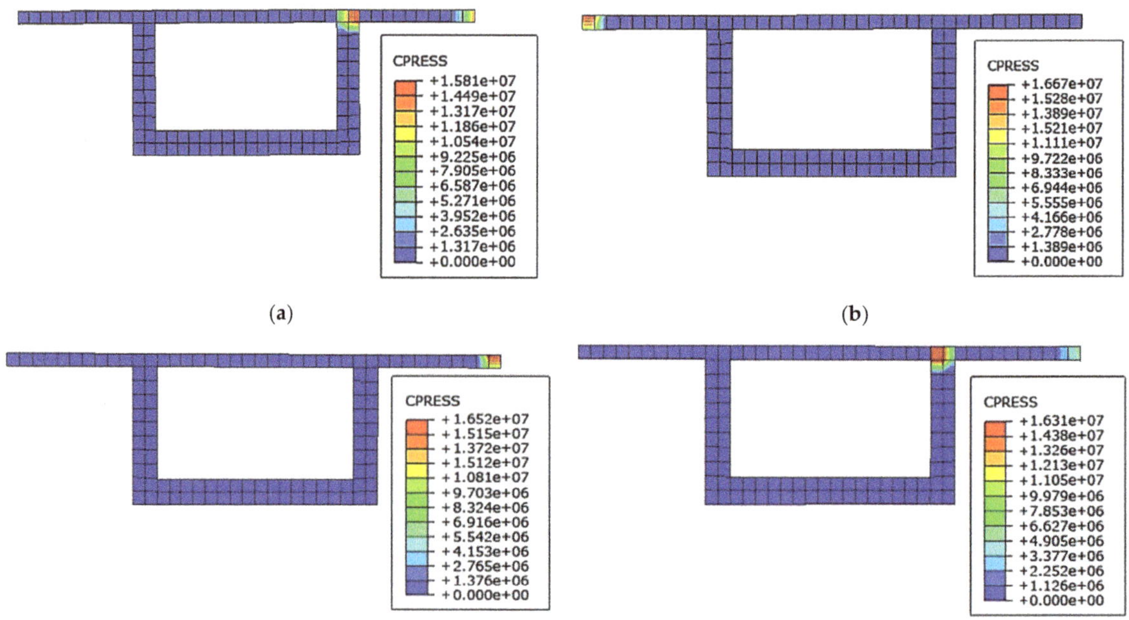

(**c**) (**d**)

Figure 14. The normal contact stress of adjacent surface at expansion joints in case G(unit: N/m^2). (**a**) The contact surface of the short unit expansion joint at the abutment position (t = 23.83 s). (**b**) The contact surface of the short unit expansion joint in the middle position (t = 20.01 s). (**c**) The contact surface of the long unit expansion joint at the abutment position in the middle position (t = 20.01 s). (**d**) The contact surface of the long unit expansion joint at the abutment position (t = 10.11 s).

According toFigure 12, these mitigation and unseating prevention devices could effectively reduce the maximum contact stress of contact surfaces. Case F (the combination of VED and SCRB) is reduced by 36.2% at the maximum; case G (the combination of VFD and SCRB) is reduced by 42.3% at the maximum; and case H (the combination of LRB and

SCRB) is reduced by 41.5% at the maximum. Cases G and H are more effectiveat seismic mitigation and unseating prevention.

In Figures 13 and 14, it is shown that the collision betweenadjacent components does not occur on the entire contact surface, but locally. The location of the maximum contact stress at different expansion joints is different. For the position of the short unit expansion joint, the collision occurs between the junction of the top plate of the box girder and the outer web and the bridge abutment in cases A and G. For the position of the middle expansion joint, the collision occurs between the outer side of the box girder top plate of the adjacent girder. For the position of the long unit expansion joint, the collision occurs between the junction of the top plate of the box girder and the inner web and the bridge abutment. After installing the pounding mitigation and unseating prevention devices, because of the installation of the rubber pad at the beam end on the side of the expansion joint, the contact stress only appears in the area with the rubber pad. Thisshows that the adjacent girders did not directly collide, but indirect collisions were caused by the force transmission of rubber pad.

4.3. Damage to Adjacent Structures at Expansion Joints

We extracted the tensile damage and compression damage at the last moment of the ground motion of each case. The maximum cumulative damage of each combined case is shown in Figure 15. The compression damage of the girder at the expansion joints of cases A and G are shown in Figures 16 and 17, respectively. The values in the figure represent the cumulative degree of plastic damage; 0 means the material is intact, 1 means complete loss of strength.

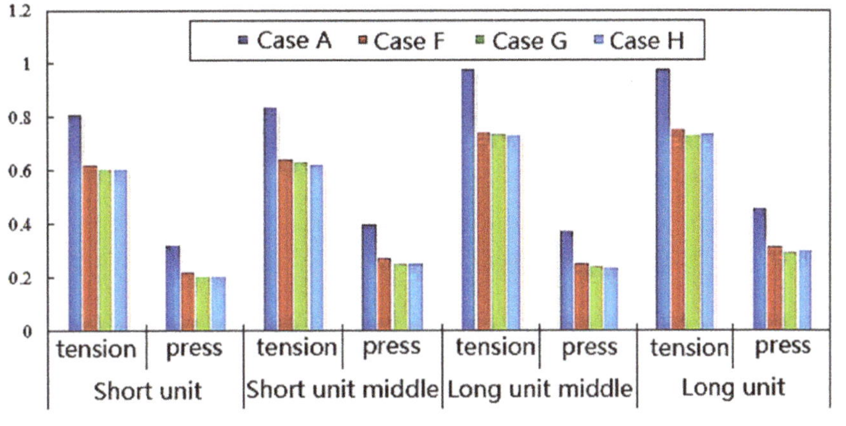

Figure 15. The cumulative damage of expansion joints adjacent surface of every case.

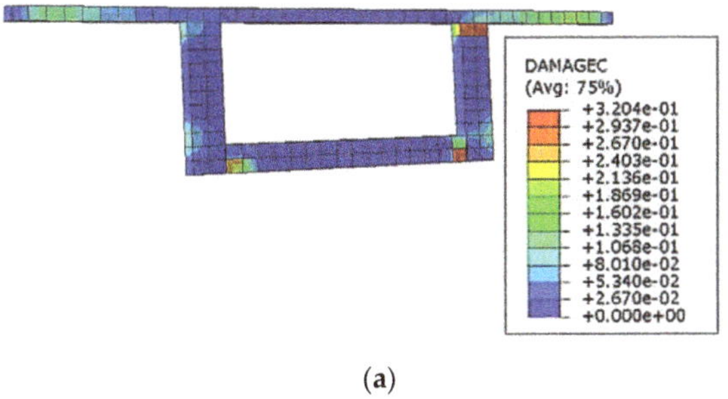

(a)

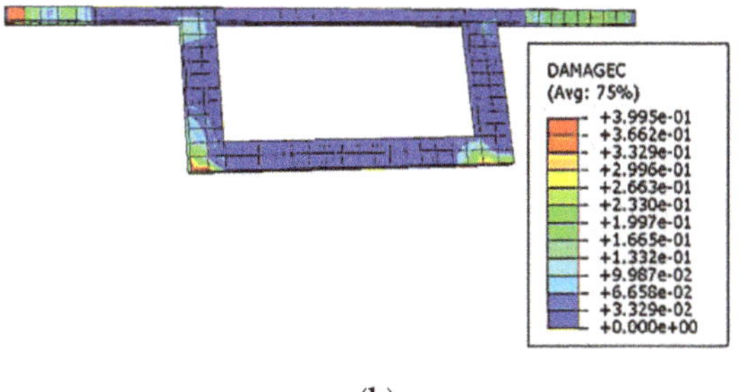

(b)

Figure 16. *Cont.*

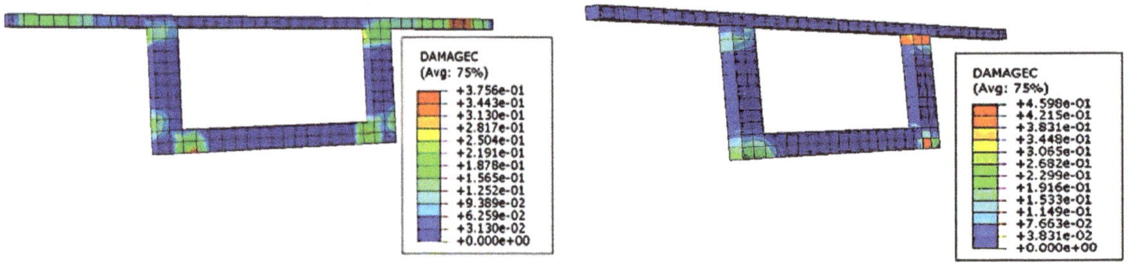

(c) (d)

Figure 16. Damage of girder in expansion joints of case A. (**a**) Compression damage at abutment position of short unit expansion joint. (**b**) Compression damage in the middle of the short unit expansion joint. (**c**) Compression damage in the middle of long unit expansion joint. (**d**) Compression damage at abutment position of the long unit expansion joint.

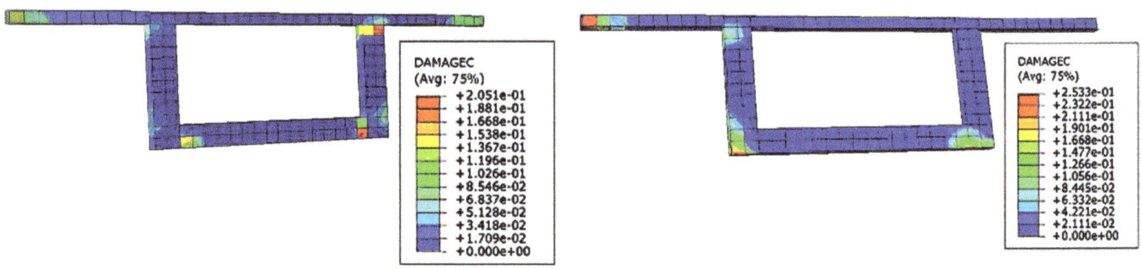

(a) (b)

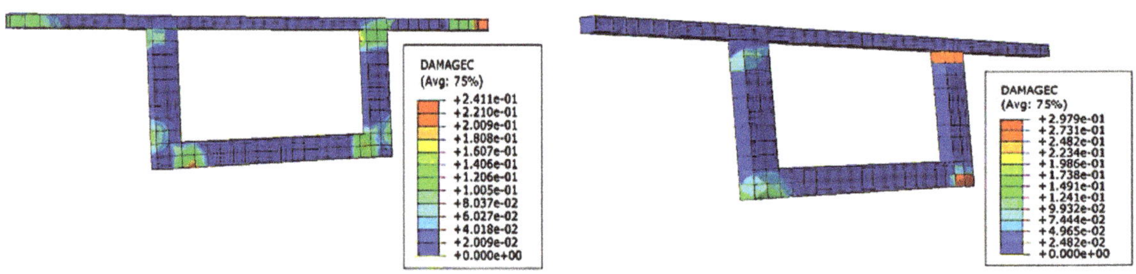

(c) (d)

Figure 17. Damage of girder in expansion joints of case G. (**a**) Compression damage at abutment position of short unit expansion joint. (**b**) Compression damage in the middle of the short unit expansion joint. (**c**) Compression damage in the middle of long unit expansion joint. (**d**) Compression damage at abutment position of the long unit expansion joint.

Figure 15 shows that the tensile damage toconcrete is more serious than the compression damage. After installing the seismic mitigation and unseating prevention device, both tension and compression damage are reduced to a certain extent, but the reduction ratio of the tensile damage is smaller and the ratio of compression damage reduction is larger. For tensile damage, the maximum reduction percentage in each combined case is about 24%. For compressive damage, case F (the combination of VED and SCRB) is reduced by 32.2% at the maximum; case G (the combination of NFVD and SCRB) is reduced by 36.6%

at the maximum; and case H (the combination of LRB and SCRB) is reduced by 36.3% at the maximum.

From the comparison of Figures 13 and 16, it can be found that the location where the maximum contact stress is generated is also the place where the compression damage is the most serious. This shows that the huge contact pressure generated by collision damages and destroysthe bridge structure.

4.4. The Number of Collisions at the Edge of Expansion Joints

The timehistory contact stress of nodes at the edges of different expansion joints in each case was extracted, and the number of collisions of each node was obtained by statistics. Figure 18 shows the number of collisions at the edge of the expansion joints in each case. Refer to Figure 5 for the location of each node.

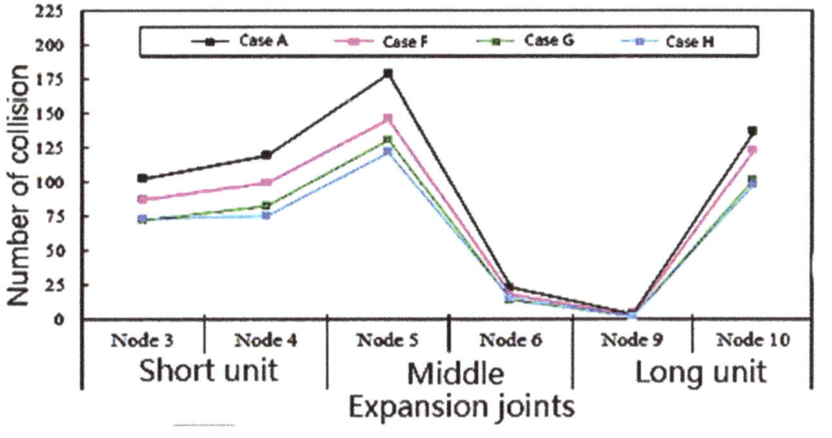

Figure 18. The contact number of node at the edge of expansion joints of every case.

Figure 18 shows that each combined case can effectively reduce the number of collisions of each edge node, and cases G and H exerta better effect on reducing the number of collisions. The collision responses between the inner node and the outer node of the same expansion joint arequite different. At the short unit expansion joint, the number of collisions on the outer side (node 3) is less than on the inner side (node 4).At the middle expansion joint, the number of collisions on the outer side (node5) is much higherthan on the inner side (node 6), and the outer contact stress is greater.At the long unit expansion joint, the number of collisions on the outside (node 9) is far lower than on the inside (node 10), and the contact stress on the inside is greater. This phenomenon shows that when the seismic wave is input along the connecting direction of piers ② and ③, the short unit mainly twists horizontally, counterclockwise, around the Z axis, while the long link mainly twists horizontally, clockwise, around the Z axis; the length of the long unit bridge is longer, and the torsion effect is obvious, which makes the numberof collisions and the collision stress between the inner and outer nodes obviously different.

4.5. Displacement Response

4.5.1. Displacement at the Short Unit Expansion Joint

We extracted the displacement of each node of the expansion joint and tookthe displacement of each node minus the displacement of the corresponding pier or abutment bottom node at each time. The displacement of the outer node was subtracted from the corresponding pier or abutment bottom, and the displacement of the inner node was subtracted from the corresponding pier or abutment bottom. Subsequently, the maximum absolute value was taken as the peak displacement of each node. We extracted the radial and tangential peak displacementof the four inner and outer nodes at the short unit expansion joint in all the cases, as shown in Figure 19.

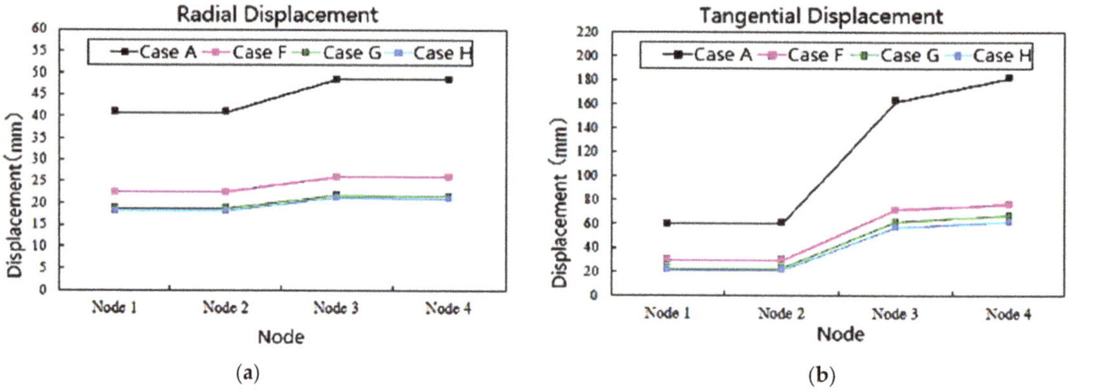

Figure 19. Nodal peak displacement of short expansion joints in every case. (**a**) Radial diaplacement (**b**) Tangential displacement.

Figure 19 shows that each combined case exertsan obvious effect on reducing the peak radial tangential displacement of each node at the edge of the short unit expansion joint, and the reduction rate reached more than 45%. Among these, cases G and H producethe greatestseismic mitigation effect. Comparing Figure 19a,b, the peak tangential displacement of the joint at the short unit expansion joint is obviously greater than the peak radial displacement of the joint. The main reason for this is that the ground motion in this paper only considers the input along the direction of piers ② and ③, and does not consider the input perpendicular to the horizontal direction. Therefore, the radial displacement of the beam is mainly caused by the torsion of the beam. The tangential displacement is composed of two parts, one of which is caused by the translation of the beam, and the other by the torsion of the beam.

Figure 19b shows that the difference in the tangential displacement ofnodes 3 and 4 in case A is mainly caused by the torsion of the girder. However, after installing the seismic mitigation and unseating prevention devices, the torsional response is effectively reduced, the inward and outward torsion is balanced and essentiallythe same, and the peak tangential displacement ofnodes 3 and 4 also become more consistent.

4.5.2. Displacement at the Middle Expansion Joint

We extracted the peak radial and tangential displacement of the four inner and outer nodes of the middle expansion joint under several combined cases, as shown in Figure 20.

Figure 20 shows that the peak seismic mitigation ratio of the radial tangential displacement of each node at the edge of the expansion joint in each combined case also reached more than 40%, and that the damping effect of cases G and H was the greatest. From Figure 20b, it can be seen that the peak tangential displacement of nodes 5,6, 7, and 8 of case A are quite different, mainly due to the torsion of the girder. However, the corresponding peak displacements of the joints in the three combined damping cases F, G and H are all effectively reduced and the difference is not significant. This shows, once again, that through the combined damping optimization design, the torsion of the girder is effectively reduced.

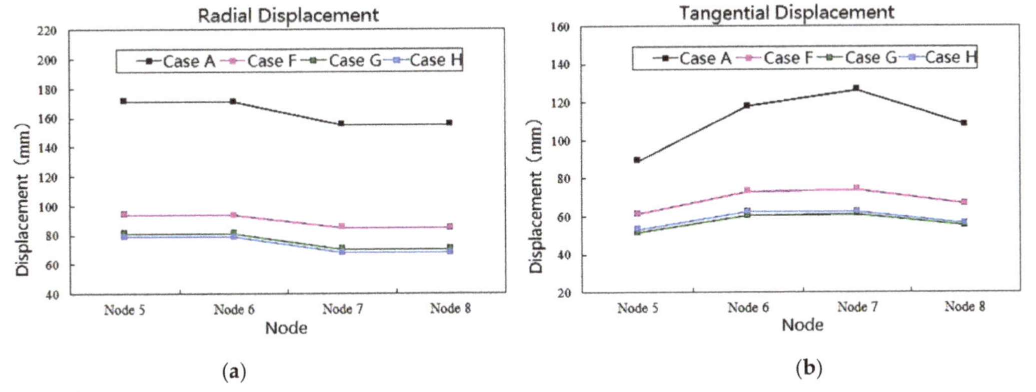

Figure 20. Nodal peak displacement of middle expansion joints of every case. (**a**) Radial diaplacement (**b**) Tangential displacement.

4.5.3. Displacement at Long Unit Expansion Joint

We extracted the peak radial and tangential displacement of the four nodes on the inner and outer edges of the long unit expansion joint, as shown in Figure 21.

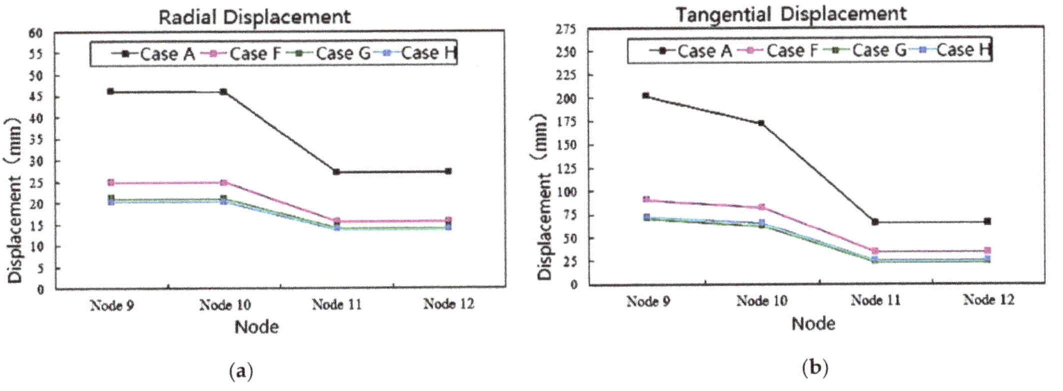

Figure 21. Nodal peak displacement of long expansion joints in every case. (**a**) Radial diaplacement (**b**) Tangential displacement.

Figure 21 shows that the long unit expansion joints are similar to the short unit expansion joints. The peak radial and tangential displacements of the joints are effectively reduced, as well as the torsion effect of the girder. The seismic mitigation ratio to the peak tangential displacement is greater than that ofthe peak radial displacement.

4.5.4. Torsion of Girder

The girder or abutment features torsion in both the tangential and vertical directions. In this paper, the torsion angle is used to indicate thedegree of torsion. The calculation formula is as follows:

$$\varphi = \frac{|u_i - u_j|_{max}}{l_{ij}} \times \frac{180}{\pi} \tag{5}$$

where u_i and u_j is the tangential displacement or vertical displacement of the inner and outer nodes of the girder or abutment,

l_{ij} is the distance between the inner and outer nodes of the main girder (bridge abutment), where the bridge deck width is 7.2 m.

Calculate the tangential and radial torsion angle of the girder or abutment in the combined case according to the above formula. The specific values are shown in Figure 22.

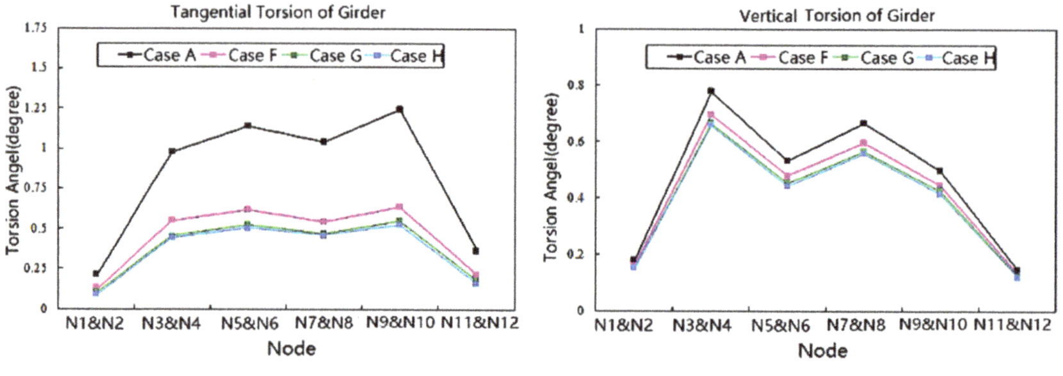

(**a**) (**b**)

Figure 22. Torsion of the girders. (**a**) Tangential torsion (**b**) Vertical torsion.

It was found that there is a torsion phenomenon in curved bridgesduring earthquakes. After installing seismic mitigation and unseating prevention devices, the tangential torsion of the curved bridge is effectively reduced, but the reduction in the vertical torsion is not large because, for vertical torsion, a certain reduction effect can only be obtained through the overall energy dissipation.

When seismic shock is input along the direction of piers ② and ③, it mainly causes the tangential torsion of the beam. When the seismic mitigation and unseating prevention device is installed, the tangential torsion is effectively reduced. To clarify this point, take the outward twist (the short unit is twisted counterclockwise around the Z axis, and the long unit is twisted clockwise around the Z axis) as an example. When the beam is twisted, for SCRB, the outer cable is in a relaxed state, and the inner cable is in a tensioned state, which can provide a large pulling force to the inside of the beam, thereby restricting the torsion of the beam.For case F (the combination of VED and SCRB), not only can the inner SCRB can provide the restraining force to pull the beam together, but the outer VED can also provide the opposite restraining force to push the beam apart.At the same time, it also dissipates part of the seismic energy; therefore, it can effectively reduce the torsion.For case G (the combination of VFD and SCRB), theprinciple is similar to VED.For CASE H (the combination of LRB and SCRB), in addition to SCRB, which can provide constraining force, the horizontal rigidity and damping of the lead rubber bearing can also provide a certain constraining force on the inside and outside, reducing the torsion of the beam. Among several damping devices, LRB and VFD exertstrongereffects on the dissipation ifseismic energy; hence, theyoffer more effective seismic mitigation and unseating prevention.

5. Conclusions

In this paper, usinga curved bridge with double-column piers as an engineering example, weestablished a three-dimensional finite element solid unit model of acurved bridge. The collision response analysis of the curved bridge was performedthrough a nonlinear dynamic time history analysis. According to the response characteristics of the curved bridge, the damping device and unseating restrainer were used for the combination optimization to analyze the seismic mitigation and unseating prevention of the curved bridge. The analysis led to the following conclusions:

(1) During earthquakes, the bending–torsion coupling phenomenon inthe girders of curved continuous girder bridges is serious, and the displacement magnitude and direction of the inside and outside of the contact surface of the expansion joints of

each bridge span are inconsistent, which can easily cause uneven local collisions at the contact surfaces. The high contact stress causes local compressive and tensile damage to reinforced concrete. At the same time, the increase in the relative radial and tangential displacement caused by the torsion of the girder could lead to the unseating of the girder.

(2) Based on the principle of energy dissipation combined with constraints, considering the combination of seismic mitigation and unseating prevention devices, three seismic mitigation cases were obtained, including the combination of VED and SCRB, the combination of VFD and SCRB, and the combination of LRB and SCRB. The results indicated that the three kinds of combined seismic mitigation cases were effective atreducing the response to collision force, stress, damage, girder torsion, and displacement, and achieved the goals of seismic mitigation and unseating prevention.

(3) The three kinds of combined device can effectively reduce the contact force, stress, and damage between adjacent structures, as well as the torsion and displacement of the beam. Among these, the pounding and unseating prevention effects of case G (the combination of VFD and SCRB) and case H (the combination of LRB and SCRB) are superior to those of case E (viscoelastic damper and steel strand cable-rubber pad combination). The performance of viscoelastic energy dampers is obviously affected by temperature, vibration frequency, and strain, while viscous dampers offer a strong energy dissipation capacity and dissipate energy under small deformations, which are also widely used in practical engineering. Therefore, it is recommended that the design of seismic mitigation and unseating prevention for curved bridges consider the combination of VFD or LRB and SCRB.

Author Contributions: Data curation and methodology, Z.L.; Formal analysis, C.Y.; Writing—review & editing, S.K. All authors have read and agreed to the published version of the manuscript.

Funding: This research was funded by the Chongqing Natural Science Foundation of China (cstc2018jcyjAX0001).

Institutional Review Board Statement: Not applicable.

Informed Consent Statement: Not applicable.

Data Availability Statement: Some data, or models used during the study are available from the corresponding author by request.

Conflicts of Interest: The authors declare no conflict of interest.

References

1. Kaleybar, R.S.; Tehrani, P. Investigating seismic behavior of horizontally curved RC bridges withdifferent types of irregularity in comparison with equivalentstraight bridges. *Structures* **2021**, *33*, 2570–2586. [CrossRef]
2. Wieser, J. Experimental and Analytical Investigation of Seismic Bridge-Abutment Interaction in a Curved Highway Bridge. Ph.D. Thesis, Department of Civil and Environmental Engineering, University of Nevada, Reno, NV, USA, 2014.
3. Wilson, T.; Mahmoud, H.; Chen, S. Seismic performance of skewed and curvedreinforced concrete bridges in mountainous states. *Eng. Struct.* **2014**, *70*, 158–167. [CrossRef]
4. Li, Z.Y.; Liu, B.; Yang, Y.W. Seismic pounding analysis of curved continuous girder bridge under strong earthquake. *J. Build. Struct.* **2016**, *37*, 349–355.
5. Jeon, J.S.; DesRoches, R.; Kim, T.; Choi, E. Geometric parameters affecting seismicfragilities of curved multi-frame concrete box-girder bridges with integralabutments. *Eng. Struct.* **2016**, *122*, 121–143. [CrossRef]
6. Jiao, C.; Liu, W.; Wu, S.; Gui, X.; Huang, J.; Long, P.; Li, W. Shake table experimental study of curved bridges with consideration ofgirder-to-girder collision. *Eng. Struct.* **2021**, *237*, 112216. [CrossRef]
7. Wang, Z.; Lee, G.C. A comparative study of bridge damage due to the Wenchuan, Northridge, Loma Prieta and San Fernando earthquakes. *Earthq. Eng. Eng. Vib.* **2009**, *8*, 251–261. [CrossRef]
8. Comartin, C.D.; Greene, M.; Tubbesing, S.K. (Eds.) *The Hyogo-ken NanbuEarthquake January 17, 1995, EERI. Preliminary Reconnaissance Report*; EERI-95-04; EERI: Oakland, CA, USA, 1995.
9. Jain, S.K.; Lettis, W.R.; Murty, C.V.R.; Barder, J.P. *Bhuj, India, Earthquake of January 26, 2001 Reconnaissance Report, Publication No. 02-01*; EERI: Oakland, CA, USA, 2002.

10. Du, X.; Han, Q.; Li, Z.X.; Li, L.Y.; Chen, S.F.; Zhang, J.F. The seismic damage of bridges in the 2008 Wenchuan earthquake and lessons from its damage. *J. Beijing Univ. Technol.* **2008**, *34*, 1270–1279.
11. Kawashima, K.; Unjoh, S.; Hoshikuma, J.; Kosa, K. Damages of bridges due to the2010 Maule, Chile, Earthquake. *J. Earthq. Eng.* **2011**, *15*, 1036–1068. [CrossRef]
12. Chouw, N.; Hao, H. Pounding damage to buildings and bridges in the 22 February2011 Christchurch earthquake. *Int. J. Protect.Struct.* **2012**, *3*, 123–140. [CrossRef]
13. Ijima, K.; Obiya, H.; Aramaki, G.; Kawasaki, N. A study on preventing the fall of skewand curved bridge decks by using rubber bearings. *Struct. Eng. Mech.* **2001**, *12*, 347–362. [CrossRef]
14. Wei, C. Seismic Analysis and Response of Highway Bridges with Hybrid Isolation. Ph.D. Thesis, Department of Civil and Environmental Engineering, University of Nevada, Reno, NV, USA, 2013.
15. Mendez Galindo, C.; Gil Belda, J.; Hayashikawa, T. Non-linear seismic dynamicresponse of curved steel bridges equipped with LRB supports. *Steel Constr.* **2010**, *3*, 34–41. [CrossRef]
16. Mohamed, W.A.; Hayashikawa, T.; Aly, A.G.; Hussien, M.H. Study on seismicresponse of curved viaduct systems with different isolation conditions undergreat earthquake ground motion. *JSCE J. Struct. Eng.* **2003**, *49*, 563–572.
17. Julian, F.D.R.; Hayashikawa, T.; Obata, T. Seismic performance of isolated curved steel viaducts equipped with deck unseating prevention cable restrainers. *J. Constr. Steel Res.* **2007**, *63*, 237–253. [CrossRef]
18. Andrawes, B.; DesRoches, R. Unseating prevention for multiple frames bridges using super elastic devices. *Smart Mater. Struct.* **2005**, *14*, 60–67. [CrossRef]
19. Raheem, S.E.A. Pounding mitigation and unseating prevention at expansion joints ofisolated multi-span bridges. *Eng. Struct.* **2009**, *31*, 2345–2356. [CrossRef]
20. Tian, Q.; Hayashikawa, T.; Ren, W.-X. Effectiveness of shock absorber device for damage mitigation of curvedviaduct with steel bearing supports. *Eng. Struct.* **2016**, *109*, 61–74. [CrossRef]
21. Jiao, C.; Liu, Y.; Wu, S.; Ma, Y.; Huang, J.; Liu, W. Influence of pounding buffer zone for mitigation of seismic response ofcurved bridges. *Structures* **2021**, *32*, 137–148. [CrossRef]
22. Shresth, B.; Hao, H.; Bi, K. Effectiveness of using rubber bumper and restrainer on mitigatingpounding and unseating damage of bridge structures subjectedto spatially varying ground motions. *Eng. Struct.* **2014**, *79*, 195–210. [CrossRef]
23. Chen, L.; Sun, L.; Xu, Y.; Di, F.; Xu, Y.; Wang, L. A comparative study of multi-mode cable vibration control using viscousand viscoelastic dampers through field tests on the Sutong Bridge. *Eng. Struct.* **2020**, *224*, 111226. [CrossRef]
24. Ruangrassamee, A.; Kawashima, K. Control of nonlinear bridge response withpounding effect by variable dampers. *Eng. Struct.* **2003**, *25*, 593–606. [CrossRef]
25. Hwang, J.-S.; Tseng, Y.-S. Design formulations for supplemental viscous dampersto highway bridges. *Earthq. Eng. Struct. Dyn.* **2005**, *34*, 1627–1642. [CrossRef]
26. Fujino, Y.; Kimura, K.; Tanaka, H. *Wind Resistant Design of Bridges in Japan: Developments and Practices*; Springer: Tokyo, Japan, 2012.
27. Symans, M.D.; Constantinou, M.C. Passive fluid viscous damping systems for seismicenergy dissipation. *ISET J. Eartheq. Tech.* **1998**, *35*, 185–206.
28. Wang, Y.Y. *Abaqus Analysis User's Guide: Prescribed Conditions, Constraints & Interactions Volume*; China Machine Press: Beijing, China, 2019.
29. Li, H.N.; Wang, D.D. Multi-scale finite element modeling and numerical analysis of reinforced concrete structure. *J. Archit. Civ. Eng.* **2014**, *31*, 20–25.
30. Zhu, B.F. *The Finite Element Method Theory and Applications*; China Water & Power Press: Beijing, China, 2018.
31. Schweizerhof, K.; Nilsson, L.; Hallquist, J.O. Crachworthiness analysis in the automotive industry. *Int. J. Comput. Appl. Technol.* **1992**, *5*, 134–156.
32. Jankowski, R. Non-linear FEM analysis of pounding-involved response of buildings under non-uniform earthquake excitation. *Eng. Struct.* **2012**, *37*, 99–105. [CrossRef]
33. Li, L.F.; Wu, W.P.; Huang, J.M.; Wang, L.H. Research on the seismic vulnerability analysis of laminated rubber bearing. *J. Hunan Univ.* **2011**, *38*, 1–6.
34. Ministry of Transport of the People's Republic of China. *Lead Rubber Bearing Isolator for Highway Bridge*; JT/T 822-2011; China Communications Press: Beijing, China, 2011.
35. Ge, S.P.; Li, A.Q. Redevelopment of velocity-depended damper elements based on ABAQUS. *Earthq. Resist. Eng. Retrofit.* **2014**, *36*, 198–217.
36. Zhou, Y. *Design of Structure with Viscous Damping*; Wuhan University of Technology Press: Wuhan, China, 2006.
37. Ma, J.X. *Advanced Structural Dynamics*; Xi'an Jiaotong University Press: Xi'an, China, 2019.

symmetry

MDPI

Article

Research on the Detection Method of Tunnel Surface Flatness Based on Point Cloud Data

Liufu Xiang [1], Yifan Ding [2], Zheng Wei [1,*], Hao Zhang [2] and Zhenguo Li [3]

1 Zhejiang Transportation Engineering Management Center, Hangzhou 311215, China; zjlysc@126.com
2 School of Civil Engineering, Zhejiang University of Technology, Hangzhou 310023, China; 2111906111@zjut.edu.cn (Y.D.); zhanghao@zjut.edu.cn (H.Z.)
3 Zhejiang Jiaotong Group Co., Ltd., Hangzhou 310051, China; zjjtgc@zjjtgc.com
* Correspondence: wz814@163.com

Abstract: The curved surface of the tunnel is symmetrical. The curved surface of the tunnel can be roughly divided into the left and right arch walls along the direction of the central axis of the tunnel. The symmetry of the tunnel needs to be analyzed when the flatness inspection of the tunnel engineering is carried out. The flatness of the initial support of the tunnel project is an important indicator of the quality inspection and acceptance of the tunnel project. The three-dimensional laser scanner (3DLS) can be used to detect its rapidity effectively. According to the points obtained by the scanner, the surface fitting method based on B-spline interpolation and the SG bar initial support value processing method are used to optimize the tunnel surface to obtain the initial degree calculation reference. Based on the method, a calculation system for the initial flatness of the tunnel based on 3DLS technology is established. At the same time, the calculation method of the overall field of view distance and the development of small blocks is proposed. Through its application and comparison with traditional methods, the analysis shows that the three-dimensional laser scanning technology is feasible in the detection of the initial branch of the tunnel, and achieves a high degree of accuracy requirements.

Keywords: three-dimensional laser scanning; surface flatness of initial support of tunnel; curved surface fitting; flatness calculation datum

Citation: Xiang, L.; Ding, Y.; Wei, Z.; Zhang, H.; Li, Z. Research on the Detection Method of Tunnel Surface Flatness Based on Point Cloud Data. *Symmetry* **2021**, *13*, 2239. https://doi.org/10.3390/sym13122239

Academic Editors: Yang Yang, Ying Lei, Xiaolin Meng and Jun Li

Received: 7 September 2021
Accepted: 16 November 2021
Published: 23 November 2021

1. Introduction

The initial support of the tunnel refers to a form of surrounding rock support used on-site to protect the safety of the construction during the excavation of the tunnel project. The flatness refers to the degree of unevenness of the sprayed concrete lining on the initial support surface during the tunnel excavation process [1]. It is one of the important criteria for the quality acceptance of the tunnel project. The results directly affect the subsequent laying of waterproof coiled materials, the paving quality of the second-stage lining, and the safety of the tunnel project; the surface level of the initial support of the tunnel is too large, which directly leads to the decrease of the construction quality of the laying of the waterproof coiled material and the second-stage lining [2]. This indirectly affects the water seepage condition of the tunnel surface, making it more prone to water seepage and leakage in the tunnel. The traditional detection methods for the flatness of the initial tunnels are low in accuracy and efficiency, and fail to meet the expectations of future construction guidance, and the current flatness detection system of the tunnel is not optimistic. The introduction of the emerging 3DLS technology and the establishment of a method system for the flatness detection of the initial support of the tunnel based on the 3DLS technology are of great significance to the improvement and development of the flatness detection system of tunnel engineering [3].

1.1. The Main Work

To establish a detection system for the surface flatness of the initial support of the tunnel based on the 3DLS technology, the tunnel engineering site (Figure 1) was firstly analyzed comprehensively, then a reasonable tunnel engineering measurement plan was formulated. The Topcon GTL-1000 ground 3DLS was used to obtain the data in tunnel engineering. For the surface point cloud data of the initial branch, the relevant parameters of the 3DLS are shown in Table 1.

Figure 1. Tunnel project site.

Table 1. Performance parameters of the 3D laser scanner.

	Country of Origin	Product Number	Scanning Speed (Point/s)	Range	Ranging Accuracy
3DLS parameters	Japan	GTL-1000	100,000	0.3~1000	2.0 mm@50 m

Take the area with a length of 2 m along the central axis of the tunnel as the analysis object, and use the SW Amberg Tunnel to preprocess the collected point cloud data. The SW Amberg Tunnel is specially designed for tunnel construction and was developed by the Swiss Amberg Technology Company (Amberg Technologies) R & D. It supports various tunnel construction methods and construction phase systems. The system is composed of software and hardware. The SW Amberg Tunnel cooperates with and guides high-precision measurement hardware equipment to efficiently and accurately complete the measurement tasks of each construction stage. Not only does it significantly improve the efficiency and accuracy of data collection and analysis, but also achieves an excellent performance in tunneling guidance, high-density scanning section over-under-excavation analysis, refined excavation volume analysis, lining thickness analysis, geotechnical engineering monitoring analysis, and tunnel completion data archiving [4].

After preprocessing by the SW Amberg Tunnel, the point cloud condition of the initial support surface of the tunnel is roughly obtained. It is believed that based on the number of point clouds after the reduction, the random sampling of the point cloud reduction method can maintain the original shape as well as the characteristics of the point cloud. Therefore, a 200×200 point cloud data set is obtained by this method and imported into MATLAB. To complete the calculation of the surface flatness of the initial support of the tunnel, the surface fitting method based on B-spline interpolation and the SG filter smoothing method based on curvature correction need to be used to optimize the surface fitting of the point cloud data and obtain the flatness calculation benchmark reference [5]. Finally, the normal vector distance from the original point cloud to the flatness calculation datum is obtained, the calculation formula of the standard deviation and the traditional detection method of the surface flatness of the initial support of the tunnel are presented, and the calculation method of flatness based on 3DLS technology is established. Draw a flatness distribution map based on the calculated flatness. It clarifies the overall and local conditions of the surface flatness of the initial support of the tunnel and improves the entire flatness detection system. It can not only determine the flatness of the surface of the initial support of the tunnel but also determine the uneven surface area of the initial support of the tunnel [6]. The measurement operation of the point cloud data by MATLAB (Commercial mathematics software produced by MathWorks, Natick, MA, USA) is shown in Figure 2.

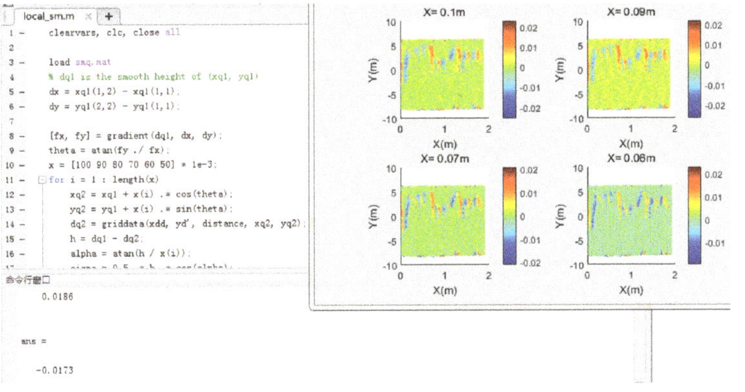

Figure 2. MATLAB operation process.

Through the comparison and analysis with traditional flatness detection methods, the surface flatness detection method of tunnel primary branches based on 3DLS technology not only meets the accuracy requirements and high accuracy, but also meets the requirements of the specification, and its detection method is feasible. The technical route of this research is shown in Figure 3.

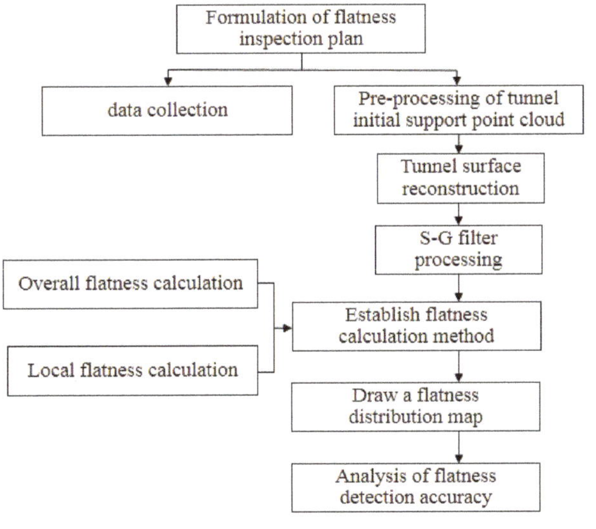

Figure 3. Technology roadmap.

1.2. Research Background

The traditional detection methods for the flatness of the initial support of the tunnel mainly include the 2 m leaning ruler combined with the tapered feeler gauge detection method and the total station detection method. Due to the randomness of the selection of detection points, the detection accuracy of this method is not high, the detection speed is slow, and the efficiency is low. When the total station is used to detect the flatness of the initial support surface of the tunnel, the randomness of the selection of detection points and the influence of the prism-free mode on the measurement accuracy will cause the fitted plane to deviate too much from the actual curved surface, which makes it impossible to obtain accurate tunnel flatness. Moreover, the selection of measurement points on the tunnel surface is subjective, so that it cannot comprehensively and truly reflect the condition of the initial support surface of the tunnel [7]. Compared with the traditional methods that have low efficiency, low accuracy, and inconvenient operation, the use of three-dimensional laser scanning technology to detect the surface flatness of the initial support of the tunnel can quickly and accurately obtain the data of the point cloud of the initial support of the tunnel. The operation is simple and does not need to touch the detected surface of the initial support of the tunnel. In addition, the accuracy of the instrument is high enough to meet the specification requirements.

At present, domestic and foreign scholars mainly apply 3D laser technology in the fields of tunnel monitoring and measurement, tunnel overall deformation analysis, etc. The application of 3D laser technology in the flatness detection of tunnel engineering is rarely studied. Duan [8] and others applied 3DLS technology to tunnel monitoring and measurement, and pointed out that when traditional detection methods are used, they are greatly affected by construction, points are easily destroyed, and data is manually recorded and inconvenient for long-term storage. Laser scanning technology has fast measurement speed and no dead ends, which effectively makes up for the shortcomings of traditional monitoring technology. Zhao [9] et al. proposed a dimensionality reduction grid deformation analysis method based on tunnel point cloud data, which can determine the tunnel deformation area and the magnitude of the deformation. Li [10] and others introduced 3DLS technology to collect 3D data of the full section of the tunnel, and qualitatively analyzed the overall deformation through chromatographic analysis, to obtain the deformation of each part of the tunnel more intuitively and automatically. Zhang [11] and others believed that the traditional method of tunnel convergence monitoring has obvious limitations and disadvantages, and clarified that the application of 3DLS technology to tunnel convergence monitoring has better advantages. The comparison with traditional methods shows the feasibility of its method. Weixing [12] and others believe that ground laser scanning technology has a huge development prospect in tunnel engineering, and clarified the advantages of ground laser scanning technology. Jong-SukYoon [13] et al. introduced a method for extracting features of tunnel concrete lining based on 3DLS technology, which provides a theoretical basis for structural health and safety inspection of tunnel construction. Manlin Xiao [14] and others introduced a tunnel surface smoothing algorithm based on mechanics correction, which can be used for 3D point cloud data collected by a 3D laser scanner. Using this algorithm, it is possible to detect and locate the damaged or deteriorated part of the inner wall of the tunnel based on the 3D laser point cloud data, effectively avoiding safety problems. Farahani [15] et al. proposed a three-dimensional laser scanning system, which can effectively obtain the contour of the 3D tunnel model, and set the deformation monitoring of the tunnel through a 3D digital image correlation system suitable for the tunnel structure. Pejić [16] proposed an optimal scheme for effectively measuring the geometry of the tunnel surface through 3DLS technology. This scheme can reliably inspect railway tunnels and achieve the purpose of optimizing the railway tunnel monitoring and measurement system. Fekete [17] and others applied the 3DLS system to the drilling and blasting tunnel operation of railway tunnel projects, which has more advantages than traditional detection methods. At present, there is relatively little research on tunnel flatness detection methods based on 3DLS technology. Therefore, research on tunnel flatness detection methods based on 3DLS technology can provide new technical ideas and methods for tunnel flatness detection.

The calculation method and analysis of the flatness are the keys to how to establish the method for detecting the flatness of the initial support surface of the tunnel based on the three-dimensional laser scanning technology. Many scholars at home and abroad have applied the 3DLS technology to flatness detection. Cheng [18] and others believe that in the engineering survey of the building facade, an electronic total station without cooperation target can be used to fit a plane and calculate the distance from the point to the fitting surface so that the fluctuation of the observation point can be observed. The overall situation can represent the flatness of the building facade. This method is feasible in the measurement of the building facade and meets the accuracy requirements. Li [19] obtained the point cloud data of the building wall according to 3DLS, obtained the accurate wall plane equation and the distance from each point to the fitting plane, and finally calculated and analyzed the wall flatness based on these distances. Based on these values, a distance statistical histogram and a distribution map of the leveling of the wall were made, and the distances were given different colors according to the threshold, which intuitively reflects the leveling of the wall. Li [20] et al. applied the three-dimensional laser scanning technology to the flatness detection of building concrete components, and also developed a color-coded deviation map to indicate the flatness of the components. At the same time, by scanning two different types of concrete components, comparative analysis shows that this method is feasible. Bosché et al. [21] proposed a new method for characterizing the flatness of building surfaces, which relies on the combination of ground laser scanning and continuous wavelet transform. It can provide accurate and extremely dense measurements on the surface of the building, while also providing a method for frequency analysis with high resolution in the spatial and frequency domains. Tang [22] et al. proposed three practical methods for evaluating the flatness of building surfaces through three-dimensional laser scanning technology. Kim [23] et al. proposed a method to detect the surface features and flatness of precast concrete components using building information modeling (BIM) and three-dimensional laser scanning technology. Based on this, a framework for evaluating the surface characteristics and flatness of concrete components is established. Fuchs [24] discussed the feasibility of introducing three-dimensional laser scanning technology into the inspection system of highway bridges, which also includes the inspection of the road smoothness of highway bridges. The research on flatness detection methods based on three-dimensional laser scanning technology is relatively complete. These studies also provide a theoretical basis and ideas for the establishment of the flatness detection system method of the primary surface of tunnel engineering.

1.3. Advantage

The establishment of a tunnel surface flatness detection system based on three-dimensional laser scanning technology is more intuitive and simple to reflect the surface flatness of tunnels. It can provide solid theoretical support and technical guidance for the tunnel engineering construction process thus has very important research significance. It solves the problem that the project in the tunnel does not meet the expectation of future construction guidance and the problem that the existing tunnel engineering initial support flatness detection system is not optimistic. The main innovation of this research is to propose a method for detecting the flatness of the initial support of the tunnel based on the three-dimensional laser scanning technology, introduce the calculation method of the flatness of the initial support of the tunnel, and introduce the concepts of overall flatness and local flatness. The comprehensive analysis of the surface flatness of the initial support of the tunnel through the combination of the overall flatness calculation method and the local flatness calculation method and the flatness distribution diagram has a good guiding role in engineering practice. It also provides research ideas for scholars who study tunnel-related fields [25].

2. Acquisition of Flatness Calculation Datum

To establish a detection system for the surface flatness of the initial support of the tunnel based on the three-dimensional laser scanning technology, the acquisition of the flatness calculation datum is very important. The point cloud of the tunnel after preprocessing (Figure 4) is very smooth, which is conducive to the reconstruction of the tunnel model. The reconstruction of the tunnel point cloud model is essentially the surface fitting of the point cloud, and the discrete point cloud is fitted to a curved surface that approximates the surface of the target object. The accuracy of the fitted surface also directly affects the calculation of the flatness of the initial support surface of the tunnel result. The flatness calculation datum surface is essentially a fitting surface suitable for flatness calculation obtained by the point cloud data through the fitting method and optimization processing. To obtain the flatness calculation reference surface, selecting a suitable fitting method can effectively improve the degree of fitting optimization.

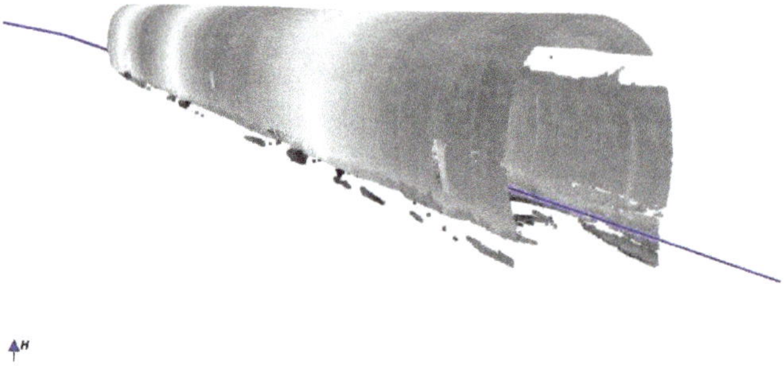

Figure 4. The point cloud of the tunnel after preprocessing.

2.1. Comparison of Fitting Methods

Common surface fitting methods include the meshing method [26], Poisson surface reconstruction method [27], Lagrangian interpolation method [28], and cubic spline interpolation method [29]. This experiment uses 3DLS The instrument obtains the surface point cloud data of the initial branch of the tunnel K109+870~K109+900 in the tunnel project, and takes the point cloud data of some areas as the analysis object, and compares and analyzes the fitting degree and fitting of the four methods when constructing the surface. Accuracy. In the schematic diagram of the degree of surface fitting by the four fitting methods, the *X* axis represents the direction along the central axis of the tunnel, the *Y* axis represents the horizontal direction of the tunnel cross-section, and the *Z* axis represents the vertical direction of the tunnel cross-section. In the schematic diagram, only the intercept Part of the area on the surface of the tunnel.

(1) The degree of fit of the four methods

After the tunnel surface is constructed by meshing (Figure 5), the surface is complete and has good continuity, but the smoothness of the surface is poor, and the details of the local area are not enough.

The tunnel surface obtained by the Poisson reconstruction method (Figure 6) is continuous and complete, which can reflect the unevenness of the tunnel surface, but the smoothness of the surface is not high, and the construction details of the point cloud cavity are insufficient.

The tunnel surface constructed by the Lagrangian interpolation method (Figure 7) can reflect the overall contour of the tunnel surface and can reflect the basic details of the tunnel surface in a local area. However, the smoothness of the surface constructed by this method is poor, and there are convex hulls. Phenomenon.

The tunnel surface constructed by the cubic B-spline interpolation method (Figure 8) is continuous and complete, with high smoothness, no local mutations, etc., and the details of the local area are rich, and the tunnel surface is better restored.

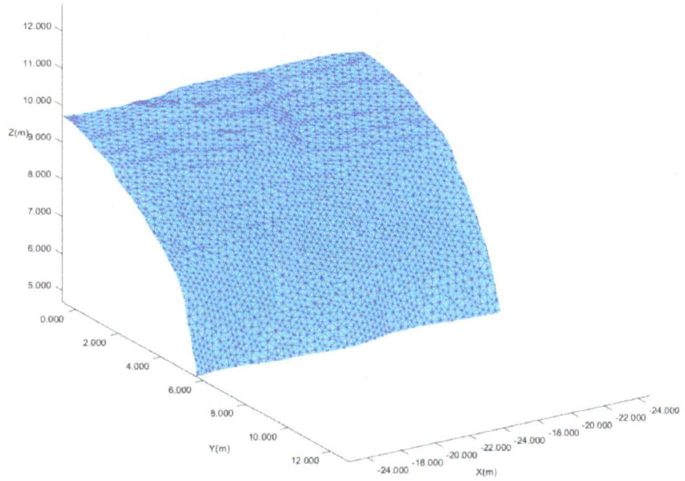

Figure 5. Partial area of tunnel surface constructed by meshing method.

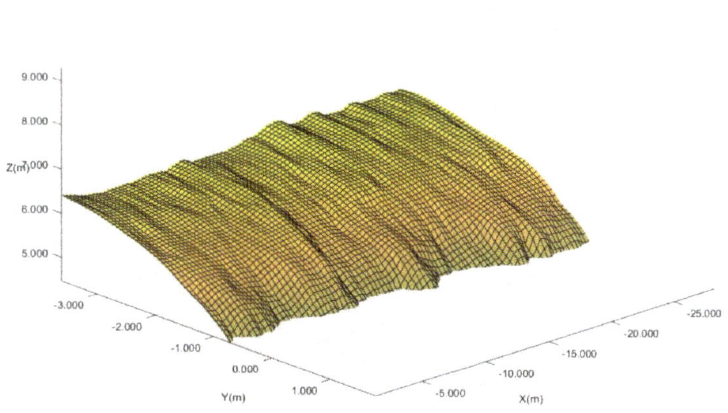

Figure 6. Partial area of tunnel surface constructed by Poisson.

(2) Analysis of the fitting accuracy of the four methods

The data this time is a total of 40,000 point clouds. The data points are extracted 5 times from the point cloud data at 5 cm intervals and 100 data points are randomly selected each time, which is divided into 5 groups. After that, the z value corresponding to each point is stored in the order of arrangement, and then the surface is constructed using four methods for the 10 cm interval point cloud data. The x and y of each stored point on the surface correspond to the corresponding z value, 5 cm interval points are compressed to get 10 cm interval point clouds, so the z value on the surface obtained by the 10 cm interval point cloud fitting is different from the corresponding z values of the points stored in the 5 cm interval point cloud. According to the x and y of the stored point, the corresponding surface can be obtained and stored, and then the difference of the corresponding point is calculated. This method is equivalent to the error calculation of the 10 cm interval point cloud, by calculating the sum of each point The fitting difference value corresponding to the fitting surface, the fitting difference value formula is

$$E = z_p - z'_p \tag{1}$$

where: $p = 1, 2, 3, \ldots, n$.

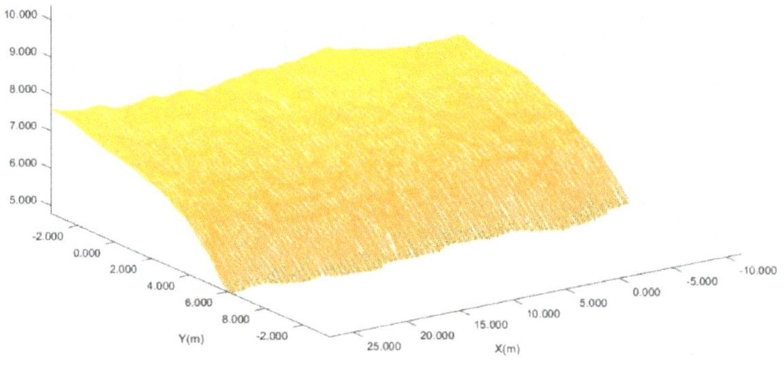

Figure 7. Local area of tunnel surface constructed by Lagrangian interpolation method.

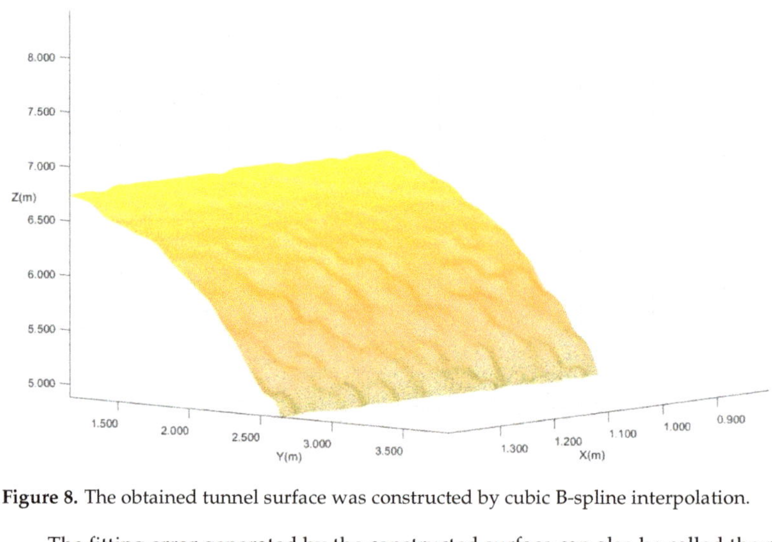

Figure 8. The obtained tunnel surface was constructed by cubic B-spline interpolation.

The fitting error generated by the constructed surface can also be called the root mean square error. By counting the root mean square error, you can get the error of the area where the surface other than the original point is located. The root mean square error formula is:

$$\sigma = \sqrt{\frac{E_1^2 + E_2^2 + \ldots\ldots + E_N^2}{n}} = \sqrt{\frac{\sum E_p^2}{n}} \tag{2}$$

According to Formula (2), four interpolation methods are used to construct the surface, and the fitting error is calculated based on 5 sets of data, as shown in Table 2.

Table 2. Fitting errors of the four sets of data corresponding to the four methods.

Fitting Error σ	Method 1 (mm)	Method 2 (mm)	Method 3 (mm)	Method 4 (mm)
Group 1	3.024	0.995	1.814	1.009
Group 2	3.071	1.015	1.775	0.984
Group 3	3.043	1.044	1.823	1.020
Group 4	3.068	1.007	1.814	1.005
Group 5	3.108	0.980	1.814	0.990

Note: Method 1 means meshing method, method 2 means Poisson reconstruction method, method 3 means Lagrangian interpolation method, method 4 means bicubic spline interpolation method.

It can be seen from the tab that the point cloud fitting errors of the Poisson reconstruction method and cubic B-spline interpolation method are kept in a small range, and the fitting accuracy is high. The fitting accuracy of the Lagrangian interpolation method is not high. The grid division method has the largest fitting error and the lowest precision. The results show that the construction of tunnel surface by the Poisson reconstruction method and cubic B-spline interpolation method has higher fitting accuracy.

By comparing and analyzing the fitting degree and fitting accuracy of the four fitting methods when constructing the surface, the following conclusions are drawn: the smoothness of the surface constructed by the meshing method is poor, and the fitting accuracy is low; Poisson reconstruction The fitting accuracy of the method is high, but the surface lacks a certain degree of smoothness; the surface fitting effect constructed by the Lagrangian interpolation method is better, but the fitting accuracy is not high; the cubic B-spline interpolation method has high fitting accuracy, compared with other methods, the surface details are complete and the smoothness is higher. This method has more advantages in the construction of the tunnel surface and has the least influence on the calculation results of the surface flatness of the initial support of the tunnel.

2.2. *Optimal Fitting of Cubic B-Spline Interpolation*

The main ideas for the optimal fitting of the original point cloud of the tunnel based on the cubic B-spline interpolation method are:

The point cloud data is divided into slices according to the x-direction, that is, the tunnel axis direction, which is equivalent to taking a tiny dx as the threshold. The x coordinate changes of the point cloud data within this range are considered to be on a 2-dimensional slice point. Then perform spline curve-fitting on this two-dimensional slice. First, convert Cartesian coordinates to polar coordinates. Since the cross-section of the tunnel is a curve similar to an arc, after changing to polar coordinates, it can be ensured that the depression angle and the polar radius can be in a one-to-one correspondence. Then use polar coordinates for interpolation and encryption based on the spline curve, and then convert the polar coordinates back to rectangular coordinates to complete the fitting of each section of the tunnel. For the same reason, perform the above processing again in the Y direction. After processing, the curve interpolation is carried out in the two orthogonal directions of the tunnel x and y, and the interpolation in the two directions is superimposed to form the fitting surface of the first branch surface of the tunnel [30].

The method of two-way slice complementary to the overall surface of the tunnel proposed in this study effectively eliminates the jagged layering effect of the one-way slice on the overall fitting surface of the tunnel, and the enlarged tunnel surface appears smoother. The cloud fitting operation speed is also faster than the overall point cloud fitting surface [31].

2.3. *S-G Filter Smoothing Based on Curvature*

The curved surface after the fitting process by cubic B-spline interpolation has a high degree of smoothness, good continuity, and more complete and rich local details, which is more consistent with the actual engineering situation. However, the fitted surface still has many point clouds that deviate from the actual surface, and its accuracy cannot meet the requirements of flatness calculation. To make the final fitting surface that can meet the requirements of flatness calculation, the fitting surface should not only be close to the actual situation but also ensure that the fitting surface has sufficient smoothness. To further limit the smoothness and authenticity of the fitted surface, this study guarantees the reliability of the fitted surface by limiting the curvature of each point on the fitted surface. Based on the fitting processing of cubic B-spline interpolation, The fitted surface is again processed by SG filtering based on curvature limitation [32]. In all tunnel projects, the design parameters of the Leicaoshan Tunnel are universal, and among many tunnels, Leicaoshan Tunnel is the most typical. The flatness detection method in this study specifies the upper and lower limits of the tunnel point cloud curvature as the curvature parameter value of the Leicaoshan Tunnel. That is, the upper limit is specified as the maximum curvature of the vault in the design parameters of Leicaoshan Tunnel, 0.395, and the lower limit is specified as the minimum curvature of the arch bottom in the design parameters of Leicaoshan Tunnel, 0.104.

The specific plan for curvature limitation is as follows:

(1) First, the curvature of any point on the curve is calculated by the curve function. The curvature of a point is calculated based on the first and second derivatives of the two points before and after. Since the fitted surface obtained by the B-spline interpolation method is fitted by the slicing method, the curvature in this step is also used in the same way. The two-dimensional curvature of each tangent surface in the previous B-spline interpolation method is calculated. Superimposed to form the curvature of the entire surface, the formula for calculating the curvature is as follows:

$$K = \frac{|y''|}{(1 + y'^2)^{3/2}} \tag{3}$$

In the formula, K is the curvature at a point on the curve, and y is the function of the corresponding curve.

(2) There are points in the calculated surface point cloud curvature that are not within the limited range. At this time, it is necessary to perform smoothing and filtering again on the surface obtained by the B-spline interpolation method. For the cross-sectional direction and the longitudinal direction, the data in the two directions is smoothed and filtered again. Taking into account the symmetry of the tunnel and the requirements for the calculation of the surface flatness of the initial support of the tunnel, Savitzky–Golay filtering is used here. The processing of S-G filtering maintains the best shape of the original data, making the processed surface closer to the actual engineering situation.

(3) After removing the unqualified points through the above method, the remaining unqualified points are only the points at the bottom of the arch. Because the curvature calculation is calculated using the curvature of the front and back slices, the curvature calculated at the two ends of the point cloud data, namely the two arch bottoms on the left and right, is meaningless in itself, and it can be directly eliminated. At this point, the final flatness calculation datum (Figure 9) is obtained surface. In the figure, the X axis direction represents the direction along the central axis of the tunnel, the Y axis represents the horizontal direction of the tunnel cross-section, and the Z axis represents the vertical direction of the tunnel cross-section. Segment fitting surface.

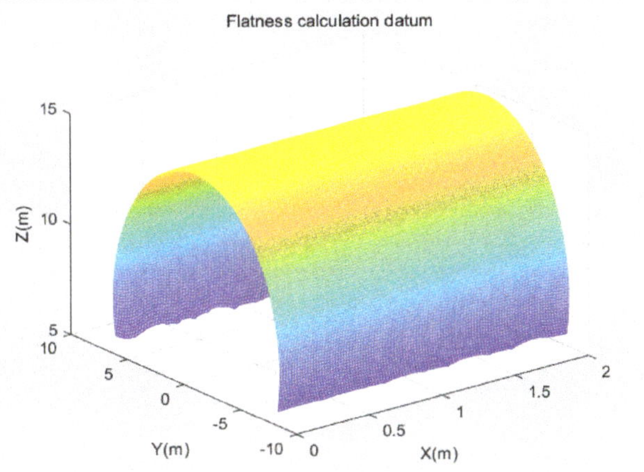

Figure 9. Flatness calculation datum.

3. Flatness Calculation

Through the above method, the flatness calculation reference surface has been obtained. The flatness calculation method in this article is based on the normal vector distance from the original point cloud of the initial support surface of the tunnel to the flatness calculation datum plane. The calculation of the normal vector distance only needs to use the original point cloud to make a normal ray perpendicular to the flatness calculation datum plane. The normal ray and the fitted surface intersect at a point, and the distance between this point and the starting point of the ray is the normal vector distance di. Flatness calculation can be divided into two parts: overall flatness and local flatness. The calculation of overall flatness is to analyze the unevenness of the overall point cloud in a region, and the calculation of local flatness is to analyze the point cloud on the local details of a region. The degree of unevenness.

3.1. Overall Flatness

The overall flatness of the initial support surface of the tunnel is mainly determined by the dispersion degree of the normal vector distance from the original point cloud to the flatness calculation datum. If the dispersion is large, it means that the original point cloud and the flatness calculation datum are quite different. The rougher the surface. If the degree of dispersion is small, it means that the difference between the original point cloud and the flatness calculation reference plane is small, and the surface is flatter.

To intuitively express the overall flatness of the surface of the initial support of the tunnel, the concept of standard deviation is introduced. Simply put, the standard deviation is a measure of the degree to which a set of values are scattered from the average. A larger standard deviation means that the difference between most of the values and its average is larger; a smaller standard deviation means that these values are closer to the mean.

To sum up, the formula for calculating the overall flatness of the initial support surface of the tunnel is as follows:

$$m_0 = \pm\sqrt{\frac{\sum d_i^2}{n-1}}(i = 1, 2, \cdots, n) \tag{4}$$

In the formula, m_0 is the overall flatness of the initial support surface of the tunnel, di is the normal vector distance, and n is the no. of point clouds collected during flatness detection.

3.2. Local Flatness

The overall flatness of the surface of the initial support of the tunnel can only reflect the overall flatness of a specific section of the tunnel. In a specific tunnel project, the overall flatness of the surface of the first support of the tunnel can only play a qualitative role, but cannot pass a quantitative one. Method to define the leveling degree of the specific local location of the tunnel surface. Therefore, the concept of local flatness is introduced, and the uneven points on the local details of the initial support surface of the tunnel are expressed through the concept of local flatness.

In the traditional method of detecting the surface flatness of the initial support of the tunnel, the 2 m ruler method is generally used to define the flatness: the maximum gap value between the reference plane of the 2 m ruler and the measuring surface. Usually, two points are measured every 200 m, and each point is continuously tested 10 times. According to the qualified rate, it is judged whether the surface flatness meets the measurement requirements. The schematic diagram for defining the flatness of the 2 m leaning rule method is shown in Figure 10.

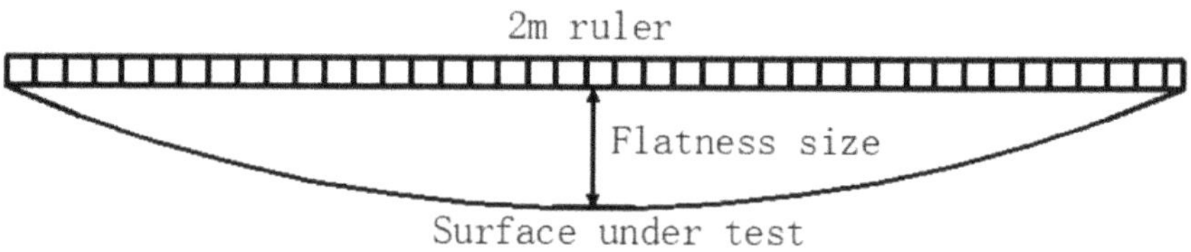

Figure 10. Schematic diagram of flatness definition of 2 m by ruler method.

By referring to the flatness definition method in the 2 m ruler method, this paper introduces the local flatness definition method based on 3DLS technology:

(1) First obtain the three-dimensional distribution map of flatness as shown in Figure 11, where the X axis represents the direction of the tunnel's central axis, the Y axis represents the direction of the tunnel cross-section, and the Z axis represents the normal vector distance of the original point cloud and flatness calculation reference plane. Flatness situation. It can be clearly seen from the three-dimensional map of flatness that the flatness of the tunnel has symmetry along the direction of the tunnel's central axis ($Y = 0$). The flatness of the left arch of the tunnel is mostly on the negative Z semi-axis, and the flatness of the right arch of the tunnel is mostly on the positive Z semi-axis.

(2) The flatness data in the Figure constitutes a 3-dimensional scalar field. First, calculate the gradient value of this scalar data point at each point. However, the size of this gradient is of no practical significance to the calculation of local flatness here, because what needs to be known here is only the direction of the gradient. Theoretically, every point on a three-dimensional surface has countless directions that can be changed, and the gradient of each point is calculated only to obtain the direction with the largest change in flatness.

(3) As it is the gradient of a 2-variable function, the value obtained corresponds to the components in the x-direction and the y-direction, which is as follows:

$$grad(f) = \frac{\partial f}{\partial x}i + \frac{\partial f}{\partial y}j \tag{5}$$

In this way, the direction of the gradient vector can be determined based on the x component and the y component, and the direction angle of the gradient can also be obtained.

(4) The flatness distribution map drawn according to the normal vector distance is similar to the topographic map. Here we draw on the principle of slope in the topographic map, slope = elevation difference/horizontal distance, in the gradient direction of each point, calculate the slope i and the inclination angle. The distance X can be adjusted according to the accuracy requirements.

(5) Once the horizontal distance is determined, the height difference between the two points can be known. It can be imagined that a certain point is the center of the circle, the horizontal distance is the radius, and the gradient direction is unique. In the gradient direction, the height changes the fastest, and the highest height difference is obtained when the horizontal distance is constant.

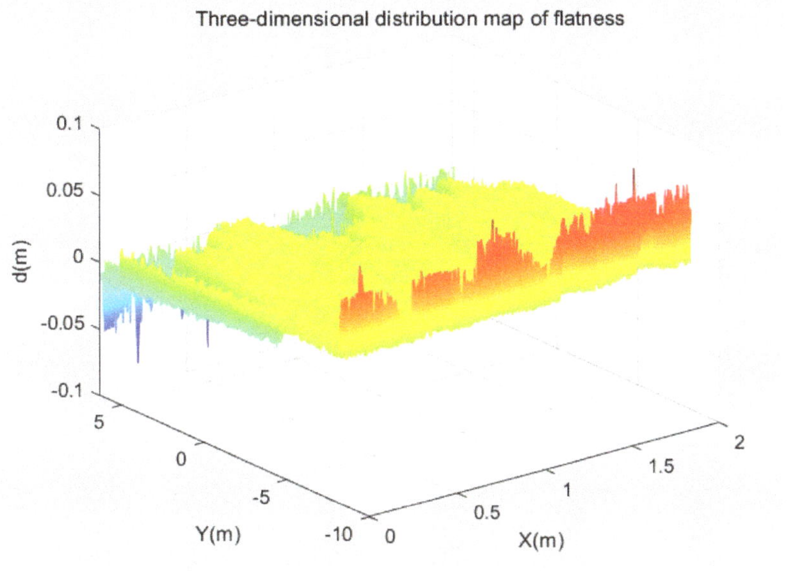

Figure 11. 3D diagram of the flatness distribution of the fitting surface after smoothing.

Note: What needs attention here is how to find the corresponding horizontal distance according to the direction of the gradient, as shown in Figure 10 below.

Assume that the red arrow is the direction angle of the calculated gradient vector, and the black dot is the normal vector distance data of a certain point, that is to say, the normal vector distance data is along the red arrow. The direction is the fastest-changing direction (gradient direction). The original coordinates of the point can be identified in Figure 10, and the length X along the gradient direction can be expressed as the green point in Figure 12. In the gradient diagram, the horizontal axis represents the direction along the central axis of the tunnel, and the vertical axis represents the direction of the cross-section of the tunnel. The endpoint of the original point gradient direction is the green point, which can be obtained in the two-dimensional coordinate plane.

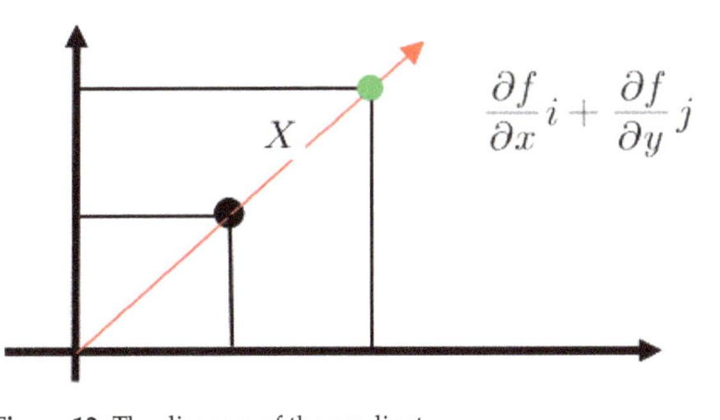

Figure 12. The diagram of the gradient.

(6) To make the definition of local flatness meet the measurement requirements as much as possible and meet the technical specifications of tunnel engineering construction, the introduced three-dimensional laser scanning technology-based tunnel engineering primary surface flatness definition formula is as follows:

$$m_1 = \frac{h\cos\alpha}{2} = \frac{x\sin\alpha}{2} = \frac{xh}{2\sqrt{h^2+x^2}} \tag{6}$$

In the formula, m_1 is the local flatness of the initial support surface of the tunnel, X represents the step distance of the original point cloud along the gradient direction, h represents the height difference between the starting point and the endpoint of the stepping direction, and α represents the inclination angle between the two points.

The schematic diagram of the local flatness definition is shown in Figure 13. The horizontal axis represents the direction of the horizontal step distance X, and the vertical axis represents the relative height h between the original point and the endpoint of the gradient direction.

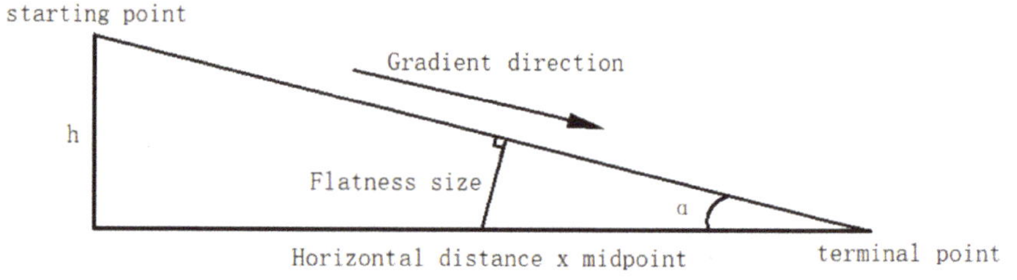

Figure 13. Schematic diagram of local flatness definition based on 3DLS technology.

4. Engineering Case Analysis

The project supported by this test is the Lushan Tunnel Project shown in Figure 14. The Lushan Tunnel is located in Fuyang, Hangzhou City, Zhejiang Province. It is a section of the newly-built Huhang Railway. The main surrounding rock grade is Grade V, and the area sections that need to collect point cloud data for flatness detection are mainly concentrated in the initial support section of the tunnel. A total of 30-m-long section areas are collected. The mileage section is DK109+870~DK109+900, whichever is selected The 2 m part is used as the analysis object of this experiment. The parameters of some regional sections of the Lushan Tunnel Project are shown in Table 3.

Figure 14. Field view of the tunnel.

The experiment uses a three-dimensional laser scanner to scan the surface of the initial support of the tunnel and collect the point cloud data of the surface of the initial support of the tunnel in a section of the area. After the point cloud data is preprocessed, the surface point cloud data of the initial support of the tunnel is curved and smoothed by the programming method to obtain the final flatness calculation datum. The normal vector distance is obtained according to the datum plane, to calculate and analyze the flatness of the initial support surface of the tunnel.

Table 3. List of regional parameters of Lushan Tunnel.

Tunnel Name	Serial No.	Scope	Surrounding Rock Grade	Length (m)	Geological Condition
Lushan Tunnel	1	DK107+468~DK107+870	V	402	Inlet section, shallow buried section
	2	DK107+870~DK107+920	IV	50	sandstone
	3	DK107+920~DK108+715	III	795	sandstone
	4	DK108+715~DK108+770	IV	55	Shallow buried section
	5	DK108+770~DK109+570	V	800	Shallow buried section
	6	DK109+570~DK109+645	IV	75	Fault affected zone
	7	DK109+645~DK109+720	IV	75	Fault affected zone
	8	DK109+720~DK110+180	III	460	Feldspar quartz sand layer

4.1. Analysis of Overall Flatness

In the experiment, the point cloud data of a 2-m-long tunnel surface cross-section was taken along the direction of the central axis of the tunnel. After preprocessing the point cloud data, a total of 40,000 points are selected by random sampling, and their normal vector distance is calculated, and a 200 × 200 normal vector distance scalar matrix is formed at the same time. Then, through the coordinates of the point cloud data, there are one-to-one correspondence in the *X*-axis central axis direction and the *Y*-axis cross-sectional direction. Finally, the normal vector distance scalar matrix is colored into a graph to analyze the overall flatness of the tunnel surface, as shown in Figure 15.

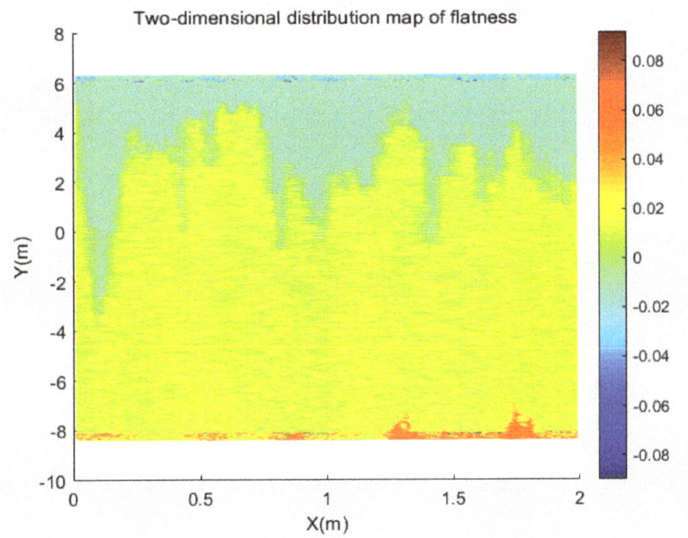

Figure 15. Overall smoothness distribution of tunnel primary branch surface.

According to the calculation formula for the overall flatness of the surface of the initial support of the tunnel, the overall flatness of the surface of the first support of the tunnel is 19.9 mm. According to the "GB-T 50299-2018 Construction Quality Acceptance Standard for Underground Railway Engineering" [33], the flatness of the sprayed concrete is allowed the deviation should be 30 mm, and the overall flatness of the initial support surface of the tunnel meets the requirements of the specification.

4.2. Analysis of Local Flatness

The analysis of local flatness first draws a flatness distribution map through the normal vector distance, and each point in the figure obtains the gradient of each point according to the gradient formula. In the gradient direction, the flatness changes the fastest. The local flatness is calculated in the gradient direction, and the local flatness is calculated and analyzed by determining the value of the horizontal distance X in the gradient direction. After the value of the horizontal distance X is determined, the height difference h between the starting point and the endpoint can be obtained, and X and h can be substituted into the local flatness calculation formula to obtain the final local flatness.

In this experiment, to analyze the influence of the value of the horizontal distance x on the local flatness results of the initial support surface of the tunnel, four-parameter values of $X = 5$ mm, $X = 10$ mm, $X = 20$ mm, and $X = 50$ mm were taken to determine the local flatness. The calculation and analysis of flatness are shown in Figure 16. In the figure, the X-axis is along the central axis of the tunnel, and the Y-axis is the direction of the cross-section of the tunnel.

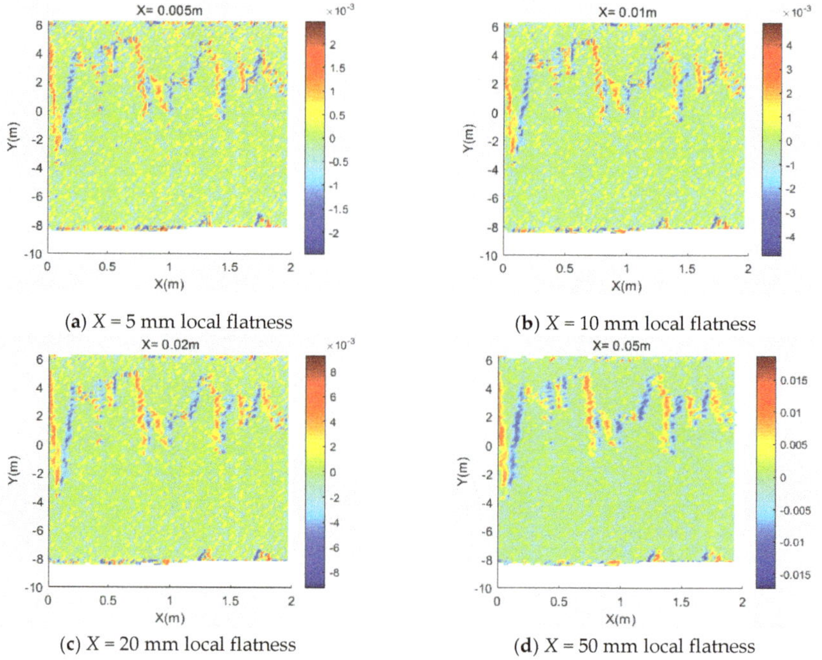

(**a**) $X = 5$ mm local flatness (**b**) $X = 10$ mm local flatness (**c**) $X = 20$ mm local flatness (**d**) $X = 50$ mm local flatness

Figure 16. Overall smoothness distribution of tunnel primary branch surface.

Through the calculation of the local flatness of the initial support surface of the tunnel, when $X = 5$ mm, the maximum local flatness is 2.5 mm and the minimum is -2.5 mm; when $X = 10$ mm, the maximum local flatness is 4.9 mm and the minimum When $X = 20$ mm, the maximum local flatness is 9.3 mm and the minimum is -9.3 mm; when $X = 50$ mm, the local flatness maximum is 18.6 mm and the minimum is -17.3 mm.

The traditional flatness detection of the initial support of the tunnel is to detect the flatness of the initial support of the tunnel by a combination of a two-meter ruler and a wedge feeler. During the inspection, place the two-meter ruler horizontally on the tunnel surface along the direction of the central axis of the tunnel. The ruler is close to the tunnel surface and finds the largest gap. Place the wedge-shaped feeler gauge here. The reading is the ruler datum plane and the initial support of the tunnel. The maximum gap distance of the protective surface is the flatness of the initial support surface of the tunnel [34]. According to the "GB-T 50299-2018 Construction Quality Acceptance Standard for Underground Railway Engineering", the allowable deviation of the flatness of the shotcrete should be 30 mm, and the local flatness of the initial support surface of the tunnel for the four sets of data meets the requirements of the specification.

When the value of X gradually increases, the local flatness value of the initial support surface of the tunnel also gradually increases. At the same time, when the value of X gradually increases, the local flatness value also tends to stabilize. To further determine when the local flatness tends to be stable, this paper adds five sets of variable test data X = 60~100 mm, as shown in Figure 17. In the figure, the *X*-axis is along the central axis of the tunnel, and the *Y*-axis is the direction of the cross-section of the tunnel.

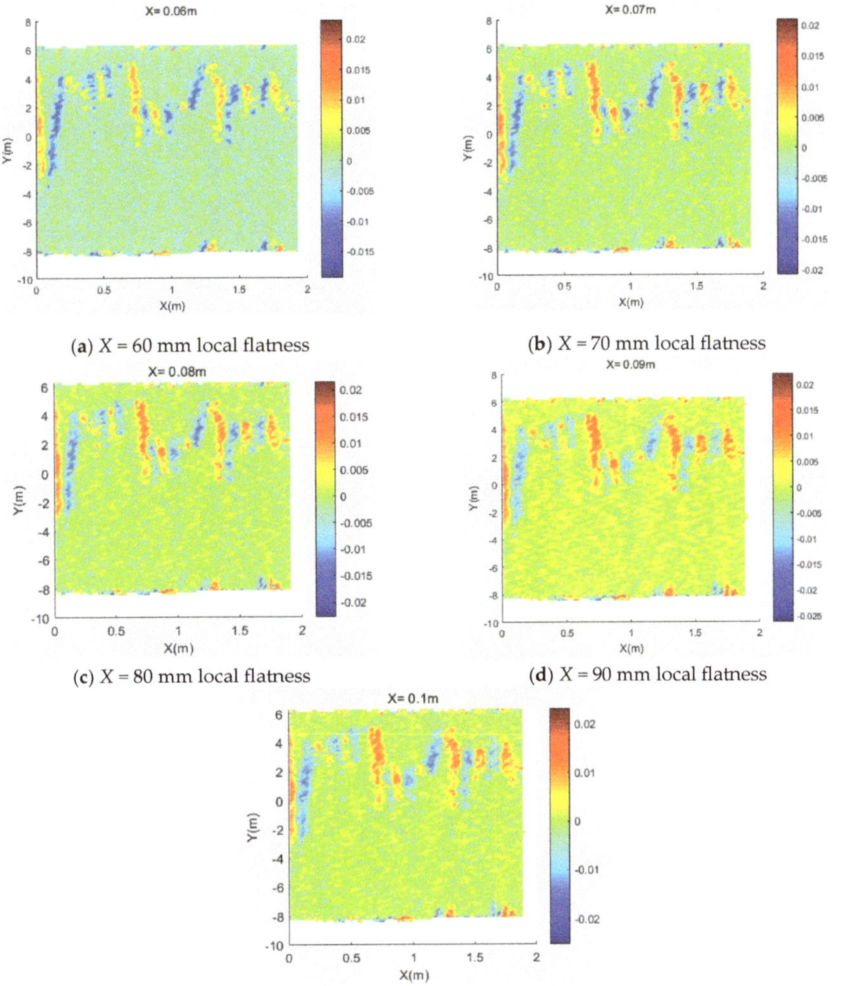

(**a**) X = 60 mm local flatness

(**b**) X = 70 mm local flatness

(**c**) X = 80 mm local flatness

(**d**) X = 90 mm local flatness

(**e**) X = 100 mm local flatness

Figure 17. Distribution diagram of local flatness of the variable set.

It can be seen from the two sets of test data that the horizontal distance X has a greater impact on the calculation of local flatness. The results of overall flatness and local flatness are summarized, as shown in Table 4 below.

Table 4. Summarization of flatness calculation results.

	Max (mm)	Minimum (mm)	Abs Max (mm)
Overall flatness	19.9		19.9
Local flatness (X = 5 mm)	2.5	−2.5	2.5
Local flatness (X = 10 mm)	4.9	−4.8	4.9
Local flatness (X = 20 mm)	9.3	−9.3	9.3
Local flatness (X = 50 mm)	18.6	−17.3	18.6
Local flatness (X = 60 mm)	19.5	−19.4	19.5
Local flatness (X = 70 mm)	19.9	−20.0	20.0
Local flatness (X = 80 mm)	20.6	−20.8	20.8
Local flatness (X = 90 mm)	20.3	−20.6	20.6
Local flatness (X = 100 mm)	20.1	−20.3	20.3

The following conclusions can be drawn from the analysis of the above table and the flatness distribution graph: (1) The calculation of local flatness depends on the program setting parameter step distance X. Choosing a suitable step distance parameter is very important for local flatness detection. When the value of X becomes larger and larger, the value of local flatness becomes larger and larger, and its value will approach the upper limit, which is similar to the calculation result of overall flatness. The approximate interval of the advance distance X is between 70 mm~90 mm, and it can be preliminarily judged that the Dangmai advance distance parameter setting between 70 mm~90 mm is suitable for local flatness detection. (2) By comparing and analyzing the flatness of the local flatness distribution map and the overall flatness distribution map, the flatness distributions of the two are basically the same. It can be explained that the detection methods of the overall flatness and the local flatness affect the flatness of the initial support of the tunnel. The representations are the same and the two have commonality. At the same time, the surface flatness of the initial support of the tunnel in this area meets the specification requirements. The local flatness calculation formula based on three-dimensional laser scanning proposed in this experiment is feasible.

4.3. Feasibility Analysis

To verify the effectiveness and feasibility of the method of detecting the surface flatness of the initial branch of the tunnel based on the 3DLS technology, this study compares and analyzes the point cloud data detected by the total station and the point cloud data detected by the 3DLS technology. The tunnel mileage K107+780, K107+785, K107+790, and K107+795 are, respectively, taken from four cross-section information, and each cross-section is taken from the left and right arch bottom, left and right arch waist, and the position of the archtop. The points were measured 4 times with a three-dimensional laser scanner at the same time interval. The thickness of the difference between the monitoring point and the design section collected by the total station is recorded as d_1, and the thickness of the difference between the monitoring point and the design section collected by 3DLS is recorded as d_2, and the statistical results of the test data collected by the two measuring instruments are summarized, as shown in Table 5 below.

It can be seen from the data in the tab that the detection value collected by 3DLS is roughly the same as the detection value collected by the total station. To further illustrate the accuracy and effectiveness of the flatness detection method based on the 3DLS technology, this study will the detection value of the total station is regarded as the most reliable value $\hat{x}$. and the detection value of the scanner is regarded as the observation value x_i. The detection difference Δd of 4 tests can be calculated respectively, and finally the median error $\hat{\sigma}$ of 3DLS can be calculated by the detection difference Δd. The error $\hat{\sigma}$ in the calculated observation value can represent the true error of 3DLS. The Medium error result is shown in Table 6.

Table 5. Section monitoring point detection value.

Section Mileage Stake	Monitoring Points Position	Class 1 Detection Value d_2/mm	Class 2 Detection Value d_2/mm	Class 3 Detection Value d_2/mm	Class 4 Detection Value d_2/mm	Total Station Detection Value d_1/mm
	Left arch	196	203	195	193	199
	Left arched waist	208	211	208	204	210
K107+780	dome	202	203	206	200	204
	Right arched waist	179	178	183	184	180
	Right arch	135	127	140	*	143
	Left arch	243	246	242	233	233
	Left arched waist	229	230	246	241	237
K107+785	dome	230	229	230	220	226
	Right arched waist	219	230	212	225	216
	Right arch	219	218	222	*	225
	Left arch	184	190	160	176	171
	Left arched waist	247	268	247	245	237
K107+790	dome	209	229	216	212	212
	Right arched waist	216	214	208	202	213
	Right arch	176	177	173	*	174
	Left arch	239	258	236	*	224
	Left arched waist	274	266	265	260	257
K107+795	dome	244	249	244	242	232
	Right arched waist	200	218	204	214	194
	Right arch	191	201	*	*	183

Note: * indicates that the location is blocked, and the data at this point has not been measured.

Table 6. The two instruments measure the difference.

Section Mileage Stake	Distance to Station/m	Monitoring Points Position	Class 1 Difference Δd/mm	Class 2 Difference Δd/mm	Class 3 Difference Δd/mm	Class 4 Difference Δd/mm
		Left arch	3	4	4	6
		Left arched waist	2	1	2	6
K107+780	0	dome	2	1	2	4
		Right arched waist	1	2	3	4
		Right arch	8	16	3	*
	Medium error $\hat{\sigma}$		5	8	3	6
		Left arch	10	13	9	0
		Left arched waist	8	7	9	4
K107+785	5	dome	4	3	4	6
		Right arched waist	3	14	4	9
		Right arch	6	7	3	*
	Medium error $\hat{\sigma}$		8	11	7	7
		Left arch	13	19	11	5
		Left arched waist	10	31	10	8
K107+790	10	dome	3	17	4	0
		Right arched waist	3	1	5	11
		Right arch	2	3	1	*
	Medium error $\hat{\sigma}$		9	20	8	8
		Left arch	15	34	12	*
		Left arched waist	7	9	8	3
K107+795	15	dome	12	17	12	10
		Right arched waist	6	24	10	20
		Right arch	8	18	*	*
	Medium error $\hat{\sigma}$		14	25	12	16

Note: * indicates that the location is blocked, and the data at this point has not been measured.

It can be seen from Table.6 that the true error of the 3DLS will increase as the distance between the section and the station increases. The cross-sectional instrument method, total station coordinate method or 3DLS method can be used, and the error in the measurement should not be greater than 25 mm [35]. Because the selection of the initial section of the tunnel in this study is a cross-sectional area of 2 m in the direction of the central axis By default, the center position of the cross-sectional area is the position of the measuring station, so the distance between the instrument and the scanned cross-section has little effect on the flatness detection results of this study. The true error of the 3DLS meets the requirements of the specification within a certain measurement range. The accuracy of the point cloud data collected by the 3DLS is improved.

5. Conclusions

This paper is based on the three-dimensional laser scanning technology to obtain the point cloud data of the initial support surface of the railway tunnel, and expounds the use of the tunnel point cloud data, through the B-spline interpolation method and the S-G smoothing method based on the curvature limitation, to obtain the flatness calculation reference plane. The normal vector distance formed by the intersection of the normal line drawn from the original point cloud and the flatness calculation datum plane is proposed, and the normal vector distance is used as the basis for flatness calculation, and two concepts based on the detection of the surface flatness of the initial support of the tunnel are introduced: the whole Flatness and local flatness. Through the analysis of the flatness distribution map and the flatness calculation results, the feasibility of the application of the three-dimensional laser scanning technology in the surface flatness detection of the initial support of the tunnel engineering is verified and discussed. The main conclusions are as follows:

(1) Compared with the traditional total station method and the two-meter ruler method in the traditional flatness detection, the efficiency is low, the accuracy is not high, and the operation is inconvenient. The use of three-dimensional laser scanning technology to detect the surface flatness of the initial support of the tunnel can quickly and accurately obtain the tunnel. A large amount of point cloud data on the surface of the initial support is easy to operate, without touching the surface of the initial support of the tunnel to be tested, and the accuracy of the instrument is high enough to make up for the problem of acquisition accuracy.

(2) Common surface fitting methods include meshing method, Poisson surface reconstruction method, Lagrangian interpolation method, and cubic B-spline interpolation method. To obtain the most suitable surface fitting method for this experiment, the experiment compares and analyzes the surface fitting degree and fitting accuracy of the four surface fitting methods. According to the comparative analysis, compared with the other three surface fitting methods, the tunnel surface constructed by the cubic B-spline interpolation method is continuous and complete, with higher smoothness, no local mutations, etc., and the details of the local area are rich and relatively The tunnel surface is restored well. At the same time, by calculating the statistical root mean square error of the four fitting methods, the fitting error of the surface area other than the original point can be obtained. The calculation result can be obtained by using the Poisson reconstruction method and the cubic B-spline interpolation method. The point cloud fitting error can be kept within a small range, and the fitting accuracy is high. Comprehensive comparative analysis shows that cubic B-spline interpolation is the most suitable fitting method for this study.

(3) After comparative analysis, a suitable surface fitting method for this experiment, namely cubic B-spline interpolation, has been obtained. On this basis, this research puts forward the method of two-way slice complementarity in the B-spline interpolation method to fit the overall surface of the tunnel and the method of SG filter smoothing based on curvature, which effectively eliminates the one-way slice to fit the whole tunnel. The jagged layering effect of the curved surface optimizes the fitting process of the tunnel curved surface. The optimized fitting surface is continuous and complete, with high smoothness,

rich and complete local details, which is consistent with the actual engineering situation, and better restores the tunnel surface. At the same time, it can also meet the requirements of the flatness calculation of the initial support of the tunnel. It can be used as a reference plane for flatness calculation.

(4) The intersection of the normal line drawn from the original point cloud and the flatness calculation datum forms the normal vector distance. Based on this, this research proposes two flatness calculation methods and draws the flatness distribution map. The calculation of overall flatness can determine the overall flatness of the initial support surface of the tunnel, and the calculation of local flatness can determine the specific location of the uneven area. Combined with the flatness distribution map, the flatness detection can be more accurate and intuitive, which can be used for tunnel engineering. The construction provides technical support and theoretical guidance.

(5) In the flatness calculation method, this research proposes a local flatness calculation method. The step distance X set by the program is the main factor affecting the local flatness. In the flatness detection of the actual tunnel engineering, it is necessary to set an appropriate stepping distance X according to actual engineering conditions. To explore and analyze the influence of step distance X on local flatness, this study set up experimental groups with different step distance X to conduct analysis. The results show that the local flatness will increase as the step distance X increases. Finally, Infinite approaches the upper limit, which is roughly stable at around 20 mm, which is roughly the same as the overall flatness calculation result. According to the analysis of the flatness distribution map obtained by the two flatness calculation methods, the flatness of the initial tunnel support surface is basically consistent, which proves that the tunnel flatness detection method proposed in this study is feasible.

(6) Through the comparative analysis with the traditional flatness detection method, the true error of 3DLS meets the specification requirements within a certain measurement range, and both are less than the 25 mm required by the specification. This shows the accuracy of the point cloud data collected by 3DLS. At the same time, the initial fitting surface fitting of the tunnel project obtained by 3DLS technology has a higher degree of optimization and is closer to the actual engineering situation. The flatness calculation method is simpler and more effective, and the flatness analysis based on the flatness distribution map is more Precise and intuitive. The method of detecting the flatness of the initial support surface of the tunnel based on the three-dimensional laser scanning technology is feasible.

In summary, compared with traditional detection methods, 3DLS are faster and more accurate, with high acquisition accuracy, wide range, and simple operation in the detection of the flatness of the initial support surface of the tunnel. The surface fitting effect is best after cubic B-spline interpolation and SG filter smoothing based on curvature limitation. In this study, the normal vector distance is formed by the intersection of the normal line drawn from the original point cloud and the flatness calculation datum surface, and based on this, the concepts of overall flatness and local flatness are proposed, and corresponding flatness distribution maps are drawn respectively. The size of the local flatness will increase as the step distance X becomes larger, and finally, approach the calculation result of the overall flatness infinitely. Comparing the measurement accuracy of 3DLS and the traditional detection instrument, and comparing the flatness distribution map and the calculation results, it can be preliminarily concluded that the tunnel flatness detection method proposed in this study is feasible.

6. Outlook

The flatness detection method in this study is mainly for curved surfaces similar to the tunnel surface. For the traditional flatness detection methods, flatness detection can only be performed on relatively flat road surfaces or building walls. Through the research of the flatness detection method in this experiment, the purpose of flatness detection on a complex curved surface is realized. However, there are still some shortcomings in the research process:

(1) Obtaining the flatness calculation datum plane in the tunnel flatness detection calculation is a key step in calculating the flatness of the initial support of the tunnel, but how to determine that the flatness calculation datum obtained by processing is the most suitable and optimal solution flatness calculation datum plane, its treatment method is still worthy of further study.

(2) At this stage, there is no system for systematic evaluation of the flatness detection of the initial branches of the tunnel using 3DLS technology, so the construction of a more complete 3DLS tunnel flatness detection and evaluation system is the next research direction.

Author Contributions: Conceptualization and methodology, Y.D., L.X. and Z.W.; data curation, Y.D.; investigation, Y.D.; writing—original draft preparation, Y.D. and L.X.; writing—review and editing, Y.D., Z.W. and H.Z.; supervision, Z.W., H.Z. and Z.L.; Validation, Z.W. and Z.L.; Project administration, Z.L. All authors have read and agreed to the published version of the manuscript.

Funding: The research was funded by the key project of Zhejiang Provincial Department of Transportation, and the grant number is 2019015.

Institutional Review Board Statement: Not applicable.

Informed Consent Statement: Not applicable.

Data Availability Statement: The data used to support the findings of this study are included in the article.

Conflicts of Interest: The authors declare no conflict of interest.

References

1. Yang, Y.; Zhang, Y.; Tan, X. Review on Vibration-Based Structural Health Monitoring Techniques and Technical Codes. *Symmetry* **2021**, *13*, 1998. [CrossRef]
2. Yang, Y.; Cheng, Q.; Zhu, Y.; Wang, L.; Jin, R. Feasibility study of tractor-test vehicle technique for practical structural condition assessment of beam-like bridge deck. *Remote Sens.* **2020**, *12*, 114. [CrossRef]
3. Yang, Y.; Li, J.L.; Zhou, C.H.; Law, S.S.; Lv, L. Damage detection of structures with parametric uncertainties based on fusion of statistical moments. *J. Sound Vib.* **2019**, *442*, 200–219. [CrossRef]
4. Yang, Y.; Xiang, C.; Jiang, M.; Li, W.; Kuang, Y. Bridge damage identification method considering road surface roughness by using indirect measurement technique. *China J. Highw. Transp.* **2019**, *32*, 99–106.
5. Yang, Y.; Li, C.; Ling, Y.; Tan, X.; Luo, K. Research on new damage detection method of frame structures based on generalized pattern search algorithm. *China J. Sci. Instrum.* **2021**, *42*, 123–131.
6. Yang, Y.; Liang, J.; Yuan, A.; Lu, H.; Luo, K.; Shen, X.; Wan, Q. Bridge element bending stiffness damage identification based on new indirect measurement method. *China J. Highw. Transp.* **2021**, *34*, 188–198.
7. Ding, K.L.; Luo, W.J.; Bao, D.D.; Yu, L.H.; Liu, M.L. Application of 3D laser scanning technology in wall flatness detection. *Eng. Surv.* **2020**, *48*, 55–59. (In Chinese)
8. Duan, L.; Luo, X.F.; Peng, X.J.; Liao, G.G.; Ling, T.; Chen, P. Application of 3D laser scanning technology in tunnel monitoring measurement. *Eng. Res.* **2020**, *5*, 238–239. (In Chinese)
9. Zhao, Q.; Wang, T. A three-dimensional laser scanning technology tunnel overall deformation analysis method. *Mapp. Sci.* **2021**, *46*, 99–105. (In Chinese)
10. Li, Y.B.; Gao, C.M.; Ma, Y.Y.; Wang, J.; Zhou, Y. Application of 3D laser scanning technology in tunnel deformation monitoring and detection. *Sci. Technol. Eng.* **2021**, *21*, 5111–5117. (In Chinese)
11. Zhang, Y.M.; Ma, Q.M.; Li, C.P.; Geng, C.L.; Li, X. Application of 3D laser scanning technology in convergence monitoring of subway tunnels. *Mapp. Bull.* **2012**, 438–440. (In Chinese) [CrossRef]
12. Wei, X.; Wang, W.S.; Vimarlund, V.; Wang, Z. Applications of terrestrial laser scanning for tunnels: A review. *J. Traffic Transp. Eng.* **2014**, *1*, 325–337.
13. Yoon, J.S.; Sagong, M.; Lee, J.S.; Lee, K.-S. Feature extraction of a concrete tunnel liner from 3D laser scanning data. *Ndt E Int.* **2009**, *42*, 97–105. [CrossRef]
14. Xiao, M.; Qi, Z.; Shi, H. The Surface Flattening based on Mechanics Revision of the Tunnel 3D Point Cloud Data from Laser Scanner. *Procedia Comput. Sci.* **2018**, *131*, 1229–1237. [CrossRef]
15. Farahani, B.V.; Barros, F.; Sousa, P.J.; Cacciari, P.P.; Tavares, P.J.; Futai, M.M.; Moreira, P. A coupled 3D laser scanning and digital image correlation system for geometry acquisition and deformation monitoring of a railway tunnel—ScienceDirect. *Tunn. Undergr. Space Technol.* **2019**, *91*, 102995. [CrossRef]

16. Peji, M. Design and optimisation of laser scanning for tunnels geometry inspection. *Tunn. Undergr. Space Technol. Inc. Trenchless Technol. Res.* **2013**, *37*, 199–206. [CrossRef]
17. Fekete, S.; Diederichs, M.; Lato, M. Geotechnical and operational applications for 3-dimensional laser scanning in drill and blast tunnels. *Tunn. Undergr. Space Technol. Inc. Trenchless Technol. Res.* **2010**, *25*, 614–628. [CrossRef]
18. Cheng, X.J.; Tang, J.B. A method of wall flatness detection based on the least square fit. *Mapp. Inf. Eng.* **2007**, *32*, 19–20. (In Chinese)
19. Li, G.; Wu, C.Y.; Feng, T.; Wang, C.X. Building Wall Flatness Detection Based on 3D Laser Scanning Technology. *Shanxi Build.* **2017**, *43*, 204–205. (In Chinese)
20. Li, D.; Liu, J.; Feng, L.; Zhou, Y.; Liu, P.; Chen, Y.F. Terrestrial Laser Scanning Assisted Flatness Quality Assessment for Two Different Types of Concrete Surfaces. *Measurement* **2020**, *154*, 107436. [CrossRef]
21. Bosche, F.; Biotteau, B. Terrestrial laser scanning and continuous wavelet transform for controlling surface flatness in construction—A first investigation. *Adv. Eng. Inform.* **2015**, *29*, 591–601. [CrossRef]
22. Tang, P.; Huber, D.; Akinci, B. Characterization of Laser Scanners and Algorithms for Detecting Flatness Defects on Concrete Surfaces. *J. Comput. Civ. Eng.* **2011**, *25*, 31–42. [CrossRef]
23. Kim, M.K.; Wang, Q.; Yoon, S.; Sohn, H. A mirror-aided laser scanning system for geometric quality inspection of side surfaces of precast concrete elements. *Measurement* **2019**, *141*, 420–428. [CrossRef]
24. Fuchs, P.A.; Chase, S.B.; Washer, G.A. Laser-Based Instrumentation for Highway Bridge Applications Structures. In Proceedings of the Structures Congress 2001, Washington, DC, USA, 21–23 May 2001.
25. Wei, Z.; Ding, Z.X.; Zhou, Z.; Huang, Y.L. Feasibility Study on Non-contact Measurement Method for Primary Support of Tunnel. *Int. J. Geosynth. Ground Eng.* **2021**, *7*, 1–12. [CrossRef]
26. Qing, W.X.; Chen, W. Improved algorithms generated by Delaunay Triangle Networks. *Comput. Sci.* **2019**, *46*, 226–229. (In Chinese)
27. Jackson, J.A. *Glossary of Geology*, 5th ed.; Springer: Berlin, Germany, 2005; 900p, ISBN 3-540-27951-2.
28. Xiong, B.S.; Wu, Z.; Yu, H.J. A surface reconstruction algorithm based on two-dimensional local Lagrange interpolation. *J. Xi'an Univ. Eng. Sci. Technol.* **2003**, *17*, 138–141. (In Chinese)
29. Su, J.K. *Research on Image Interpolation Algorithm Based on B-Spline*; Guangdong University of Technology: Guangzhou, China, 2014. (In Chinese)
30. Liu, S.Y.; Han, X.; Jia, C.Q. Grid stitching and fusion of cubic B-spline interpolation. *J. Image Graph.* **2018**, *23*, 1901–1909. (In Chinese)
31. Wei, Z.; Yao, T.; Shi, C. Research on the Construction of 3D Laser Scanning Tunnel Point Cloud Based on B-Spline Interpolation. 2021. Available online: https://link.springer.com/chapter/10.1007/978-3-030-79672-3_8 (accessed on 6 September 2021).
32. Liu, Z.F.; Jiang, L.; Zhang, H. An Application Study of an S-G Filtering Adaptive Method in Atmospheric Echo Signals. *Electron. World* **2020**, *19*, 42–43. (In Chinese)
33. GB/T 50299-2018. Quality Acceptance Standard for Underground Railway Works. Available online: https://www.gongbiaoku.com/book/am917777413 (accessed on 6 September 2021).
34. Ji, Z.Y. Domestic road flatness detection technology. *Architecture* **2009**, *20*, 86–87. (In Chinese)
35. China Railway Corporation Enterprise Standards. *TB 10101-2018 Railway Engineering Measurement Specifications*; China Railway Press: Beijing, China, 2018. (In Chinese)

symmetry

MDPI

Article

Dynamic Response Analysis of a Multiple Square Loops-String Dome under Seismic Excitation

Zhenwei Lin [1], Chao Zhang [2,*], Jucan Dong [3], Jianliang Ou [1] and Li Yu [4]

[1] College of Civil Engineering, Yango University, Fuzhou 350017, China; zweilin@ygu.edu.cn (Z.L.); jlou@ygu.edu.cn (J.O.)
[2] College of Civil Engineering, Fuzhou University, Fuzhou 350108, China
[3] Shenzhen Expressway Engineering Consultant Co., Ltd., Shenzhen 518029, China; dongjucan@szewec.com
[4] Gemdale Properties & Investment, Kunshan 215399, China; yuli3@gemdalepi.com
* Correspondence: zhangchao1985@fzu.edu.cn

Abstract: The interaction between multiple loops and string cables complicates the dynamic response of triple square loops-string dome structures under seismic excitation. The internal connection between the multiple square loops-string cables and the grid beams was studies to provide a favorable reference for an anti-seismic structure. With a finite element model of the Fuzhou Strait Olympic Sports Center Gymnasium, established by SAP2000 software, the structural dynamic characteristic parameters were obtained first, and then this study adopted a time-history analysis method to study the internal force response of the cables and the roof grid beams of the multiple square loops-string dome (MSLSD) under three types of seismic array excitation. The influence of two factors, namely the seismic pulse and the near and far seismic fields, on the dynamic response of this structure was analyzed by three groups of different types of seismic excitation (PNF, NNF, PFF). As shown from the results, the first three-order vibration modes were torsional deformations caused by cables, the last five were mainly the overall roof plane vibration and antisymmetric vibration. Under the excitation of the three seismic arrays, the internal force responses of stay cables, square cables in the outer ring and the string cables were largest, while the maximum internal force response of the struts changed with the direction of seismic excitation. The largest internal force response of the roof grid beams occurred in local components such as BX3, BX7 and BY7, and the largest deformation of the beam nodes occurred in JX7, JX12 and JY4. In general, the seismic pulse and the near seismic field weakened the internal force response of the struts and cables but increased the internal force response and deformation of the dome beams, while the near and far seismic fields outweighed the seismic pulse. All the above provides an important reference for structural monitoring and seismic resistance.

Keywords: multiple square loops (MSL)-string; seismic excitation; dynamic response; seismic pulse; near and far field

Citation: Lin, Z.; Zhang, C.; Dong, J.; Ou, J.; Yu, L. Dynamic Response Analysis of a Multiple Square Loops-String Dome under Seismic Excitation. *Symmetry* **2021**, *13*, 2062. https://doi.org/10.3390/sym13112062

Academic Editors: Yang Yang, Ying Lei, Xiaolin Meng and Jun Li

Received: 26 September 2021
Accepted: 21 October 2021
Published: 1 November 2021

1. Introduction

With the recent continuous development of long-span spatial structures, the string dome, as a roof structure boasting light weight, high rigidity, good stability and strong spanning capability, has been widely adopted in large gymnasiums, opera houses and other public buildings such as the Tianbao Center Lobby, the 2008 Olympic Badminton Stadium, and the Jinan Olympic Sports Center Gymnasium [1,2]. The new square loops-string (SL-S) structure is one with pre-tensioned square ring cables and stay cables, in which struts bear axial compression and form a reliable support to arch up the lattice cylindrical shell. Compared with the traditional large-scale string dome structure, the square loops-string structure is simpler in lines, clearer in force transmission paths and better in force-bearing and antideformation performance of the roof grid beam structure. Such a structure has been integrated into the canopies of Shenzhen North Railway Station and the Fuzhou Strait Olympic Sports Center Gymnasium [3,4].

In addition to exhibition competitions and large-scale mass activities, the dome-structure gymnasium can also serve as an earthquake refuge. Collapse of buildings during earthquakes causes serious casualties and economic losses. Much research has been carried out on recent years carried out by Chinese and foreign experts on the mechanical properties and stability of the string dome structure, such as seismic response and anti-seismic analysis [5–8], buckling and dynamic response under friction [9], model optimization design based on artificial neural networks [10], elastoplastic dynamic response [11], structural optimization algorithm and design [12–14], long-span discontinuous mechanical properties [15], initial geometric defect analysis [16], cable tension estimation [17], static and the dynamic analysis by finite element method [18,19], and thrust line analysis of masonry domes [20]. Gong, S.Y. [21] summarized the research on shape selection, statics and stability, dynamic and seismic resistance, prestress and optimization of the string dome. Li, X.Y. [22] used a simplified soil model through a modified SR method to analyze the dynamic characteristics and seismic response of a string dome structure under soil-structure interaction (SSI). Jiang, Z.R. [23] studied the nonlinear dynamic buckling of a long-span elliptic paraboloid string dome by introducing such factors as geometric nonlinearity, initial geometric defects, material elastoplasticity, and half-span live load distribution. Ruggieri, S. [24,25] developed a new numerical practical procedure to investigate the evaluation of floor deformability in the performance of a simple linear analysis on 3D numerical models by removing the rigid floor hypothesis and adopting a simple floor model, and defined a new approach for predicting the fundamental period of vibration for reinforced concrete buildings through regression analysis procedures of 40 new buildings using a numerical model with elastic dynamic parameters.

However, most of the current seismic response analyses on string structures are limited to the traditional string dome structure, without enough research on the dynamic characteristics of the latest string dome structure, i.e., the Multiple Square Loops-String (MSL-S) structure. Thus, a project adopting the MSL-S dome structure with the Fuzhou Strait Olympic Sports Center Gymnasium as the object of study, we used the time history method to analyze the internal force of cables and grid beams and the displacement response of grid beams under seismic excitation to provide a favorable reference for antiseismic responses of the MSL-S. The influence of two factors, namely, the near and far seismic fields and the seismic pulse, on the dynamic response of this structure was analyzed.

2. Project Overview

The Fuzhou Strait Olympic Sports Center Gymnasium covers about 42,000 m^2. The roof has an elliptical MSL-S structure with a span of 116 m $\times$ 97 m and the roof structure is mainly composed of grid frame beams, triple square loops and the surrounding concrete columns, in which the frame beams include outer steel concrete ring beams, grid steel beams and the roof support, while the string structure consists of struts, stay cables, square ring cables and string beams. The dome is symmetrical in the east-west direction but asymmetrical in the north-south direction, with the north top higher than the south. The model structure is shown in Figure 1a. The grid beams are made from rectangular steel and the struts have circular steel pipes. The grid beams are rigidly connected with reinforced bases and the struts are articulated with cables and roof grid beams, the nodes of which are shown in Figure 1b,c.

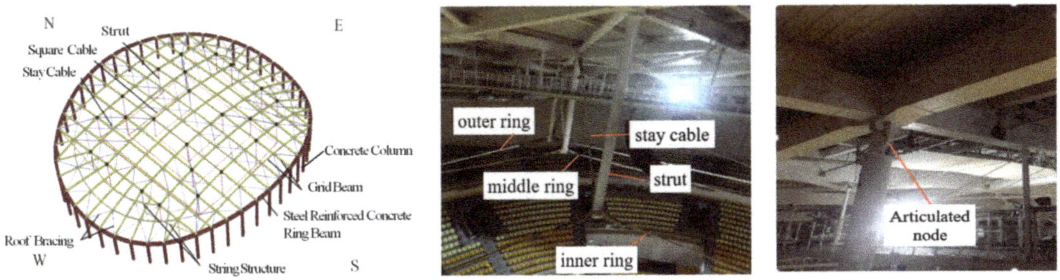

(a) **(b)** **(c)**

Figure 1. String dome structure of the Fuzhou Strait Olympic Sports Center Gymnasium. (**a**) Finite element model; (**b**) Cable connection of each ring; (**c**) Articulation node.

3. Numerical Simulation and Model Validation

3.1. Modelling Parameters

3.1.1. Geometric Dimension

The elevation of the dome reaches 39.7 m, the span of the square loop-string structure is 106.5 m, the span of the main truss of grid beams peaks at 97.5 m with the distance between main trusses ranging from 6.3 m to 7.6 m and the distance between secondary trusses is 9.5 m. The four sides of the dome have a relatively large load area with long-span horizontal cables. To increase the vertical rigidity of the structure, two middle struts (SC13~SC16) are added on the north and south sides of the outer square ring cables and independent string beams (SH13~SH14) are erected both on the east and west sides, the maximum span of which is 75.9 m. The struts and cables of the MSL-S structure system are shown in Figure 2.

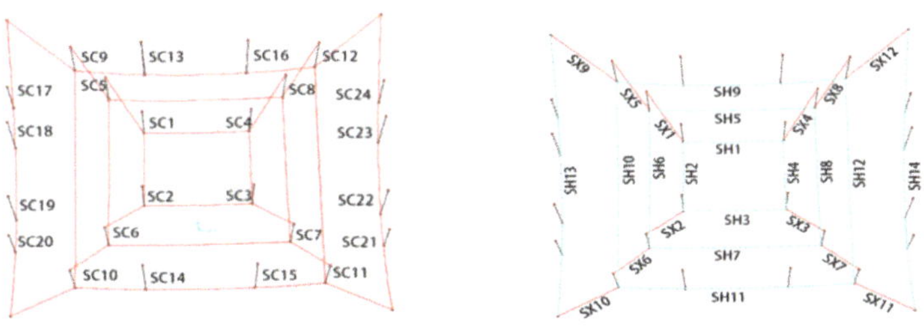

(a) **(b)**

Figure 2. Schematic diagram for the number of struts and cables. (**a**) Strut numbers; (**b**) Number of square ring cables and stay cables.

3.1.2. Material Characteristics

The grid beams, struts and additional roof support of the MSL-S roof structure are all made of the Q345 steel with an elasticity modulus of 2.06×10^5 N/mm^2 and a volume density of 7.85×10^3 kg/m^3. While the cables have PE finished products composed of $\Phi5$ or $\Phi7$ high-strength steel wires with an elasticity modulus of 1.95×10^5 N/mm^2 and a tensile strength of 1.67×10^3 MPa.

3.1.3. Finite Element Model Creation

General finite element software SAP2000 V19 was used to establish the three-dimensional finite element model of the string dome structure of the gymnasium. The roof grid beams used the steel box-sections simulated by the Frame unit, the prestressed cables were simulated by the Cable unit and the struts were simulated by the Link unit. The struts nodes at both ends were simulated as articulated to ensure the smooth rotation during the prestressing process. The berm under the roof, an auxiliary structure of grid structure system, was applied to the corresponding struts in the form of a line load (1.5 kN/m) after simplification; Given the additional load of the cable connection device, PE pipes and fillers, the weight of the cables was multiplied by a coefficient of 1.05. With the ring beams outside the grid beams connected with the concrete column nodes, the degrees of freedom in six directions were coupled, and the bottom of concrete column was a fixed end constraint. The three-dimensional finite element model is shown in Figure 1a.

3.2. Model Validation

To verify the accuracy and reliability of the finite element numerical model, the stress monitoring data and displacement monitoring data of the key roof nodes during the construction process were determined [26]. Stress monitoring distributions of roof grid beams was conducted at 16 different places with strain gauges installed on both the upper and lower surfaces of the beam spans, as shown in Figure 3a; while deformation monitoring was conducted in nine places with the measuring points set in the middle of the grid beams, as shown in Figure 3c.

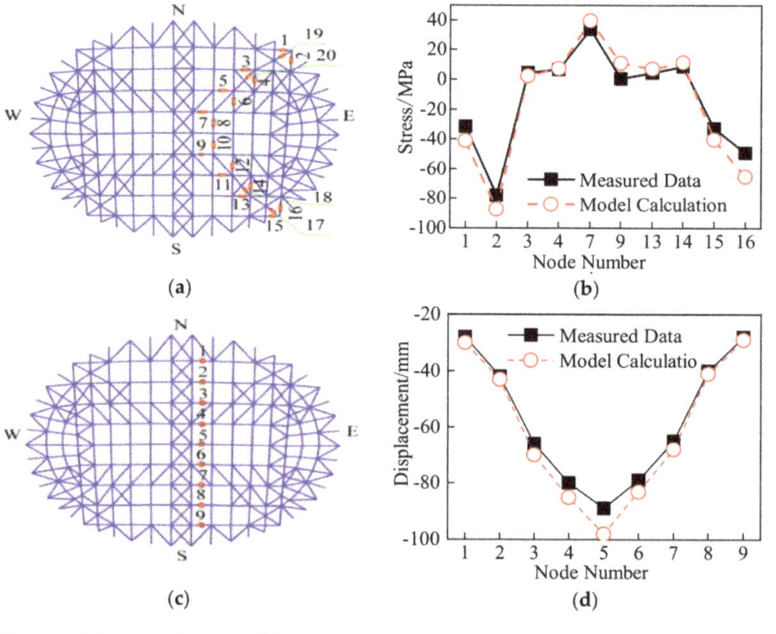

Figure 3. Schematic diagram of the monitoring points of roof grid beams. (**a**) Schematic diagram for stress monitoring points; (**b**) stress diagram of grid beams; (**c**) schematic diagram of displacement monitoring points; (**d**) vertical displacement of roof.

A comparison of the results between the measured data and the model calculation of grid beam stress and node vertical displacement are, respectively, displayed in Figure 3b,d, from which we can see that the overall difference between the two data sets is small. Among them, No. 16 beam is the one with the biggest stress error, which is 16.1 MPa between the actual measurement and theoretical calculation, and No. 5 beam has the biggest displacement error, which is 10.1 mm. Given the good agreement between the simulation and experiment results in each construction stage with acceptable errors, the numerical model was used as the simulation analysis model.

3.3. Structural Dynamic Characteristics

The first eight orders of vibration modes simulated and analyzed by the finite element model are shown in Figure 4. The natural frequency and participating mass ratios in the main directions are shown in Table 1. It can be seen from the model analysis that the first-order vibration mode demonstrates the torsion of the middle ring cable, the second one as the torsion of the inner and outer ring cables, the third as the torsion of the inner ring cable, the fourth as the antisymmetric vertical vibration of the whole roof along the short span, the fifth as the antisymmetric vertical vibration of the whole roof along the long span, the sixth as the symmetric vertical vibration of the whole roof along 45 degree, the seventh as the outer vibration of the south outer ring cables, and the eighth as the coupling of the translational motion and antisymmetric vibration of roof along the long-span direction. Therefore, the first three vibration modes are dominated by the torsional deformation of square ring cables while the last five ones are mainly the overall roof plane vibration and antisymmetric vibration.

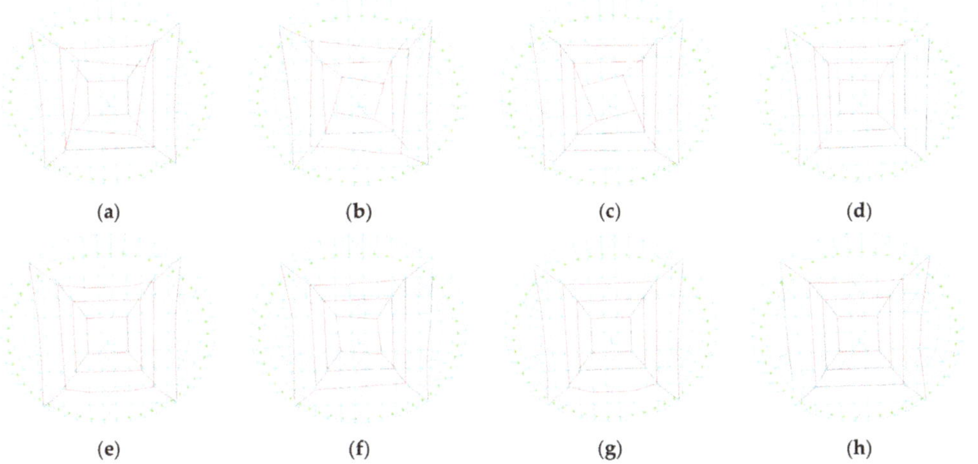

Figure 4. Schematic diagram of the first eight-order vibration modes. (**a**) First-order; (**b**) second-order; (**c**) third-order; (**d**) fourth-order; (**e**)fifth-order; (**f**) sixth-order; (**g**) seventh-order; (**h**) eighth-order.

Table 1. Structural natural frequency.

Order	1	2	3	4	5	6	7	8
Natural Frequency (Hz)	0.76	0.86	0.86	0.91	1.13	1.17	1.28	1.29
Main Direction	RZ	RZ	RZ	UX	UY	UX	UY	UX
Partecipating Masses Ratio	1.342×10^{-3}	7.96×10^{-3}	5.452×10^{-4}	0.07886	0.02809	0.09835	0.03773	0.63

4. Seismic Response Law of Roof Structure under Different Seismic Arrays

To facilitate the analysis of the seismic excitation response of the roof frame within the ring-cable structure of the gymnasium, the key structural components were numbered. As the central grid beams cover the longest span with the largest internal force value and displacement amount, they were selected as the research object. The numbers of grid beams and beam nodes are shown in Figure 5, where B represents the grid frame beam, J represents the node, X and Y, respectively, stand for the EW direction and SN direction. The numbers of struts and cables are shown in Figure 2.

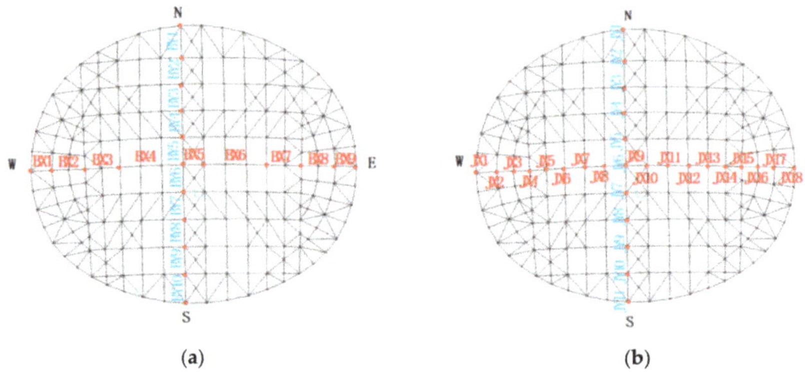

Figure 5. Schematic diagram of roof grid beams and tip nodes. (**a**) Grid beam numbers; (**b**) number of grid beam nodes.

4.1. Seismic Arrays Input

The degree of seismic fortification intensity of the gymnasium is seven, and falls into the third seismic group. In light of the rules of seismic grouping, the Imperial Valley-06 far-field seismic wave files, DLT-352 and DLT-DWN, were collected from Pacific Earthquake Engineering Research Center (PEERC). The seismic acceleration curve and response spectrum curve are showed in Figure 6a–d. The horizontal seismic waves (X or Y direction) of DLT-352 were combined with the vertical seismic waves (Z direction) of DLT-DWN, and the seismic excitation effect partial coefficients, γ_{Eh} and γ_{Ev}, were, respectively, set as 1.3 and 0.5 under two components, and set as 1.0, 0.85 and 0.65 under three components [27]. The combination formulas are as follows.

$$\text{No. 1 Seismic Array: } K1 = 1.3 \times X + 0.5 \times Z, \tag{1}$$

$$\text{No. 2 Seismic Array: } K2 = 1.3 \times Y + 0.5 \times Z, \tag{2}$$

$$\text{No. 3 Seismic Array: } K3 = 1.0 \times X + 0.85 \times Y + 0.65 \times Z, \tag{3}$$

To better analyze the response of the structure under seismic excitation, the peak ground acceleration (PGA) was adjusted to 0.22 g in consideration of the impact of earthquakes that rarely occur. Meanwhile, when analyzing the elastoplasticity of dome structure in the context of rarely occurring earthquakes, the damping ratio was set as 0.05 and the nonlinear time history analysis method based on "self-weight" was used to analyze the seismic response.

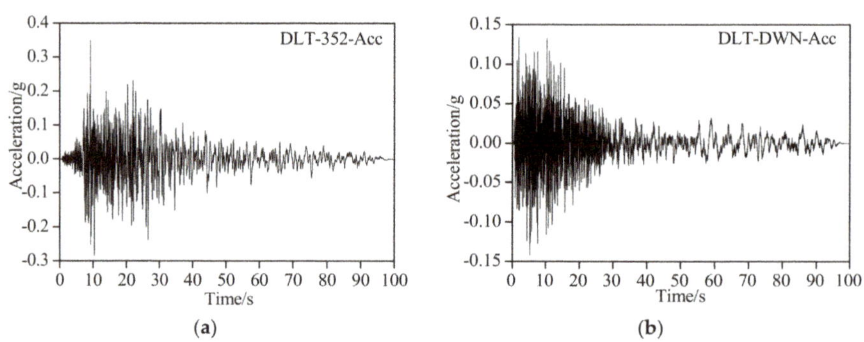

Figure 6. *Cont.*

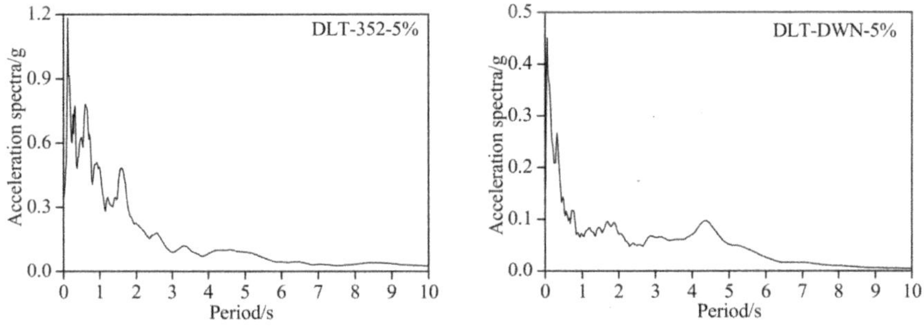

(c) (d)

Figure 6. Acceleration curve and response spectrum curve of seismic wave. (**a**) DLT-352 acceleration time history; (**b**) DLT-DWN acceleration time history; (**c**) DLT-352 acceleration response spectrum; (**d**) DLT-DWN acceleration response spectrum.

4.2. Dynamic Response under the Excitation of No. 1 Seismic Array

4.2.1. Analysis of the Internal Force Response of Roof Square Ring Cables

As the stay cables and square ring cables unite to mainly transfer tensions, and the struts units transfer compressions, the internal force response of cables and struts of the ring-cable structure under No. 1 seismic array excitation is shown in Figure 7, where the structural dynamics remain relatively small due to the equal altitudes at both ends of the EW (X) direction, as well as great rigidity. It can be seen from the figure that struts (SC1 to SC16) tend to be more average in force and smaller in internal force response with comparable values, and the smallest internal force response occurs in the additional struts (SC1 to SC16) at the north and south sides of the outer ring. The inner ring and middle ring stay cables (SX1 to SX8) and ring cables (SH1 to SH8) share a relatively small internal force response within 90 kN, while the outer ring stay cables (SX9 to SX12) and ring cables (SH9 to SH12) experience a significantly increased internal force response, almost double that of the inner ring and middle ring. String cables are long in span and unilaterally connected, and the cables (SH13 to SH14) and struts (SC17 to SC24) have the largest internal force response.

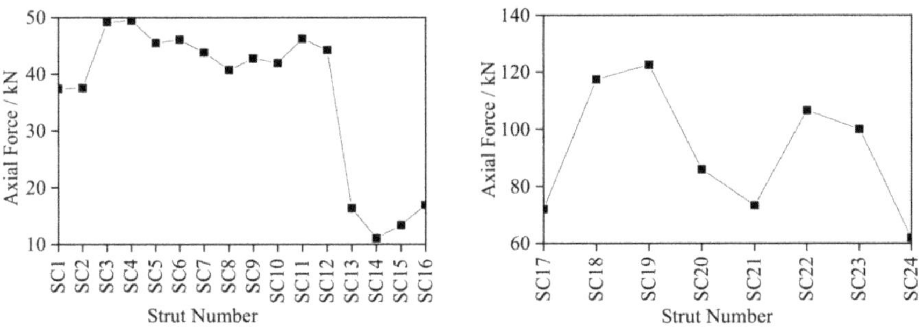

(a) (b)

Figure 7. *Cont.*

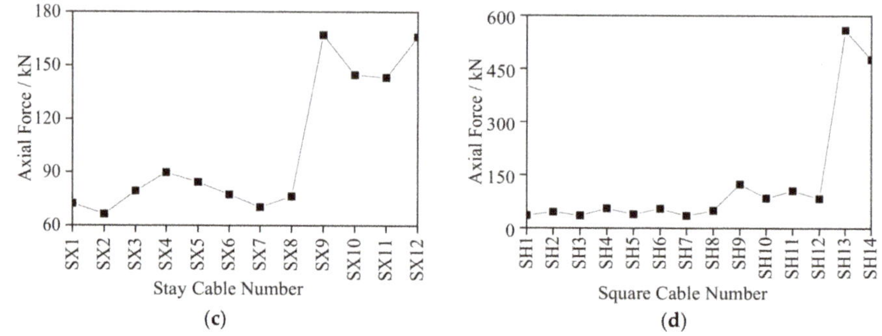

Figure 7. Internal force diagram for square ring cable structure. (**a**) Internal force of struts in the square ring cable; (**b**) internal force of struts in string cables; (**c**) internal force of stay cables; (**d**) internal force of square cables.

4.2.2. Analysis of the Internal Force Response of Roof Grid Beams

The internal force response of the grid beams under No. 1 seismic array excitation is shown in Figure 8. The ring-cable structural frame system is irregular, and the internal force of the roof frame (BX1 to BX9) in the EW direction is distributed axisymmetrically with the mid-span beam BX5, which has the smallest internal force response. The X-shaped steel pipes set in the SN direction in the mid-span grid beam strengthen the local roof stiffness. As there is a long span between BX3 and BX7, their internal force response is the largest. The internal force response of the roof grid beams (BY1 to BY10) are irregularly distributed and generally greater than that in the EW direction. As the input direction of combined seismic waves is perpendicular to the roof framing elements, a larger internal force response occurs in out-of-plane vibration, maximizing at BY7 about 1440 kN.

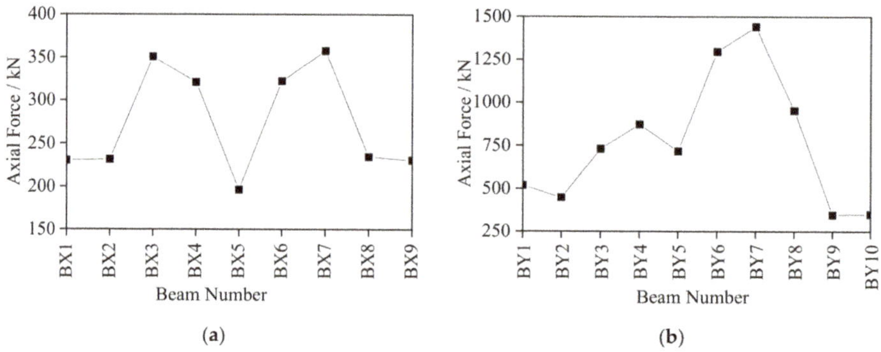

Figure 8. Internal force diagram for roof grid beams. (**a**) EW direction; (**b**) SN direction.

4.2.3. Analysis of the Roof Node Deformation

The vertical displacement of roof nodes moving along the positive direction of the Z axis is represented as positive displacement, and vice versa. It can be seen from Figure 9 that the displacement and deformation of the roof grid beam nodes (JX1 to JX18) in the EW direction are basically symmetrically distributed. While the maximal displacement is reached at nodes JX7 and JX12, the increased local stiffness at the middle span greatly decreases the displacement of JX9 and JX10. The vertical displacement of roof grid beam nodes (JY1 to JY11) in the SN direction is distributed asymmetrically, with the greatest displacement appearing at node JY7, which is about 45 mm. Comparing Figure 8a with Figure 8b, the difference between the node displacement and internal force response of the roof grid beams can be seen. The reason why the overall grid beam node displacement in the SN direction is smaller than that in the EW direction is that the long-span arch structure facilitates the antisymmetrical vibration mode under combined seismic excitation, leading to a large vertical displacement.

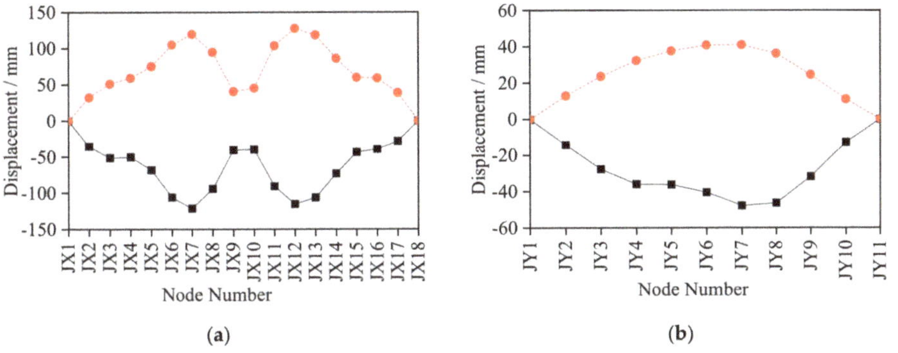

Figure 9. Vertical displacement diagram for roof grid beams nodes. (**a**) EW direction; (**b**) SN direction.

4.3. Dynamic Response under Excitation of No. 2 and No. 3 Seismic Arrays

4.3.1. Analysis on the Internal Force Response of Roof Square Ring Cables

The structural stiffness of the gymnasium in the SN direction is relatively weak, and the structural dynamic response is larger and similar under the Excitation of No. 2 and No. 3 seismic arrays. Therefore, these two seismic arrays were considered simultaneously for better comparison and analysis.

As seen in Figure 10, the internal force response of the struts and the cables under No. 2 seismic array excitation was larger than that under the No. 3 seismic array and was about four to six times larger than that under the No. 1 seismic array. As an additional vertical acceleration is generated by the altitude difference between the north and south sides when the seismic waves move along the Y direction, the small rise span of the roof structure in this direction, coupled with less out-of-plane stiffness than in-plane stiffness, resulting in a greater structural dynamic response. In addition, the internal force response of the struts (SC1 to SC12) shows a downward trend from the inner ring to the outer ring. The additional struts (SC13 to SC16) of the outer ring have the least internal force response, followed by the struts in the string cables (SC17 to SC24), with an average value of about 80 kN. The internal force response of the middle ring stay cables (SX5 to SX8) and loop cables (SH5 to SH8) are slightly smaller than those of the inner ring. The outer ring stay cables (SX9 to SX12), loop cables and string cables (SH9 to SH14) have the largest internal force response.

4.3.2. Analysis of the Internal Force Response of Roof Grid Beams

As seen in Figure 11, the internal force response of the roof grid beams (BX1 to BX9) in the EW direction under No. 2 seismic array is larger than that under No. 3 seismic array, which is contrary to that of the beams from BY1 to BY10. The internal force response of the roof grid beams from BX1 to BX9 under No. 2 seismic array is symmetrically distributed, in which the grid beams BX3 and BX7 have the largest internal force response that is obviously greater than that in the SN direction, thus serving as the main stress component of the structural system. The internal force response of the grid beam in the SN direction is irregularly distributed with an average value of about 200 kN under No. 2 seismic array, and with the largest value of 1120 kN under No. 3 seismic array.

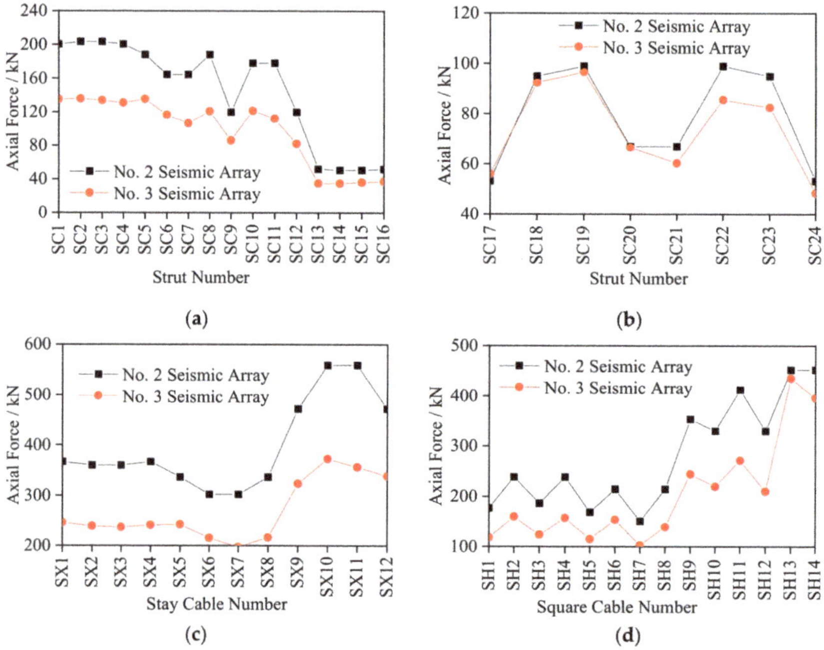

Figure 10. Internal force diagram of the square ring cable structure. (**a**) Internal force of struts in square ring cable; (**b**) internal force of struts in string cable; (**c**) internal force of stay cables; (**d**) internal force of square cables.

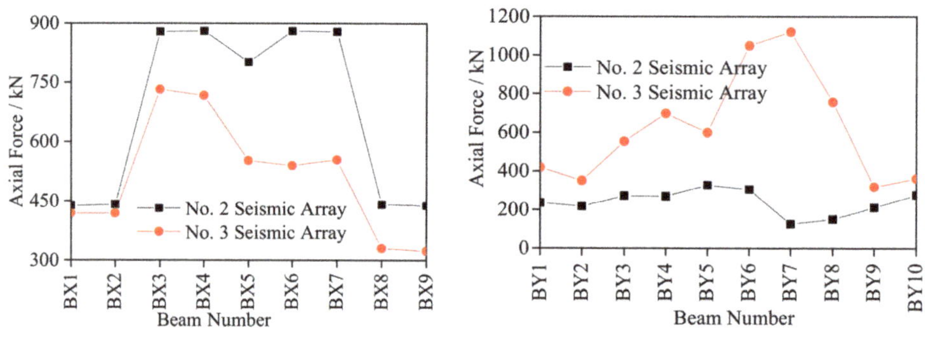

Figure 11. Internal force diagram of roof grid beams. (**a**) EW direction; (**b**) SN direction.

Compared with No. 1 seismic array excitation, the internal force response of the roof grid beams in the EW direction under No. 2 seismic array excitation has a significant increase, the maximum value of which is about 2.5 times that of No. 1, while a substantial decrease occurs in the internal force response of the roof grid beams in the SN direction, the maximum value of which only accounts for about 25% of that in the EW direction. The reason is that when the seismic wave moves in the Y direction, the grid beams in the EW direction become perpendicular to the moving direction of combined seismic waves and thus generate out-of-plane vibration, greatly increasing the internal force of grid beams. As the grid beams in the SN direction are parallel to the moving direction of the seismic wave, they suffer less an effect.

4.3.3. Analysis of the Roof Nodes Deformation

The displacement response of roof grid beam nodes under No. 2 and No. 3 seismic array excitation is displayed in Figure 12. The displacement response in the EW direction under No. 2 seismic array excitation is weaker than that under No. 3 seismic array excitation, in addition to nodes JX9 and JX10, while the displacement response in the SN direction is different. The displacement response is basically symmetrically arranged in the EW direction (JX1 to JX18), with the displacement maximizing at the midspan nodes of JX9 and JX10 under No. 2 seismic array, and at the nodes of JX7 and JX13 under No. 3 seismic array excitation. The displacement peak in the SN direction appears at JY4, and the displacement of the north nodes (JY2 to JY5) under seismic excitation exceeds that of the south nodes (JY7 to JY10), because of weak lateral rigidity caused by the longer steel pipe columns on the north side. Compared with No. 1 seismic array, the displacement of the roof grid beam nodes under No. 2 seismic array excitation experiences a decrease in the EW direction, while an increase is seen in the SN direction, with the maximum value almost doubled.

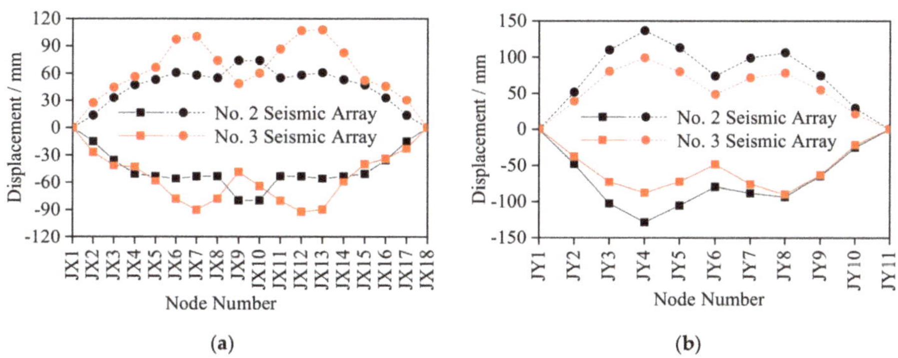

Figure 12. Vertical displacement diagram of roof grid beams nodes. (**a**) EW direction; (**b**) SN direction.

5. Effect Law of Different Types of Ground Motions on the Seismic Response of Roof Structures

Seismic action comes with strong randomness and uncertainty. Located at the junction of the Pacific plate and the Asia-Europe plate, Fuzhou has witnessed the occurrence of different types of earthquakes from time to time. Therefore, further research on the dynamic response of the gymnasium structure under different seismic fields and seismic impulses is of great significance for seismic monitoring and emergency response tasks related to the gymnasium.

5.1. Input of Different Types of Seismic Waves

Imperial Valley-06 far-field seismic wave files from the PEERC were selected, including two groups of pulse near field(PNF) seismic waves, two groups of nonpulse near field (NNF) seismic waves and two groups of pulse far field(PFF) seismic waves. The spectrum curve of seismic wave acceleration response (ARS) is displayed in Figure 13. In consideration of rarely occurring earthquakes, the PGA was adjusted to 0.22 g and the horizontal direction (X-direction) and vertical direction (Z-direction) of the six sets of seismic waves were chosen as the earthquake inputs according to the partial coefficients. The seismic parameter information is detailed in Table 2, where the nonlinear time history analysis method based on "self-weight" is used. The damping ratio was set as 0.05 and the average value of the response under various types of seismic array conditions was selected as the structural dynamic response.

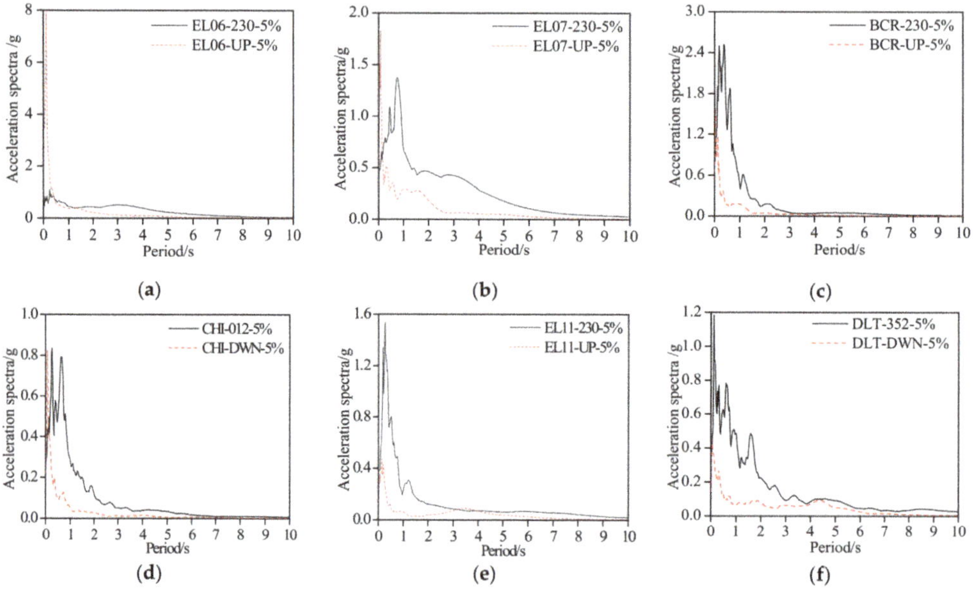

Figure 13. Seismic wave response spectrum curves under load case. (**a**) EL06-ARS; (**b**) EL07-ARS; (**c**) BCR-ARS; (**d**) CHI-ARS; (**e**) EL11-ARS; (**f**) DLT-ARS.

Table 2. Table of Seismic Load Cases.

ID	Type	Load Case	Seismic Wave Measurement	PGA (g)	Partial Factor (Direction)	Input Proportionality Factor
F1	Pulse Near Field (PNF)	EL06	EL06-230 EL06-UP	0.449 1.895	1.3(Ux) 0.5(Uz)	6242 569
F2		EL07	EL07-230 EL07-UP	0.469 0.578	1.3(Ux) 0.5(Uz)	5976 1865
F3	Nonpulse Near Field (NNF)	BCR	BCR-230 BCR-UP	0.777 0.532	1.3(Ux) 0.5(Uz)	3607 2026
F4		CHI	CHI-012 CHI-DWN	0.270 0.216	1.3(Ux) 0.5(Uz)	10381 4991
F5	Pulse Far Field (PFF)	EL11	EL11-230 EL11-UP	0.379 0.144	1.3(Ux) 0.5(Uz)	7395 7486
F6		DLT	DLT-352 DLT-DWN	0.350 0.142	1.3(Ux) 0.5(Uz)	8008 7592

5.2. Analysis of the Internal Force Response of Roof Square Ring Cables

The internal force responses of the structural components of the multiple square ring cables under the excitation of the three types of seismic waves are shown in Figure 14. It can be seen from Figure 14a,b that the internal force response law of the struts (SC1 to SC24) remains basically the same with comparable values. Except for some struts (such as SC1, SC11, SC21, etc.), the internal force response under the excitation of the PNF combined seismic wave is the smallest, indicating that, in general, seismic impulses weaken the internal force response of the struts, while the near and far seismic fields equip the struts with greater internal force response.

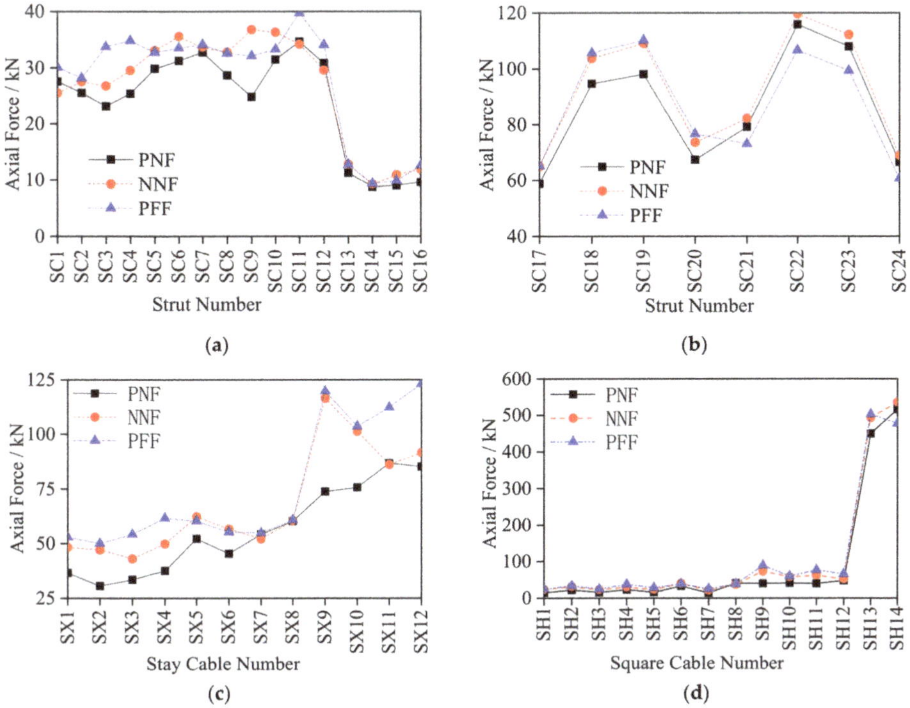

Figure 14. Internal force diagram of the square cable structure. (**a**) Internal force of struts in square ring cable; (**b**) internal force of struts in string cable; (**c**) internal force of stay cables; (**d**) internal force of square cables.

The internal force response of stay cables and ring cables are shown in Figure 14c,d. In general, the internal force response under the PFF combined seismic wave is the biggest, while that under the PNF combined seismic wave is the smallest. The outer ring stay cables (SX9 to SX12) and the ring cables (SH9 to SH14) undergo the largest increase in the internal force response under the NNF and PFF seismic excitation, and the smallest under the PFF seismic excitation. SH13, under the PFF seismic excitation, has the largest internal force response with an axial force value of 503 kN, while SH14, under the NNF seismic excitation, has the largest internal force response with an axial force value of 535 kN and an increase of 0.203. That proves that the seismic impulse fails to cause larger internal force responses of stay cables and ring cables. Instead, it weakens their internal force response. Other than SH14, the seismic excitation in the far field generates a greater internal force response than the near field.

5.3. Analysis of the Internal Force Response of Roof Grid Beams

Figure 15 shows the internal force response of the roof grid beams under the three types of seismic excitation. It can be seen that the change law of the structure system under different seismic excitation remains basically the same. The PNF seismic array has the biggest excitation, followed by NNF and then PFF, which indicates that the seismic impulse generates a larger structural dynamic response to the roof grid frame beams, the near seismic field produces a greater internal force response than that of the far seismic field under the same impulse, and the influence of the near and far seismic fields exceed that of the impulse. In addition, as the frame beam BX5 of the roof vault is rigid enough and the axial forces under different seismic conditions are equivalent, with the smallest value, the aforementioned two factors exert little effect on this cross-grid beam.

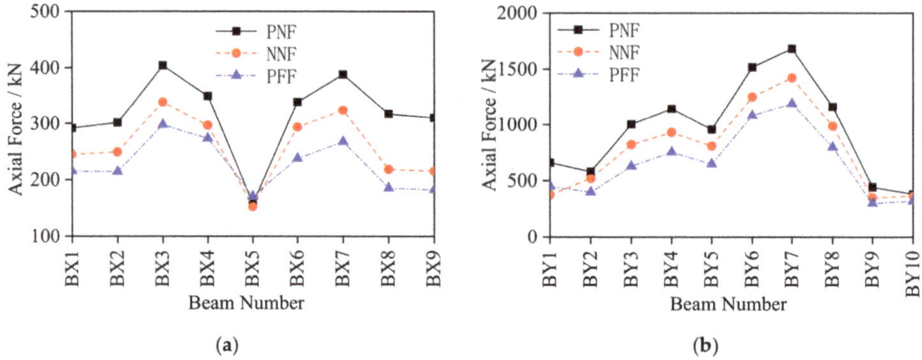

Figure 15. Internal force diagram for the midspan frame beam. (**a**) EW direction; (**b**) SN direction.

5.4. Analysis of the Roof Nodes Deformation

The vertical displacement response of the roof grid beam nodes under three types of seismic excitation is shown in Figure 16, in which the overall distribution law is consistent with that of the grid beam internal force response. PNF combined seismic excitation has the largest vertical displacement, followed by NNF and then PFF. In addition, the positive deformation of nodes JX11–JX17 and the negative deformation of JX2–JX8 are distributed alternately under the excitation of NNF and PFF, with little difference. This indicates that the seismic impulse excitation further increases the displacement response of the roof grid beam nodes, and the seismic excitation of the near field produces a greater nodal displacement response than the far field. Overall, the near field factor exceeds the impulse factor.

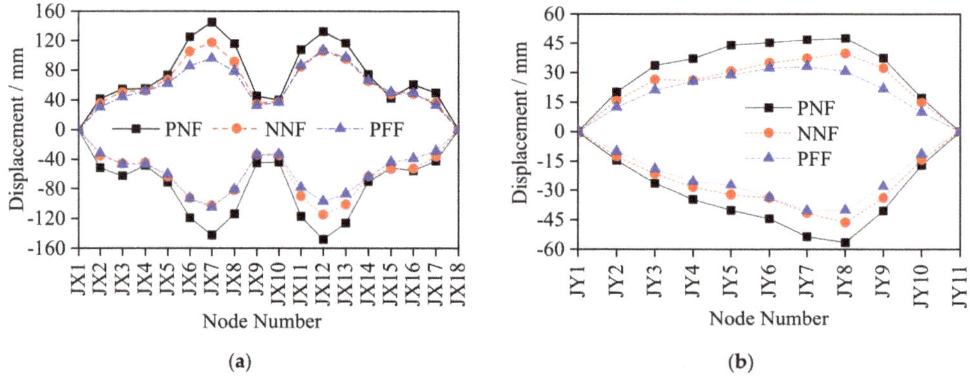

Figure 16. Vertical displacement diagram of roof span nodes. (**a**) EW direction; (**b**) SN direction.

6. Conclusions and Future Plans

With the establishment of a finite element numerical model of the MSL-S roof structure of the Fuzhou Strait Olympic Sports Center Gymnasium, this study analyzed the dynamic characteristics of string structures. The structural dynamic response under combined seismic excitation was studied through the nonlinear time history analysis method based on "self-weight" and the influence of different seismic impulses and near and far seismic fields on the dynamic response of the structure was studied. The conclusions are as follows.

1. The natural frequency of the first eight-order vibration modes gradually increased from 0.762 to 1.292, with the main vibration mode being the torsional deformation

caused by the cables, while the last five mainly related to the overall roof plane vibration and antisymmetric vibration. Therefore, attention must be paid to the cable vibration of the structure.

2. Under excitation of the No. 1 seismic array, the internal force responses of structural struts, stay cables and ring cables were relatively small. The internal force of the struts of each ring remained equivalent within 50 kN, and weaker than that of the string cables between 60 kN to 125 kN. The largest internal force response of the stay cables was in outer ring between 140 kN to 170 kN, and the internal force of the string cables was the largest between 450 kN to 600 kN. The internal force response of the roof grid beam was symmetrically arranged in the EW direction with the largest internal force response occurring in BX3 and BX7 (about 350 kN). The response in the SN direction was generally larger, among which BY7 had the largest response of 1441.31 kN. The vertical displacement of the grid beam in the EW direction exceeded that in the SN direction. Therefore, emphasis must be laid on monitoring the stress and deformation of the local components of BX3, BX7 and BY7 and the internal force of the string cables.
3. The structural dynamic response under excitation of No. 2 seismic array was larger than that under the excitation of No. 3 seismic array except for the internal force response of the roof grid beams in the SN direction and the displacement response of the roof grid beam nodes in the EW direction. The internal force response of struts showed a downward trend from the inner ring to the outer ring, with the largest internal force response appearing in stay cables, ring cables and string cables of the outer ring. Under excitation of No. 2 and No. 3 seismic arrays, the internal force response of the struts of each ring was within 200 kN, while the internal force of the outer ring stay cables was the largest with a peak value of 559.71 kN, and the peak value of the internal force of the square cables was 452.22 kN (SH13). The internal force response of the roof grid beam was relatively large, with the peak value in the EW direction of 880.64 kN and in the SN direction of 1119.97 kN. The maximum value of displacement of the grid beam nodes in the SN direction was about twice that in the EW direction, with the largest displacement in node JY4 (136.72 mm). Thus, emphasis must be placed on monitoring the struts in the inner ring and the stay cables, ring cables and string cables in the outer ring, the stress of the roof mid-span grid beam and the deformation of local nodes such as JX7, JX12 and JY4.
4. The seismic impulse factor weakened the internal force response of cables and struts. The internal force response of the cables under the seismic excitation in the far field was larger than that in the near field. The seismic impulse increased the internal force response and displacement deformation of the roof grid beam and the seismic excitation in the near field produced greater internal force response and displacement deformation of the grid beam than the far field. The influence of near and far fields exceeded that of the impulse. Thus, attention must be paid to the internal force response and displacement deformation of the roof grid beam when the seismic excitation contains an impulse factor, especially in the near field.

The research focuses on the internal connection of the structural dynamic response between the cables, strut and roofs in the structural system of MSLSD, which predicted possible damage of the structure under different types of seismic excitation in advance and provided an important reference for structural seismic resistance. In addition, the research is relevant to the operation and maintenance of the building, including regular inspections. Due to the limitation of monitoring equipment, the existing monitoring system cannot collect corresponding data for all the key roof nodes to improve the monitoring database; the actual structure and the finite element model needs to be further optimized to analyze the structure dynamic response more accurately. At the same time, compared with the traditional string dome structure, the advantages and disadvantages of this new structure system (MSLSD) need to be further studied to provide a reference for promotion and improvement.

Author Contributions: Conceptualization, Z.L. and C.Z.; methodology, C.Z.; software, Z.L. and L.Y.; validation, Z.L., J.D. and C.Z.; formal analysis, J.O.; investigation, J.O.; data curation, L.Y.; writing—original draft preparation, Z.L.; writing—review and editing, Z.L. and J.D. All authors have read and agreed to the published version of the manuscript.

Funding: This research is funded by the National Natural Science Foundation of China (Project No. E51508102). Natural Science Foundation of Fujian Province (Project No. 2019J05128). Natural Science Foundation of Fujian Province (Project No. 2019J01233).

Institutional Review Board Statement: Not applicable.

Informed Consent Statement: Not applicable.

Data Availability Statement: Not applicable.

Conflicts of Interest: The authors declare no conflict of interest.

References

1. Chen, Z.H.; Liu, H.B.; Wang, X.D.; Zhou, T. Research Review of String Dome Structure. *J. Build. Struct.* **2010**, *31*, 210–215.
2. Olofin, I.O.; Liu, R.G. Suspen-Dome System: A Fascinating Space Structure. *Open Civ. Eng. J.* **2017**, *11*, 131–142. [CrossRef]
3. Fu, X.Y.; Sun, C.; Wu, B.; Meng, M.L.; Feng, Y.W. New String Structure System Design of Shenzhen South Railway Station Canopy. *Build. Struct.* **2015**, *45*, 47–52.
4. Tuo, M.B.; Wang, Y.; Wu, P.C. Construction Technology of Square Loop-String Structure of Fuzhou Strait Olympic Sports Center Gymnasium. *Constr. Technol.* **2015**, *44*, 41–44.
5. Rao, P.X. Seismic Response and Dynamic Stability Analysis of Long-Span Cavit-Square String Dome Structure under Multi-dimensional and Multi-Point Excitation. Marster's Thesis, Nanchang University, Nanchang, China, 2016.
6. Yu, L. Study on Anti-Seismic Performance and Cable Breakage of Multiple Square Loop-String Dome Structure. Marster's Thesis, Fuzhou University, Fuzhou, China, 2017.
7. Zhang, Q.W.; Zhang, Y.; Yao, L.; Fan, F.; Shen, S. Finite element Analysis of the Static Properties and Stability of a 800 m Kiewitt Type Mega-latticed Structure. *J. Constr. Steel Res.* **2017**, *137*, 201–210. [CrossRef]
8. Zhao, R.X. Static Stability and Dynamic Response Analysis of String Dome Structure. Marster's Thesis, Hebei University, Baoding, China, 2020.
9. Zhao, Z.W.; Wu, J.J.; Liu, H.Q.; Liang, B. Influence of Friction on Buckling and Dynamic Behavior of Suspen-Dome Structures. *Struct. Eng. Int.* **2020**, *30*, 262–269.
10. Guo, J.M.; Zhao, X.X.; Guo, J.H.; Yuan, X.F.; Dong, S.L.; Xiong, Z.L. Model updating of suspended-dome using artificial neural networks. *Adv. Struct. Eng.* **2017**, *20*, 1727–1743. [CrossRef]
11. Liu, H.J.; Feng, Y.L.; Luo, Y.F. Dynamic Stability of Long-span Suspen-Domessubjected to Seismic Excitations. *Adv. Mater. Res.* **2012**, *378*, 209–212. [CrossRef]
12. Kaveh, A.; Rezaei, M.; Shiravand, M.R. Optimal Design of Nonlinear Large-scale Suspen Dome Using Cascade Optimization. *Int. J. Space Struct.* **2018**, *33*, 3–18. [CrossRef]
13. Liu, X.C.; Zhang, A.L.; Zhang, X. Particle Swarm Optimization Algorithm for Suspendome Structure under Multiple Loading Cases. *Eng. Comput.* **2016**, *33*, 767–788. [CrossRef]
14. Shen, X.H.; Zhang, Q.; Lee, D.S.H. Static Behavior of a Retractable Suspen-Dome Structure. *Symmetry* **2021**, *13*, 1105. [CrossRef]
15. Yu, J.H.; Leng, M.; Zhang, Z.Y.; Jiang, Z.Y.; Wang, Z.K. Mechanical Properties Research on Suspended-dome Structure with Discontinuous Support under Different Parameters. *Open Civ. Eng. J.* **2017**, *11*, 303–314.
16. He, S.; Jiang, Z.R.; Cai, J. Investigation on Simulation Methods of Initial Geometric Imperfection Distribution in Elasto-plastic Stability Analysis of Single-layer Reticulated Shells. *KSCE J. Civ. Eng.* **2018**, *22*, 1193–1202. [CrossRef]
17. Yang, Z.J.; Liang, B.; Liu, H.Q.; Zhao, Z.W. Cable Tension Estimation for Suspen-Dome Structures Based on Numerical Method. *Iran J. Sci. Technol. Trans. Civ. Eng.* **2020**, *44*, 115–126. [CrossRef]
18. Itu, C.; Bratu, P.; Borza, P.N.; Vlase, S.; Lixandroiu, D. Design and analysis of inertial platform insulation of the eli-np project of laser and gamma beam systems. *Symmetry* **2020**, *12*, 1972. [CrossRef]
19. Chinnuraj, S.; Thyla, P.R.; Elango, S.; Venugopal, P.R.; Mohanram, P.V.; Nataraj, M.; Mohanraj, S.; Manojkumar, K.N.; Ayyasamy, S. Static and dynamic behavior of steel-reinforced epoxy granite CNC lathe bed using finite element analysis. *Proc. Inst. Mech. Eng. Part L* **2020**, *234*, 595–609. [CrossRef]
20. Varma, M.; Ghosh, S.; Milani, G. Finite element thrust line analysis of cracked axisymmetric masonry domes reinforced with tension rings. *Int. J. Mason. Res. Innov.* **2018**, *3*, 72–87. [CrossRef]
21. Gong, S.Y. The research of Suspen-dome Structure. *IOP Conf. Ser. Mater. Sci. Eng.* **2017**, *242*, 012050. [CrossRef]
22. Li, X.Y.; Wang, G.X.; Xue, X.D. Dynamic Performance Analysis of String Dome Structure under Soil-structure Interaction. *Ind. Constr.* **2015**, *45*, 30–36.
23. Jiang, Z.R.; Shi, K.R.; Gao, X.N. Analysis of Nonlinear Buckling of a Long-Span Elliptic Paraboloid Suspended Dome Structure. *Adv. Mater. Res.* **2013**, *639*, 191–197.

24. Ruggieri, S.; Porco, F.; Uva, G. A practical approach for estimating the floor deformability in existing RC buildings: Evaluation of the effects in the structural response and seismic fragility. *Bull. Earthq. Eng.* **2020**, *18*, 2083–2113. [CrossRef]
25. Ruggieri, S.; Fiore, A.; Uva, G. A New Approach to Predict the Fundamental Period of Vibration for Newly-designed Reinforced Concrete Buildings. *J. Earthq. Eng.* **2021**, *4*, 1–26. [CrossRef]
26. Zhang, C.; Yu, L.; Liu, T. Monitoring and Numerical Simulation Analysis of Multiple Square Loop-String Roof Cable Tension Construction Process. *J. Wuhan Univ. Technol.* **2016**, *40*, 797–802.
27. GB 50011-2010. *Code for Seismic Design of Buildings*; China Architectural Engineering Press: Beijing, China, 2016; pp. 42–43.

symmetry

MDPI

Article

Review on Vibration-Based Structural Health Monitoring Techniques and Technical Codes

Yang Yang [1,*], Yao Zhang [2,*] and Xiaokun Tan [1]

1 School of Civil Engineering, Chongqing University, Chongqing 400044, China; 20164896@cqu.edu.cn
2 School of Architecture and Civil Engineering, Xiamen University, Xiamen 361005, China
* Correspondence: yangyangcqu@cqu.edu.cn (Y.Y.); zhangyao@xmu.edu.cn (Y.Z.)

Abstract: Structural damages occur in modern structures during operations due to environmental and human factors. The damages accumulating with time may lead to a significant decrease in structure performance or even destruction; natural symmetry is broken, resulting in an unexpected life and economic loss. Therefore, it is necessary to monitor the structural response to detect the damage in an early stage, evaluate the health condition of structures, and ensure the operation safety of structures. In fact, the structure and the evaluation can be considered as a special symmetry. Among several SHM methods, vibration-based SHM techniques have been widely adopted recently. Hence, this paper reviews the vibration-based SHM methods in terms of the vibrational parameters used. In addition, the technical codes on vibration based SHM system have also been reviewed, since they are more important in engineering applications. Several related ISO standards and national codes have been developed and implemented, while more specific technical codes are still required to provide more detailed guidelines in practice to maintain structure safety and natural symmetry.

Keywords: structural health monitoring (SHM); vibration; frequency domain; time domain; time-frequency domain; technical codes

Citation: Yang, Y.; Zhang, Y.; Tan, X. Review on Vibration-Based Structural Health Monitoring Techniques and Technical Codes. *Symmetry* **2021**, *13*, 1998. https://doi.org/10.3390/sym13111998

Academic Editor: Jan Awrejcewicz

Received: 19 September 2021
Accepted: 18 October 2021
Published: 22 October 2021

1. Introduction

Civil engineering structures are usually designed to serve for 50 years to 100 years, during which they are expected to maintain structural integrity. Unpredicted and unexpected structure failure due to accumulated damages during design life may cause significant life and economic losses; therefore, structural health monitoring (SHM) is very important since it can monitor the structural response, evaluate the structural safety in real time and maintain structure safety and symmetry in nature.

SHM techniques have been developed for many years [1–10]. Generally, the vibration based SHM methods are the most widely adopted. The objectives of these SHM approaches are determining the existence of structural damages, identifying the location and severity of structural damages, evaluating structure safety, predicting the remaining service life of the structure, and making decision of the maintenance strategy, if possible. In fact, the vibration characteristics of a structure are a function of its physical parameters. Structural damage causes change in physical parameters of the structure, and change in physical parameters therefore can be used as an indicator of structure health condition. Through the signal monitored by the sensors installed on the structure, the vibration characteristics can be extracted and the corresponding change can be detected and analyzed. In addition, from the change of vibration characteristics, one can further obtain the change in the physical parameters of the structure to diagnose the structure health condition. Recently, with the rapid development of modern computer technology and the progress of sensor technology and signal processing technology, test signals can be accurately and quickly analyzed and processed. Therefore, vibration-based structural health monitoring technology has become a research hotspot at home and abroad.

Based on these techniques and recently developed IT and sensing techniques, many technical standards and codes on vibration-based SHM have been developed and implemented for engineering applications. For a SHM system that can be used in practice, it usually contains four parts or sub-systems: sensing sub-system, including both a fixed sensor system and a portable sensor, data acquisition and transmission sub-system composing of a data acquisition unit, data transmission network and corresponding software system, data management and control sub-system containing of data management software and control server, and structure performance evaluation sub-system including structural health assessment server, structural health assessment workstation, and corresponding software system.

Although SHM has become an important field in the development of civil engineering disciplines, related technical methods still need to be improved, and there is still a lack of complete technical standards and specifications for vibration-based structural health monitoring. It is difficult for engineers to design suitable SHM system for a given structure based on the existing codes because there is no explicit answer of what kind of sensors to be used, where the sensors to be installed, and how to evaluate the structure health condition by using monitored data. Therefore, this work aims to review both vibration-based SHM techniques and technical codes, and provide a certain reference for the application of technical methods and standard specifications for vibration-based structural health monitoring. In Section 2, the vibration-based SHM approaches are reviewed, and attention will be paid on the more recently developed ones due to the length limit. In addition, the advantages and drawbacks of each approach has been summarized. In Section 3, the developed technical codes including both ISO standards and national codes are reviewed, and the suitable situations that each code can be applied have been reviewed, which may help to find out most suitable SHM method for a given structure. In Section 4, the challenges and future development are discussed.

2. Vibration-Based Structural Health Monitoring Techniques

Generally, in terms of vibration parameters used, vibration-based structural health monitoring techniques can be classified into three categories: frequency domain, time domain and time-frequency domain approaches.

2.1. Frequency Domain Methods for Vibration-Based SHM

Modal parameters including frequency, mode shape, and damping are usually used in frequency domain methods for vibration-based SHM [11–15]. In addition, several Frequency Response Function (FRF) related parameters have also been widely adopted. Compared to time domain methods and time-frequency domain methods, the frequency domain methods can be used in more situations because the frequency domain properties of structure are more stable. Even if the structure is subjected to greatly different loading conditions, the extracted dynamic properties are almost the same, and they only depend on the structure itself. However, it should be noted that the algorithms to extract these properties are not easy to implement and sometimes time consuming.

2.1.1. Frequencies and Mode Shapes

Frequencies and mode shapes are the most frequently used modal parameters in SHM methods. When a structure is damaged, its frequencies usually drop accordingly. In particular, the lower frequencies drop slightly while higher frequencies drop a little bit significantly [16]. In addition, temperature change also affects the natural frequency of the structure. The importance of temperature effects for a damage detection method based on relative frequency shift of several weak-axis bending vibration modes of beam-like structures was investigated by Gillich et al. [17]. However, frequencies alone are not usually adopted to identify the local damage since they are global indicator and they do not contain location information. Therefore, they are usually used together with mode shapes, which contain location information and more sensitive to local damages. The extracted mode shapes during monitoring were compared to those measured at the undamaged stage, and the damage indices of MAC and COMAC were proposed to locate and evaluate the local damages:

$$\mathrm{MAC}\left(\varphi_i^{\mathrm{u}},\varphi_i^{\mathrm{d}}\right)=\frac{\left[\left(\varphi_i^{\mathrm{u}}\right)^{\mathrm{T}}\varphi_i^{\mathrm{d}}\right]^2}{\left[\left(\varphi_i^{\mathrm{u}}\right)^{\mathrm{T}}\varphi_i^{\mathrm{u}}\right]\left[\left(\varphi_i^{\mathrm{d}}\right)^{\mathrm{T}}\varphi_i^{\mathrm{d}}\right]} \tag{1}$$

$$\mathrm{COMAC}\left(\varphi_i^{\mathrm{u}}(x_j),\varphi_i^{\mathrm{d}}(x_j)\right)=\frac{\left(\varphi_i^{\mathrm{u}}(x_j)\varphi_i^{\mathrm{d}}(x_j)\right)^2}{\left(\varphi_i^{\mathrm{u}}(x_j)\right)^2\left(\varphi_i^{\mathrm{d}}(x_j)\right)^2} \tag{2}$$

where φ_i^{u} and φ_i^{d} are the ith undamaged and damaged mode shape, x_j is the coordinate of jth point. It is observed that when φ_i^{d} exactly matches φ_i^{u}, the MAC value should be 1, hence a MAC value close to 1 indicates that the structure is still in good condition, but a MAC value greatly less than 1 means that the structure is damaged. Compared to MAC, COMAC has location information, the COMAC value at x_j close to 1 indicates that the structure is still intact at x_j and the COMAC value at x_j greatly less than 1 means that the structure has damage at x_j.

A lot of research [18–24] has been conducted to identify local damages by using both frequencies and mode shapes since they contain both global and local information of structures, and some improvements have also been proposed so that they can be applied successfully in practice. One direction to improve is to construct the baseline of structure mode shapes more accurately. Finite element (FE) model updating has been widely used for this purpose [25–27]. Conventional FE model updating was constructed for regenerating of baseline of frequencies and mode shapes. The frequencies and mode shapes obtained by FE model were compared to those measured by monitoring system to check the existence of local damages in the building. Then the stiffness matrix (usually, the mass matrix is not included) can be updated so that the updated frequencies and mode shapes can match the measured ones. Finally, the location and severity (stiffness loss) can be obtained by the updated FE model. In fact, the FE model updating can be generalized as a constrained optimization problem:

$$\min_{x_k}\|\sum_i w_i(\lambda_{FE,i}(x_k)-\lambda_i)\|_2^2\ s.t.\ x_{lk}\le x_k\le x_{uk} \tag{3}$$

where $\lambda_{FE,i}(x_k)$ is the ith frequency or mode shape obtained by FE model using design parameters x_k, λ_i is the measured ith frequency or mode shape, w_i in the range of 0 to 1 is the weight factor, x_{lk} and x_{uk} are the upper and lower bounds on the kth design variable. There are several standard procedures to solve this kind of constrained optimization problem.

To reduce iteration times and increase computation efficiency, substructure techniques have been developed [28–35]. It divided the whole structures into several small substructures, each of which was treated independently. Then the substructures were assembled to regenerate the global structure by imposing interface constraints. Weng et al. [28] proposed a new iterative substructuring method, which can accurately obtain the eigen-solutions and eigen-sensitivities of structures. Li et al. [30] proposed a sub-structure damage identification method based on frequency domain dynamic response reconstruction, and verified it numerically and experimentally. Papadimitriou et al. [31] proposed the component mode synthesis technology, which can effectively re-analyze in the generalized coordinate space of the accurate component model calculated by using the reference finite element model and the characteristic interface mode. The substructure techniques are usually more effective than conventional FE model updating method since substructure is more sensitive to local damage. The FE model updating methods including substructure techniques are considered as a typical inverse problem in mathematics, where restraint and optimization algorithm are very important.

Constraint is important since FE model updating is generally ill conditioned due to less measurements than unknown parameters to be determined. Hence, the target function to be optimized should include an additional term leading to a convex error function, and the selection of regularization parameters should be determined by specific structures and experience. The Tikhonov regularization is frequently adopted [36–38], and it is proven effective for a lot of practical scenarios, but the identified size of damage is usually larger than expect.

In addition to conventional optimization algorithms, several advanced optimization algorithms proposed for artificial intelligence and pattern recognition have been adopted in vibration-based SHM approaches, such as genetic algorithm [39–43], artificial neural network [44,45], and particle swarm optimization [46,47] and Artificial bee colony algorithm [48,49], etc. Unlike conventional optimization techniques which require established model to optimize parameters, these advanced ones are model-independent. This is actually very helpful in vibration-based SHM system since the measured and monitored data are usually insufficient and contains significant uncertainties, which brings great difficulties in convergence when identifying parameters by conventional optimization methods. However, it should also be admitted that these advanced approaches have their own drawbacks, for example, the computation load of genetic algorithm is very high since it is a global optimizer. In fact, for different types of structures, different optimization techniques should be considered due to various degree-of-freedoms; unfortunately, there is no common sense on how to select the optimization algorithms based on the type of structures.

Recently, machine learning methods become more popular due to the quick development of artificial intelligence [50–58]. It can definitely help to improve the reconstruction of structural model, but the model is a data-driven model rather than the physics-based model in FE model updating methods. Generally, the machine learning methods contains three steps, data acquisition, feature extraction, and feature classification, which are also the most important steps in pattern recognition. Frequencies and mode shapes are usually obtained during data acquisition and pre-processing as input of these algorithms. Feature extraction may depend on "model", which means that the features of undamaged "model" and damaged "model" should be labeled artificially during the training process. Then the algorithms are trained by the labeled data to generate the classifier. This is also known as supervised learning, and artificial neural network, convolutional neural network, and supported vector machine, etc., are the most typical ones. Actually, these methods are quite useful in real vibration-based SHM systems since it does not require regeneration of physics-based models of structures; therefore, it has great potential in the future application. However, it should be noted that there is a huge amount of data to be labeled during the training process, which costs a lot of manpower.

Bayesian methods [59–63] have been proposed and developed to reduce the influence of measurement noise and model errors on identifying local damages, since deterministic methods may fail when the change of frequencies and mode shapes due to damage is concealed by measurement noise or model errors. Bayesian methods use prior information from experiments and experience to construct the posterior probability of uncertainties and identified and evaluated damages accordingly. They are typical probabilistic methods, and they can even help on ill-conditioned inverse problems since they introduce a regularization term by using the probability distributions of uncertainties. However, it is noteworthy that prior information is very important in Bayesian methods, if the prior information is not accurate enough, the damage identification may fail even though the measurement is noiseless and model is perfect. In fact, the Bayesian probability of parameters θ under a given structure response $\mathbf{R}$ is as follows:

$$p(\theta\,|\mathbf{R}\,) = \frac{p(\mathbf{R}\,|\,\theta\,)p(\theta)}{p(\mathbf{R})} \tag{4}$$

where $p(\mathbf{R}\,|\,\theta\,)$ is the posterior joint probability distribution of the structure response under the condition of θ, $p(\theta)$ is the prior probability distribution of θ, $p(\mathbf{R})$ is a standardized constant, and $p(\theta\,|\mathbf{R}\,)$ is the posterior joint probability distribution of θ under the condition of $\mathbf{R}$. It should also be noted that the integral value of $p(\theta\,|\mathbf{R}\,)$ equals to 1.

In practice, the sampling data for generating prior distribution is usually sparse which makes the task difficult; therefore, sparse Bayesian learning [64–69] has been proposed to construct parameterized prior which can accurately construct the prior distribution based on sparse data. Several investigations have been conducted to show the feasibility of Bayesian methods on SHM by using frequencies and mode shapes, however most of which used lab-scale experiments and numerical simulations. Further studies are expected to show the applicability of Bayesian methods on real structures.

The other direction to improve is elimination of the dependency of the baseline or the undamaged model of structures. An assumption was proposed for this purpose: the mode shapes and mode shape curvatures are smooth and no sudden change with respect to location can be found for undamaged structures [70–73]. Once the sudden change of mode shapes or mode shape curvatures is observed, it is believed that the local damage occurs there. It is also proven by some studies that the mode shape curvatures are more sensitive than mode shapes on local damages, especially for early-stage damages. However, the extraction of mode shapes from accelerations or displacements is usually polluted by noise, and the mode shape curvatures obtained by central difference on mode shapes are even less accurate, resulting that the sudden change of mode shape curvatures due to local damages are covered by numerical error. Hence, how to improve the accuracy of mode shape curvature during monitoring should be investigated. On the other hand, the machine learning methods may also be independent on "model", which means that the features from undamaged "model" and damaged "model" do not need to be completely labeled, and the algorithms themselves can identify can classify the features. This is what is called "semi-supervised learning" [58,74] and "unsupervised learning" [75,76]. It is attractive but the identification accuracy needs to improve significantly, otherwise the false alarm will be issued unexpectedly and frequently.

2.1.2. Damping

Although damping can also be used for SHM system to monitor the health condition of structures [14,77–81], it is less frequently observed in practice than frequencies and mode shapes since it is more difficult to measure. Frizzarin et al. [77] analyzed the damping by using ambient vibration data to detect damage without baseline, and demonstrated the proposed method by a large-scale concrete bridge model with seismic damage. Mustafa et al. [78] introduced an energy based damping evaluate approach to evaluate the health condition of a truss bridge by numerical simulations. Cao et al. [79] compared damping based damage detection methods by using reinforced concrete structures and fiber reinforced composites, and clarified the factors that influenced the capability of damping on damage detection. Recently, Liu et al. [14] proposed a novel complex eigen-parameter identification method to evaluate the stiffness reduction and damping defect simultaneously on a non-classically damped shear building.

Ideally, the damping change due to local damage can be observed because the cracks may increase the frictions between interfaces. However, the measurement is vulnerable to noise, especially for structures subject to ambient environmental vibrations, so that the change of damping due to local damage is concealed by the measurement error. On the other hand, the damping model is difficult to select or construct whereas which is important in identification of damping. Classical Rayleigh damping which is a combination of mass and stiffness is frequently adopted in practice because it is the simplest damping model. However, it cannot be applied to many structures; therefore, some more advanced damping models have been proposed. It should be noted that for different types of structure, different damping models should be considered. Moreover, damping is a global property for a structure, similar to frequency, so damping itself can hardly be used to identify the location of local damages.

2.1.3. FRFs and Related Variants

FRFs are actually an extension of conventional modal parameters, because they contain the information over the entire frequency range. For different types of structure, the optimal frequency range may be various, which is highly dependent on experience and trial experiments. There is lack of theoretical analysis and numerical simulation investigations of how to select the sensitive frequency range to local damages for different structures. Operational deflection shapes [82–85] and their curvatures, power spectral density [86–88], frequency shift curve, and its curvature [89,90] are the most frequently used FRFs related variants.

FE updating methods can be applied to FRFs and related variants [91]. Conventional FE updating methods are effective in identifying local damages but have lower computational efficiency. Unfortunately, there are less investigations on applying substructure techniques, advanced regulation algorithms and optimization algorithms to FRFs and related variants, because it is difficult to select the sensitive frequency range for a given structure and it is also difficult to converge due to uncertainties in the measurement of FRFs.

Machine learning methods can also be applied to FRFs and related variants [92–94]. Conventional FRFs are curves which can be represented as one-column vectors, hence artificial neural network is suitable for identifying local damages by using FRFs. Usually, principal component analysis is applied to FRFs first to extract the most important components, which are then used as input to artificial neural network. Fourier amplitude spectra is a 2D surface FRFs related variant [95], therefore, convolutional neural network can be applied to it to construct the SHM system. In addition to neural networks, the Dirichlet process clustering [96] can be applied to SHM system to identify early-stage damages on bridges by using FRFs. However, these investigations have been conducted through numerical simulations and lab-scale experimental studies. Whether these machine learning methods based on FRFs and related variants are still effective should be further examined by field measurement. It should also be noted that since the quality of dataset for training is crucial for machine learning methods, therefore, the performance of these methods should be further examined when more FRFs data are available.

2.2. Time Domain Methods for Vibration-Based SHM

Instead of extracting the frequency related properties from the time history of dynamic responses of a structure, the dynamic responses of a structure in time domain can be used for SHM directly. Among them, acceleration and displacement are the most frequently used. The time domain methods usually do not require much calculation resources and therefore are timesaving, but they are used for the structures subject to stabilize environmental excitations because different excitation may cause quite different dynamic response and may cause the methods to fail to identify damages.

2.2.1. Accelerations

FE updating methods can be applied to accelerations for SHM [97], similar to frequencies and mode shapes. Tikhonov regularization [98,99], adaptive Tikhonov regularization [100], and L1 regularization [101] were successfully used to identify local damages based on accelerations. However, the dataset of time history of accelerations is much larger than the dataset of frequencies and mode shapes, hence the convergency is difficult to achieve. Moreover, there is no proven procedures on how to select the certain time history of accelerations, which is now generally dependent on experience.

Machine learning methods can also be applied to accelerations and variances for SHM [102–108], including both supervised and unsupervised methods. Although research showed that the machine learning methods can locate and evaluate local damages successfully in lad-scale experiments and numerical benchmark studies, no evidence have been provided that they can also be applied to real structure in practice. On the other hand, unlike frequencies and mode shapes which are only dependent on the structure itself, the accelerations are highly dependent on environmental excitations. The environmental excitations are always varying with respect to time, therefore it is quite difficult for the training algorithms in machine learning methods to differentiate the change of accelerations due to local damage and that due to environmental excitations.

Bayesian methods are also applicable when accelerations are used [109,110], and they show great potential in application to real structures. Research work has been conducted through experimental and numerical study. It should be noted that since the prior information is important to the Bayesian methods, hence the change of operational conditions should be considered carefully. For example, the traffic load of a bridge may increase with the economy development, therefore the prior information constructed previously may change with respect to time.

Statistical time series methods [111,112] are proposed especially for time history of dynamic responses, which usually fit time series models such as autoregressive model, autoregressive with exogenous model, and Mahalanobis squared distance, etc. All of them show their distinguishing advantages, but they still have their own limitations. For autoregressive model and autoregressive with exogenous model, it is difficult to determine the model order, which is currently highly dependent on experience. For Mahalanobis squared distance method, the data from the undamaged structures under various conditions is required, which is almost impossible for old building and structures. Statistical moment of accelerations can also be used to identify local damages of structure, Yang et al. [113,114] proposed a fusion of statistical moments by combining the fourth-order statistical moment of displacement with the eighth-order statistical moment of acceleration for the damage identification of structures. However, the order of moment to be selected is highly dependent on experience since different structures may have various statistical moments sensitive to damages, which limits the widely application of statistical moment. Temperature also plays an important role in SHM approach in time domain. Hios et al. [115] proposed a new stochastic global model method based on statistical hypothesis testing, and determined a functional hybrid model that can describe temperature-dependent dynamics. OBrien et al. [116] used temperature data to validate damage indicators based on measured data collected under uncontrolled traffic conditions, and showed that temperature can be used as a proxy for damage since stiffness of concrete structure is dependent on temperature.

2.2.2. Displacements

Generally, the methods that can be applied on accelerations can also be applied to displacement [117]. However, the displacement is usually not directly measured during monitoring [118]. In principle, acceleration measurement can be doubly integrated to give displacement, but this process is notoriously error-prone due to unknown initial conditions such as integration constants and low frequency noise of measurement that is amplified in an inverse square manner. In reality, the displacement at the measured location can only be recovered from field measured acceleration in an approximate sense, depending on the frequency characteristics of the contributing activities.

In the SHM system, one strategy to resolve the issue of unknown initial conditions makes use of the basic fact in structural dynamics that initial condition effects decay exponentially with time. Thus, one can start the numerical integration process before the main event to be captured, so as to allow a 'burn-in' time for the (unknown) initial condition effect to die down to negligible level during the main event that matters. On the other hand, the presence of noise especially in the low frequency regime presents a major difficulty. In addition to amplification of low frequency noise during numerical integration, data acquisition hardware typically has 'pink' noise in the low frequencies, i.e., with PSD inversely proportional to frequency. Integrating the 'raw' measured acceleration will often lead to significant systematic over-estimation of displacement, in many cases a 'flying off' trace of time history. In particular, a constant error in the acceleration gives a linear trend in the velocity and a quadratic trend in the displacement. One basic strategy is to suppress the noise by a causal filter with parameters designed to significantly attenuate the frequency components in the data below and above specified cut-off frequencies. Acceleration data is filtered before numerically integrated to give velocity data, which is filtered again and then integrated to give displacement data. Filtering produces distortion in the acceleration data and hence the integrated displacement. This will need to be controlled and verified in the development of SHM system.

2.3. Time-Frequency Domain Methods for Vibration-Based SHM

In addition to time domain and frequency domain dynamic properties, the properties in both time domain and frequency domain can also be used for SHM due to the development of advanced time-frequency analysis. Compared to time domain and frequency domain methods, the amount of time-frequency domain methods is much fewer. The time-frequency domain methods are more powerful because it contains the information of stable frequency domain properties and can further show the change with time. However, it is admitted that they require a lot of calculation resources and space for data storage.

Short time Fourier Transform, Wavelet Transform [119], and Hilbert-Huang Transform [120] including empirical mode decomposition are the most widely used time-frequency analysis methods. Usually, only the measured accelerations or dynamic strains are required for these methods, and the high frequency components may change significantly once the local damage occurs. Therefore, it is not necessary for these methods to construct the undamaged model for the monitored structure. This is a very attractive advantage of these methods; however, it is noteworthy that the time-frequency methods can only locate the local damages but cannot evaluate the severity of local damages. Further investigations of applying time-frequency domain methods on real structures are expected in the near future.

3. Current Technical Codes Related to Vibration-Based Structural Health Monitoring

In this section, the current technical standards and codes related to vibration-based structural health monitoring are reviewed, including both ISO standards and national codes.

ISO has four technical committees related to building and construction: TC 59 (Committee on Architecture and Civil Engineering), TC 98 (Committee on Fundamentals of Structural Design), TC135 (Committee on Nondestructive Testing), and TC268 (Committee on Urban and Community Sustainable Development). TC 59 including its SCs has published 124 ISO standards of which 33 under the direct responsibility of ISO/TC 59 (Table 1), TC 98 including its SCs has published 23 ISO standards (Table 2), TC 135 including its SCs has published 97 ISO standards of which 1 under the direct responsibility of ISO/TC 135 (Table 3), and TC 268 (including its SCs) has published 26 ISO standards of which 10 under the direct responsibility of ISO/TC 268 (Table 4). The ISO standards published by ISO/TC135 are all about non-destructive testing, including detailed procedures of different non-destructive testing methods. They are not reviewed herein since they are more relevant to damage detection rather than SHM system.

In fact, the codes published by ISO TC 59 do not only specify the general principles to determine requirements of structural performance, but also provides a general approach to assess the structural safety based on structural performance. They are a very important framework of a SHM system, but they lack details on how to implement in real engineering projects. The standards published by ISO TC 98 focus more on reliability and show the requirements and procedures to assess structure health condition based on structural reliability. They also provide approaches and procedures to prepare national and organization codes. However, structural reliability is more abstract, and it contains more complicated mathematical models, which is difficult to be applied in real SHM projects. The regulations published by ISO TC 268 focus on smart building and sustainable development. They provide the foundation on how to construct smart community infrastructures. Of course, the SHM system is helpful to construct smart community infrastructures and maintain sustainable development. Therefore, they provide the future work scope for current SHM systems, but they lack more details on how to construct a SHM system for an existing building.

Table 1. Published ISO standards by TC 59's SCs.

Subcommittee	Subcommittee Title	Published Standards	Standards under Development
ISO/TC 59/SC 2	Terminology and harmonization of languages	4	2
ISO/TC 59/SC 8	Sealants	30	14
ISO/TC 59/SC 13	Organization and digitization of information about buildings and civil engineering works, including building information modelling (BIM)	18	5
ISO/TC 59/SC 14	Design life	10	1
ISO/TC 59/SC 15	Framework for the description of housing performance	8	2
ISO/TC 59/SC 16	Accessibility and usability of the built environment	1	1
ISO/TC 59/SC 17	Sustainability in buildings and civil engineering works	12	3
ISO/TC 59/SC 18	Construction procurement	8	3

Table 2. Published ISO standards by TC 98's SCs.

Subcommittee	Subcommittee Title	Published Standards	Standards under Development
ISO/TC 98/SC 1	Terminology and symbols	2	0
ISO/TC 98/SC 2	Reliability of structures	8	2
ISO/TC 98/SC 3	Loads, forces, and other actions	13	0

Table 3. Published ISO standards by TC 135's SCs.

Subcommittee	Subcommittee Title	Published Standards	Standards under Development
ISO/TC 135/SC 2	Surface methods	14	2
ISO/TC 135/SC 3	Ultrasonic testing	24	3
ISO/TC 135/SC 4	Eddy current testing	7	0
ISO/TC 135/SC 5	Radiographic testing	26	0
ISO/TC 135/SC 6	Leak testing	4	0
ISO/TC 135/SC 7	Personnel qualification	7	1
ISO/TC 135/SC 8	Thermographic testing	4	2
ISO/TC 135/SC 9	Acoustic emission testing	10	3

Table 4. Published ISO standards by TC 268's SCs.

Subcommittee	Subcommittee Title	Published Standards	Standards under Development
ISO/TC 268/SC1	Smart community infrastructures	16	15

3.1. Standards Published by ISO/TC59

3.1.1. ISO 11863:2011

ISO 11863:2011 (ISO/TC 59/SC 15) [121], specifies the basic requirements and principles to determine and check the basic requirements of structural performance. This is very important to a SHM system since it can help to restrain the scope of the SHM system and select proper vibration parameters to be monitored. It also specifies the thresholds for capability, which is in fact essential for a SHM system since automated alert algorithm highly depends on the pre-defined thresholds. In addition, it provides guidelines on assessing the difference between designed and measured capabilities, which is helpful to generate the maintenance strategy in SHM system. However, it does not provide any detailed feasible procedures for any specific structures. Therefore, it is difficult to construct a SHM system for a specific building or structure by just following ISO 11863. For example, it requires "the threshold level is a minimum level of demand", but the statement is quite general because for different buildings or structural components the requirements may be quite different, hence it is difficult to be applied in real application directly.

3.1.2. ISO 15928-1:2015

ISO 15928-1:2015 [122] shows a general method on assessment of the structural safety performance of buildings, in which the principles on how to evaluate the design and construction of buildings are provided. It mainly focuses on the design and construction stage of buildings; therefore, it is helpful for generating SHM systems during construction. Although the idea of evaluating the structural safety performance of building during design can be shared in constructing SHM systems for operation stage, it should also be noted that design of building and design of monitoring system for building are quite different. Therefore, the evaluation processes outlined in this code cannot be directly used for design of monitoring system during operation. Moreover, it only shows that the evaluation "may be carried out by analysis, testing, service experience or a combination of the above" [120] without any details. The details of design of building provided in Eurocode 1-9, or other national standards are important supplementary standards, however they are not covering the design of monitoring system.

3.2. Standards Published by ISO/TC98

3.2.1. ISO 4356:1977

ISO 4356:1977 [123] establishes the basic principles that should be adopted when setting up national standards, regulations and recommendations for the deformation of buildings at the limit states of serviceability. Traditionally, measurement of deformation is usually considered as static measurement rather than dynamic measurement. However, recently developed signal processing techniques can be used to reconstruct the time history of displacement by integrating accelerations twice. In fact, deformation of structural components is quite important in monitoring; when the deformation approaches the threshold at limit states of serviceability, an alert should be issued by the SHM system. Although this code was drafted for the purpose of building design, it can also be used as guideline for design of monitoring system since it provides the basic principles to determine the deformations of buildings at the serviceability limit states, which can also be considered as the base for answering the critical question that how to evaluate the health condition of the building based on the monitored data.

3.2.2. ISO 13822:2010

ISO 13822:2010 [124] provides general requirements and procedures for the assessment of existing structures based on the principles of structural reliability and consequences of failure. Although it is applicable to the assessment of any type of existing structure of any material that was originally designed, analyzed, and specified based on accepted engineering principles and design rules, it only provides very general requirements and procedures without any specified methods. In addition, it mainly focuses on routine visual inspections, including visible deformations and surface defects like cracks and spalling, but limited information of vibration parameters is required in this code. There is no doubt that it is helpful to generate SHM systems of buildings. However, it is noteworthy that the regular inspection and real time monitoring is a little bit different, e.g., the former requires experienced technician while the latter depends on pre-installed sensors.

3.2.3. ISO 2394:2015

ISO 2394:2015 [125] presents a risk- and reliability-informed foundation for decision making of maintenance strategy by considering design and assessment of structures for the purpose of developing code. It is certainly helpful for building SHM systems by answering the critical question that how to evaluate the health condition of buildings based on the monitored data and how to plan maintenance strategy and build automated alert system based on evaluated condition. However, there is only the basic idea without any detailed procedures for specific buildings and the vibration parameters to be monitored.

3.3. *Standards Published by ISO/TC268*

3.3.1. ISO 37104:2019

ISO 37104:2019 (ISO/TC 98/SC 2) [126] provides general guidance on how to implement and maintain sustainable development management systems in accordance with ISO 37101 principles, which can be applied to cities and other forms of settlement. It should be noted that SHM is important for sustainable development since it can help buildings to extend their life. Therefore, this code should be paid attention to when designing the SHM system, so that the designed SHM system can fulfill the requirements of sustainable developments.

3.3.2. ISO 37105:2019

ISO 37105:2019 [127] specifies a descriptive framework for a city, including a structurally related basis for a city or community. When more buildings have installed SHM systems, it may help to generate sustainable cities. Therefore, this code should also be followed to achieve the objective.

3.3.3. ISO/TS 37107:2019

ISO/TS 37107:2019 [128], provides a top-level maturity model for Smart Sustainable Communities (MMSSC) which can be used for self-assessment of individual cities and communities. A simple way to assess community's maturity in adopting the good practices is sketched. In fact, with the development of IoT, the SHM systems can also be connected to form a network which may perform as an important of smart sustainable communities. Hence, this code is helpful for integrate multiple SHM systems in the future.

3.4. *National Codes*

Several national codes on SHM have been published in the past two decades. In North America, Intelligent Sensing for Innovative Structures of Canada published the first guideline for SHM, "Guidelines for Structural Health Monitoring" [129] in 2001, in which the techniques of both static and dynamic structural testing, periodic regular inspection, and continuous monitoring were presented and summarized. The Federal Highway Administration, U.S., published guidelines for SHM of bridges and tunnels, "Development of a Model Health Monitoring Guide for Major Bridges" [130] and "Tunnel Operations, Maintenance, Inspection, and Evaluation Manual" [131], where the regular visual inspection is the most important method. The International Federation for Structural Concrete also published "Monitoring and Safety Evaluation of Existing Concrete Structures" as a guideline for SHM of existing concrete structures [132], whereas the vibration-based SHM methods are less important than the quality and durability evaluation of concrete.

In Europe, the Structural Assessment, Monitoring and Control of European Union developed "Guideline for Structural Health Monitoring" [133] in 2006 to present the basic regulations and procedures of SHM, including determination of actions, structural condition analysis, design and operation of monitoring, numerical analysis and general damage identification. It is comprehensive and provides a framework for the following standards and codes. The Russian Federation also published its national code, GOSTR 53778-2010, "Building and structures, technical inspections and monitoring regulations" [134], where regular visual inspections, modal testing methods, condition classification, and grading system were presented. Similar to the Federal Highway Administration, U.S., the German administration also published the codes for SHM of bridges and tunnels, "Quality assurance for structural maintenance, surveillance, checking and assessment of bridges and tunnels, monitoring of bridges and other engineering structures", [135] where checking procedures and methods were presented. However, it was drafted in German, so it is not so easy for engineers in other countries to understand.

In Asia, "GB50982-2014 Technical Code for Monitoring of Buildings and Bridge Structures" [136], was published by Ministry of Housing and Urban-Rural Development of China. It has nine chapters and covers the basic requirements for SHM systems and general monitoring methods, and specific methods for high rise buildings, long-span spatial building, bridge structures, and other structures. For each type of structure, the requirements of SHM system for both construction stage and operation stage are provided. The monitored data are mostly vibration related, in addition to temperature and humidity. This is a comprehensive technical code on SHM, and its supplementary codes "Application and Analysis of Technical Code for Monitoring of Buildings and Bridge Structures" has also been published. However, it should be admitted that it can be improved by including more advanced vibration-based SHM techniques reviewed in Section 2 and more details about the sensor selection and arrangement.

4. Challenges and Future Development

Many vibration based SHM techniques have been proposed and developed recently, which have been reviewed in Section 2, however, their real applications in practical buildings and structures are rare. Some challenges are summarized as follows:

(1) Although various damage indicators and damage indexes based on vibration parameters have been proposed, it should be admitted that the sensitivities of them are not high enough to detect damage at early stage. Usually, the vibration parameters related to lower vibration mode can be measured more easily and accurately, but unfortunately those related to higher vibration mode are more sensitive to minor local damages. Considering that higher vibration modes can be hardly extracted if only ambient environmental excitation exists, damage index more sensitive to local damage at early stage by using lower vibration modes should be investigated in the future.

(2) The uncertainties of damage detection and evaluation in a SHM system are usually inevitable due to measurement noise, non-ideal boundary conditions, and ambient environmental vibrations. It increases the difficulty in extracting modal properties and calculating damage indicators and sometime the damaged signal can be concealed by the uncertainties. The statistical signal or statistical damage index may be investigated to reduce the uncertainties during monitoring.

(3) Data transmission, processing, and storage should also be paid attention to although it is usually be ignored in many research works. In fact, it is very important for a practical SHM system in real applications. The collection of the same type of data should be simultaneous, and they should be transmitted to the local server or cloud server smoothly. The requirements of hardware should be investigated in the future so that the proposed vibration based SHM methods can be applied better in practice.

(4) Currently, the benchmarks of SHM systems are lab studies and numerical studies, which are quite different from actual buildings and structures. Therefore, it is necessary to generate a benchmark study by using real building or structure. In fact, a data sharing platform is desired to examine the proposed SHM approaches, which may be helpful for the development and improvement of vibration based SHM methods in the future.

On the other hand, it can be concluded that although there are some ISO standards and national codes relevant to vibration based SHM system, unfortunately they cannot answer the following critical questions well:

(1) What monitoring methods should be used for a given building?
(2) What types of sensors should be used and where are they installed?
(3) How can the health condition of the building be assessed based on the collected data?

The above issues are important and unavoidable in SHM, since a properly designed SHM system should provide early warning to potential structural collapse to ensure the safety of building and residents' lives. Therefore, it is important to fill the gap in ISO standards and national codes to clearly specify which monitoring methods should be adopted for a specific building, where to install the sensors, how to evaluate the health condition of structures based on the collected data, and how to plan maintenance strategy and predict the remaining life of the building or structure.

5. Conclusions

The vibration-based SHM techniques and related ISO standards and national codes have been reviewed. The advantages and drawbacks of each method as well as the applicability of each standard or code have been presented. For different types of structure, different vibration-based SHM techniques should be selected. There is no universal approach for all types of structures and all kinds of damages. Although the standards, codes, and regulations have provide basic requirements and principles of a SHM system, it is still difficult for engineers to answer questions such as what types of sensors are to be used, where to install them, and how to use the monitored data to evaluate the structure health condition and predict the remaining life for a given structure. Therefore, it is necessary to develop such a code of SHM system construction that can be applied to real civil engineering structures.

Author Contributions: Conceptualization, Y.Y.; methodology, Y.Y. and Y.Z.; draft writing, Y.Z. and X.K.T.; final writing, Y.Z. and Y.Y. All authors have read and agreed to the published version of the manuscript.

Funding: We are grateful to the following agencies for their supports in this study: The National Key Research and Development Program of China(Grant No. 2020YFF0217802), National Natural Science Foundation of China (Grant No. 51778090, 51911530244), Science and technology plan project of Chongqing Science and Technology Bure (Grant No. cstc2020jscx-msxm0907), Graduate Research and Innovation Foundation of Chongqing, China (Grant No. CYS19004, CYS20022), Science and Technology Project of China Power Construction Co., Ltd. (Grant No. KJ-2020-117).

Institutional Review Board Statement: Not applicable.

Informed Consent Statement: Not applicable.

Data Availability Statement: Not applicable.

Conflicts of Interest: The authors declare no conflict of interest.

References

1. Doebling, S.W.; Farrar, C.R.; Prime, M.B.; Shevitz, D.W. *Damage Identification and Health Monitoring of Structural and Mechanical Systems from Changes in Their Vibration Characteristics: A Literature Review*; Los Alamos National Laboratory Report LA-13070-MS; Los Alamos National Laboratory: Los Alamos, NM, USA, 1996.
2. Farrar, C.R.; Doebling, S.W.; Nix, D.A. Vibration–based structural damage identification. *Philos. Trans. R. Soc. A Phys. Eng. Sci.* **2001**, *359*, 131–149. [CrossRef]
3. Sohn, H.; Farrar, C.R.; Hemez, F.M.; Shunk, D.D.; Stinemates, D.W.; Nadler, B.R. *A Review of Structural Health Monitoring Literature: 1996–2001*; Los Alamos National Laboratory Report; Los Alamos National Laboratory: Los Alamos, NM, USA, 2003.
4. Carden, E.P.; Fanning, P. Vibration based condition monitoring: A review. *Struct. Health Monit.* **2004**, *3*, 355–377. [CrossRef]
5. Fan, W.; Qiao, P.Z. Vibration-based damage identification methods: A review and comparative study. *Struct. Health Monit.* **2011**, *10*, 83–111. [CrossRef]
6. Li, H.N.; Yi, T.H.; Ren, L.; Huo, L.S. Reviews on innovations and applications in structural health monitoring for infrastructures. *Struct. Monit. Maint.* **2014**, *1*, 1–45. [CrossRef]
7. Kong, X.; Cai, C.S.; Hu, J.X. The state-of-the-art on framework of vibration-based structural damage identification for decision making. *Appl. Sci.* **2017**, *7*, 497. [CrossRef]
8. Sony, S.; Laventure, S.; Sadhu, A. A literature review of next-generation smart sensing technology in structural health monitoring. *Struct. Health Monit.* **2019**, *26*, e2321.1–e2321.22. [CrossRef]
9. Han, Q.H.; Ma, Q.; Xu, J.; Liu, M. Structural health monitoring research under varying temperature condition: A review. *J. Civ. Struct. Health* **2021**, *11*, 149–173. [CrossRef]

10. Dong, C.Z.; Catbas, F.N. A review of computer vision–based structural health monitoring at local and global levels. *Struct. Health Monit.* **2021**, *20*, 692–743. [CrossRef]
11. Liu, P.L. Identification and damage detection of trusses using modal data. *J. Struct. Eng.* **1995**, *121*, 599–608. [CrossRef]
12. Radzieński, M.; Krawczuk, M.; Palacz, M. Improvement of damage detection methods based on experimental modal parameters. *Mech. Syst. Signal Process.* **2011**, *25*, 2169–2190. [CrossRef]
13. Zhao, J.H.; Zhang, L. Structural damage identification based on the modal data change. *Int. J. Eng. Manuf.* **2012**, *4*, 59–66. [CrossRef]
14. Liu, J.; Lu, Z.R.; Yu, M.L. Damage identification of non-classically damped shear building by sensitivity analysis of complex modal parameter. *J. Sound Vib.* **2019**, *483*, 457–475. [CrossRef]
15. Han, J.P.; Zheng, P.J.; Wang, H.T. Structural modal parameter identification and damage diagnosis based on Hilbert-Huang transform. *Earthq. Eng. Eng. Vib.* **2013**, *13*, 101–111. [CrossRef]
16. Salawu, O.S. Detection of structural damage through changes in frequency: A review. *Eng. Struct.* **1997**, *19*, 718–723. [CrossRef]
17. Gillich, G.R.; Furdui, H.; Wahab, M.A.; Korka, Z.I. A robust damage detection method based on multi-modal analysis in variable temperature conditions. *Mech. Syst. Signal Process.* **2019**, *115*, 361–379. [CrossRef]
18. Ratcliffe, C.P. Damage detection using a modified Laplacian operator on mode shape data. *J. Sound Vib.* **1997**, *204*, 505–517. [CrossRef]
19. Zhang, Y.; Wang, L.Q.; Xiang, Z.H. Damage detection by mode shape squares extracted from a passing vehicle. *J. Sound Vib.* **2012**, *331*, 291–307. [CrossRef]
20. Khiem, N.T.; Tran, H.T. A procedure for multiple crack identification in beam-like structures from natural vibration mode. *J. Sound Vib.* **2014**, *20*, 1417–1427. [CrossRef]
21. Capecchi, D.; Ciambella, J.; Pau, A.; Vestroni, F. Damage identification in a parabolic arch by means of natural frequencies, modal shapes and curvatures. *Meccanica* **2016**, *51*, 2847–2859. [CrossRef]
22. Yang, Y.; Cheng, Q.; Zhu, Y.; Wang, L.; Jin, R. Feasibility study of tractor-test vehicle technique for practical structural condition assessment of beam-like bridge deck. *Remote Sens.* **2020**, *12*, 114. [CrossRef]
23. Yang, Y.; Xiang, C.; Jiang, M.; Li, W.; Kuang, Y. Bridge damage identification method considering road surface roughness by using indirect measurement technique. *China J. Highw. Transp.* **2019**, *32*, 99–106.
24. Yang, Y.; Liang, J.; Yuan, A.; Lu, H.; Luo, K.; Shen, X.; Wan, Q. Bridge element bending stiffness damage identification based on new indirect measurement method. *China J. Highw. Transp.* **2021**, *34*, 188–198.
25. Friswell, M.; Mottershead, J.E. *Finite Element Model Updating in Structural Dynamics*; Kluwer Academic Publishers: Dordrecht, The Netherlands, 1995.
26. Sanayei, M.; AliKhaloo, A.; Gul, M.; Catbas, F.N. Automated finite element model updating of a scale bridge model using measured static and modal test data. *Eng. Struct.* **2015**, *102*, 66–79. [CrossRef]
27. Suzuki, A.; Kurata, M.; Li, X.H.; Shimmoto, S. Residual structural capacity evaluation of steel moment-resisting frames with dynamic-strain-based model updating method. *Earthq. Eng. Struct. Dyn.* **2017**, *46*, 1791–1810. [CrossRef]
28. Weng, S.; Xia, Y.; Xu, Y.L.; Zhu, H.P. An iterative substructuring approach to the calculation of eigen-solution and eigen-sensitivity. *J. Sound Vib.* **2011**, *330*, 3368–3380. [CrossRef]
29. Weng, S.; Xia, Y.; Zhou, X.Q.; Xu, Y.L.; Zhu, H.P. Inverse substructure method for model updating of structures. *J. Sound Vib.* **2012**, *331*, 5449–5468. [CrossRef]
30. Li, J.; Law, S.S.; Ding, Y. Substructure damage identification based on response reconstruction in frequency domain and model updating. *Eng. Struct.* **2012**, *41*, 270–284. [CrossRef]
31. Papadimitriou, C.; Papadioti, D.C. Component mode synthesis techniques for finite element model updating. *Comput. Struct.* **2013**, *126*, 15–28. [CrossRef]
32. Liu, Y.; Sun, H.; Wang, D.J. Updating the finite element model of large-scaled structures using component mode synthesis technique. *Intell. Autom. Soft Comput.* **2013**, *19*, 11–21. [CrossRef]
33. Yu, J.X.; Xia, Y.; Lin, W.; Zhou, X.Q. Element-by-element model updating of large-scale structures based on component mode synthesis method. *J. Sound Vib.* **2016**, *362*, 72–84. [CrossRef]
34. Wang, T.; He, H.; Yan, W.; Chen, G.P. A model-updating approach based on the component mode synthesis method and perturbation analysis. *J. Sound Vib.* **2018**, *433*, 349–365. [CrossRef]
35. Weng, S.; Zhu, H.P.; Xia, Y.; Li, J.J.; Tian, W. A review on dynamic substructuring methods for model updating and damage detection of large-scale structures. *Adv. Struct. Eng.* **2020**, *23*, 584–600. [CrossRef]
36. Rucevskis, S.; Sumbatyan, M.A.; Akishin, P.; Chate, A. Tikhonov's regularization approach in mode shape curvature analysis applied to damage detection. *Mech. Res. Commun.* **2015**, *65*, 9–16. [CrossRef]
37. Wang, S.Q.; Xu, M.Q.; Xia, Z.P.; Li, Y.C. A novel Tikhonov regularization-based iterative method for structural damage identification of offshore platforms. *J. Mar. Sci. Technol.* **2019**, *24*, 575–592. [CrossRef]
38. Bao, Y.Q.; Li, H.; Ou, J.P. Emerging data technology in structural health monitoring: Compressive sensing technology. *J. Civ. Struct. Health Monit.* **2014**, *4*, 77–90. [CrossRef]
39. Meruane, V.; Heylen, W. An hybrid real genetic algorithm to detect structural damage using modal properties. *Mech. Syst. Signal Process.* **2011**, *25*, 1559–1573. [CrossRef]

40. Liu, H.B.; Jiao, Y.B. Application of genetic algorithm-support vector machine (GA-SVM) for damage identification of bridge. *Int. J. Comput. Intell. Appl.* **2011**, *10*, 383–397. [CrossRef]
41. Amiri, G.; Seyed Razzaghi, S.A.; Bagheri, A. Damage detection in plates based on pattern search and Genetic algorithms. *Smart Struct. Syst.* **2011**, *7*, 117–132. [CrossRef]
42. Beygzadeh, S.; Salajegheh, E.; Torkzadeh, P.; Salajegheh, J.; Naseralavi, S.S. An improved genetic algorithm for optimal sensor placement in space struc- tures damage detection. *Int. J. Space Struct.* **2014**, *29*, 121–136. [CrossRef]
43. Hou, R.R.; Xia, Y.; Xia, Q.; Zhou, X.Q. Genetic algorithm based optimal sensor placement for L_1-regularized damage detection. *Struct. Control Health Monit.* **2019**, *26*, e2274. [CrossRef]
44. Saeed, R.A.; Galybin, A.N.; Popov, V. Crack identification in curvilinear beams by using ANN and ANFIS based on natural frequencies and frequency response functions. *Neural Comput. Appl.* **2012**, *21*, 1629–1645. [CrossRef]
45. Neves, A.C.; González, I.; Leander, J.; Karoumi, R. Structural health monitoring of bridges: A model-free ANN-based approach to damage detection. *J. Civil. Struct. Health Monit.* **2017**, *7*, 689–702. [CrossRef]
46. Guo, H.Y.; Li, Z.L. Structural damage identification based on evidence fusion and improved particle swarm optimization. *J. Vib. Control.* **2014**, *20*, 1279–1292. [CrossRef]
47. Chen, Z.P.; Yu, L. A novel PSO-based algorithm for structural damage detection using Bayesian multi-sample objective function. *Struct. Eng. Mech.* **2017**, *63*, 825–835.
48. Ding, Z.H.; Yao, R.Z.; Li, J.; Lu, Z.R. Structural damage identification based on modified Artificial Bee Colony algorithm using modal data. *Inverse Probl. Sci. Eng.* **2017**, *26*, 422–442. [CrossRef]
49. Ding, Z.H.; Lu, Z.R.; Huang, M.; Liu, J. Improved artificial bee colony algorithm for crack identification in beam using natural frequencies only. *Inverse Probl. Sci. Eng.* **2017**, *25*, 218–238. [CrossRef]
50. Bishop, C.M. *Pattern Recognition and Machine Learning*; Springer: Berlin/Heidelberg, Germany, 2006.
51. Bakhary, N.; Hao, H.; Deeks, A.J. Structure damage detection using neural network with multi-stage substructuring. *Adv. Struct. Eng.* **2010**, *13*, 95–110. [CrossRef]
52. Jiang, S.F.; Zhang, C.M.; Zhang, S. Two-stage structural damage detection using fuzzy neural networks and data fusion techniques. *Expert Syst. Appl.* **2011**, *38*, 511–519. [CrossRef]
53. Hakim, S.J.S.; Razak, H.A. Structural damage detection of steel bridge girder using artificial neural networks and finite element models. *Steel Compos. Struct.* **2013**, *14*, 367–377. [CrossRef]
54. Hakim, S.J.S.; Razak, H.A. Adaptive Neuro Fuzzy Inference System (ANFIS) and Artificial neural networks (ANNs) for structural damage identification. *Struct. Eng. Mech.* **2013**, *45*, 779–802. [CrossRef]
55. Bandara, R.P.; Chan, T.H.T.; Thambiratnam, D.P. Frequency response function based damage identification using principal component analysis and pattern recognition technique. *Eng. Struct.* **2014**, *66*, 116–128. [CrossRef]
56. Kourehli, S.S. LS-SVM regression for structural damage diagnosis using the iterated improved reduction system. *Int. J. Struct. Stab. Dyn.* **2016**, *16*, 1550018. [CrossRef]
57. Gui, G.Q.; Pan, H.; Lin, Z.B.; Li, Y.H.; Yuan, Z.J. Data-driven support vector machine with optimization techniques for structural health monitoring and damage detection. *KSCE J. Civ. Eng.* **2017**, *21*, 523–534. [CrossRef]
58. Ye, X.W.; Jin, T.; Yun, C.B. A review on deep learning based structural health monitoring of civil infrastructures. *Smart Struct. Syst.* **2019**, *24*, 567–586.
59. Figueiredo, E.; Radu, L.; Worden, K.; Farrar, C.R. A Bayesian approach based on a Markov-chain Monte Carlo method for damage detection under unknown sources of variability. *Eng. Struct.* **2014**, *80*, 1–10. [CrossRef]
60. Lam, H.F.; Hu, Q.; Wong, M.T. The Bayesian methodology for the detection of railway ballast damage under a concrete sleeper. *Eng. Struct.* **2014**, *81*, 289–301. [CrossRef]
61. Behmanesh, I.; Moaveni, B. Probabilistic identification of simulated damage on the Dowling Hall footbridge through Bayesian finite element model updating. *Struct. Control Health Monit.* **2015**, *22*, 463–483. [CrossRef]
62. Behmanesh, I.; Moaveni, B.; Papadimitriou, C. Probabilistic damage identification of a designed 9-story building using modal data in the presence of modeling errors. *Eng. Struct.* **2017**, *131*, 542–552. [CrossRef]
63. Yin, T.; Jiang, Q.H.; Yuen, K.V. Vibration-based damage detection for structural connections using incomplete modal data by Bayesian approach and model reduction technique. *Eng. Struct.* **2017**, *132*, 260–277. [CrossRef]
64. Tipping, M.E. Sparse Bayesian learning and the relevance vector machine. *J. Mach. Learn. Res.* **2001**, *1*, 211–244.
65. Wipf, D.P.; Rao, B.D. Sparse Bayesian learning for basis selection. *IEEE Trans. Signal Process.* **2004**, *52*, 2153–2164. [CrossRef]
66. Williams, O.; Blake, A.; Cipolla, R. Sparse Bayesian learning for efficient visual tracking. *IEEE Trans. Pattern Anal. Mach. Intell.* **2005**, *27*, 1292–1304. [CrossRef]
67. Ji, S.; Xue, Y.; Carin, L. Bayesian compressive sensing. *IEEE Trans. Signal Process.* **2008**, *56*, 2346–2356. [CrossRef]
68. Zhang, Z.; Rao, B.D. Sparse signal recovery with temporally correlated source vectors using sparse Bayesian learning. *IEEE J. Sel. Top. Signal Process.* **2011**, *5*, 912–926. [CrossRef]
69. Lin, J.; Nassar, M.; Evans, B.L. Impulsive noise mitigation in powerline communications using sparse Bayesian learning. *IEEE J. Sel. Areas Commun.* **2013**, *31*, 1172–1183. [CrossRef]
70. Yoon, M.K.; Heider, D.; Gillespie, J.W., Jr.; Ratcliffe, C.P.; Crane, R.M. Local damage detection with the global fitting method using mode shape data in notched beams. *J. Nondestruct. Eval.* **2009**, *28*, 63–74. [CrossRef]

71. Yoon, M.K.; Heider, D.; Gillespie, J.W., Jr.; Ratcliffe, C.P.; Crane, R.M. Local damage detection with the global fitting method using operating deflection shape data. *J. Nondestruct. Eval.* **2010**, *29*, 25–37. [CrossRef]
72. Cao, M.S.; Radzieński, M.; Xu, W.; Ostachowicz, W. Identification of multiple damage in beams based on robust curvature mode shapes. *Mech. Syst. Signal Process.* **2014**, *46*, 468–480. [CrossRef]
73. Feng, D.M.; Feng, M.Q. Output-only damage detection using vehicle-induced displacement response and mode shape curvature index. *Struct. Control Health Monit.* **2016**, *23*, 1088–1107. [CrossRef]
74. Chen, S.; Cerda, F.; Rizzo, P.; Bielak, J.; Garrett, J.H.; Kovacevic, J. Semi-supervised multiresolution classification using adaptive graph filtering with appli- cation to indirect bridge structural health monitoring. *IEEE Trans. Signal Process.* **2014**, *62*, 2879–2893. [CrossRef]
75. Rafiei, M.H.; Adeli, H. A novel unsupervised deep learning model for global and local health condition assessment of structures. *Eng. Struct.* **2018**, *156*, 598–607. [CrossRef]
76. Cha, Y.J.; Wang, Z.L. Unsupervised novelty detection–based structural damage localization using a density peaks-based fast clustering algorithm. *Struct. Health Monit.* **2018**, *17*, 313–324. [CrossRef]
77. Frizzarin, M.; Feng, M.Q.; Franchetti, P.; Soyoz, S.; Modena, C. Damage detection based on damping analysis of ambient vibration data. *Struct. Control Health Monit.* **2010**, *17*, 368–385. [CrossRef]
78. Mustafa, S.; Matsumoto, Y.; Yamaguchi, H. Vibration- based health monitoring of an existing truss bridge using energy-based damping evaluation. *J. Bridge Eng.* **2017**, *23*, 04017114. [CrossRef]
79. Cao, M.S.; Sha, G.G.; Gao, Y.F.; Ostachowicz, W. Structural damage identification using damping: A compendium of uses and features. *Smart Mater. Struct.* **2017**, *26*, 043001. [CrossRef]
80. Ay, A.M.; Khoo, S.; Wang, Y. Probability distribution of decay rate: A statistical time-domain damping parameter for structural damage identification. *Struct. Health Monit.* **2019**, *18*, 66–86. [CrossRef]
81. Adhikari, S. *Structural Dynamic Analysis with Generalized Damping Models: Analysis*; Wiley-ISTE: London, UK, 2014.
82. Schwarz, B.J.; Richardson, M.H. *Introduction to Operating Deflection Shapes*; CSI Reliability Week: Orlando, FL, USA, 1999.
83. Pai, P.F.; Young, L.G. Damage detection of beams using operational deflection shapes. *Int. J. Solids. Struct.* **2001**, *38*, 3161–3192. [CrossRef]
84. Waldron, K.; Ghoshal, A.; Schulz, M.J.; Sundaresan, M.J.; Ferguson, F.; Pai, P.F.; Chung, J.H. Damage detection using finite element and laser operational deflection shapes. *Finite Elem. Anal. Des.* **2002**, *38*, 193–226. [CrossRef]
85. Zhang, Y.; Lie, S.T.; Xiang, Z.H. Damage detection method based on operating deflection shape curvature extracted from dynamic response of a passing vehicle. *Mech. Syst. Signal. Process.* **2013**, *35*, 238–254. [CrossRef]
86. Fang, S.E.; Perera, R. Power mode shapes for early damage detection in linear structures. *J. Sound Vib.* **2009**, *324*, 40–56. [CrossRef]
87. Li, J.; Hao, H. Damage detection of shear connectors based on power spectral density transmissibility. *Key Eng. Mater.* **2013**, *569*, 1241–1248. [CrossRef]
88. Pedram, M.; Esfandiari, A.; Khedmati, M.R. Damage detection by a FE model updating method using power spectral density: Numerical and experimental investigation. *J. Sound Vib.* **2017**, *397*, 51–76. [CrossRef]
89. Zhang, Y.; Wang, L.Q.; Lie, S.T.; Xiang, Z.H. Damage detection in plates structures based on frequency shift surface curvature. *J. Sound Vib.* **2013**, *332*, 6665–6684. [CrossRef]
90. Zhang, Y.; Lie, S.T.; Xiang, Z.H.; Lu, Q.H. A frequency shift curve based damage detection method for cylindrical shell structures. *J. Sound Vib.* **2014**, *333*, 1671–1683. [CrossRef]
91. Sipple, J.D.; Sanayei, M. Finite element model updating using frequency response functions and numerical sensitivities. *Struct. Control Health Monit.* **2014**, *21*, 784–802. [CrossRef]
92. Li, J.C.; Li, U.; Xu, Y.L.; Samali, B. Damage identification in civil engineering structures utilizing PCA-compressed residual frequency response functions and neural network ensembles. *Struct. Control Health Monit.* **2011**, *18*, 207–226. [CrossRef]
93. Samali, B.; Dackermann, U.; Li, J. Location and severity identification of notch-type damage in a two-storey steel framed structure utilising frequency response functions and artificial neural network. *Adv. Struct. Eng.* **2012**, *15*, 743–757. [CrossRef]
94. Dackermann, U.; Li, J.C.; Samali, B.J. Identification of member connectivity and mass changes on a two-storey framed structure using frequency response functions and artificial neural networks. *J. Sound Vib.* **2013**, *332*, 3636–3653. [CrossRef]
95. Duan, Y.F.; Chen, Q.Y.; Zhang, H.M.; Yun, C.B.; Wu, S.K.; Zhu, Q. CNN-based damage identification method of tied-arch bridge using spatial-spectral information. *Smart Mate. Struct.* **2019**, *23*, 507–520.
96. Rogers, T.J.; Worden, K.; Fuentes, R.; Dervilis, N.; Tygesen, U.T.; Cross, E.J. A Bayesian non-parametric clustering approach for semi-supervised Structural Health Monitoring. *Mech. Syst. Signal Process.* **2019**, *119*, 100–119. [CrossRef]
97. Jafarkhani, R.; Masri, S.F. Finite element model updating using evolutionary strategy for damage detection. *Comput-Aided Civ. Inf.* **2011**, *26*, 207–224. [CrossRef]
98. Wang, J.; Yang, Q.S. Modified Tikhonov regularization in model updating for damage identification. *Struct. Eng. Mech.* **2012**, *44*, 585–600. [CrossRef]
99. Zhu, H.P.; Mao, L.; Weng, S. A sensitivity-based structural damage identification method with unknown input excitation using transmissibility concept. *J. Sound Vib.* **2014**, *333*, 7135–7150. [CrossRef]
100. Li, X.Y.; Law, S.S. Adaptive Tikhonov regularization for damage detection based on nonlinear model updating. *Mech. Syst. Signal Process.* **2010**, *24*, 1646–1664. [CrossRef]

101. Zhang, C.D.; Xu, Y.L. Comparative studies on damage identification with Tikhonov regularization and sparse regularization. *Struct. Control Health Monit.* **2016**, *23*, 560–579. [CrossRef]
102. Abdeljaber, O.; Avci, O.; Kiranya, S.; Gabbouj, M.; Inman, D.J. Real-time vibration-based structural damage detection using one-dimensional convolutional neural networks. *J. Sound Vib.* **2017**, *388*, 154–170. [CrossRef]
103. Bao, Y.Q.; Tang, Z.Y.; Li, H.; Zhang, Y.F. Computer vision and deep learning–based data anomaly detection method for structural health monitoring. *Struct. Health Monit.* **2019**, *18*, 401–421. [CrossRef]
104. Ghiasi, R.; Torkzadeh, P.; Noori, M. A machine-learning approach for structural damage detection using least square support vector machine based on a new combinational kernel function. *Struct. Health Monit.* **2016**, *15*, 302–316. [CrossRef]
105. Zhou, Q.F.; Ning, Y.P.; Zhou, Q.Q.; Luo, L.K.; Lei, J.Y. Structural damage detection method based on random forests and data fusion. *Struct. Health Monit.* **2012**, *12*, 48–58. [CrossRef]
106. Lai, Z.; Nagarajaiah, S. Semi-supervised structural linear/nonlinear damage detection and characterization using sparse identification. *Struct. Control Health Monit.* **2019**, *26*, e2306. [CrossRef]
107. Santos, J.P.; Crémona, C.; Calado, L.; Silveira, P.; Orcesi, A.D. On-line unsupervised detection of early damage. *Struct. Control Health Monit.* **2016**, *23*, 1047–1069. [CrossRef]
108. Avci, O.; Abdeljaber, O. Self-organizing maps for structural damage detection: A novel unsupervised vibration-based algorithm. *J. Perform. Constr. Facil.* **2016**, *30*, 04015043. [CrossRef]
109. Arangio, S.; Beck, J.L. Bayesian neural networks for bridge integrity assessment. *Struct. Control Health Monit.* **2012**, *19*, 3–21. [CrossRef]
110. Arangio, S.; Bontempi, F. Structural health monitoring of a cable-stayed bridge with Bayesian neural networks. *Struct. Inf. Eng.* **2015**, *11*, 575–587. [CrossRef]
111. Gul, M.; Catbas, F.N. Structural health monitoring and damage assessment using a novel time series analysis methodology with sensor clustering. *J. Sound Vib.* **2011**, *330*, 1196–1210. [CrossRef]
112. Mosavi, A.A.; Dickey, D.; Seracino, R.; Rizkalla, S. Identifying damage locations under ambient vibrations utilizing vector autoregressive models and Mahalanobis distances. *Mech. Syst. Signal Process.* **2012**, *26*, 254–267. [CrossRef]
113. Yang, Y.; Li, J.L.; Zhou, C.H.; Law, S.S.; Lv, L. Damage detection of structures with parametric uncertainties based on fusion of statistical moments. *J. Sound Vib.* **2019**, *442*, 200–219. [CrossRef]
114. Yang, Y.; Li, C.; Ling, Y.; Tan, X.; Luo, K. Research on new damage detection method of frame structures based on generalized pattern search algorithm. *China J. Sci. Instrum.* **2021**, *42*, 123–131.
115. Hios, J.D.; Fassois, S.D. A global statistical model based approach for vibration response-only damage detection under various temperatures: A proof-of-concept study. *Mech. Syst. Signal Process.* **2014**, *49*, 77–94. [CrossRef]
116. Brien, E.J.O.; Heitner, B.; Žnidarič, A.; Schoefs, F.; Causse, G.; Yalamas, T. Validation of bridge health monitoring system using temperature as a proxy for damage. *Struct. Control Health Monit.* **2020**, *27*, e2588.
117. Xu, B.; Song, G.; Masri, S.F. Damage detection for a frame structure model using vibration displacement measurement. *Struct. Health Monit.* **2011**, *11*, 281–292. [CrossRef]
118. Li, J.; Hao, H.; Fan, K.; Brownjohn, J. Development and application of a relative displacement sensor for structural health monitoring of composite bridges. *Struct. Control Health Monit.* **2015**, *22*, 726–742. [CrossRef]
119. Yang, Y.C.; Nagarajaiah, S. Blind identification of damage in time-varying systems using independent component analysis with wavelet transform. *Mech. Syst. Signal Process.* **2014**, *47*, 3–25. [CrossRef]
120. Huang, N.E. *The Hilbert-Huang Transform in Engineering*; Taylor and Francis Group: New York, NY, USA, 2005.
121. ISO 11863:2011. *Buildings and building-Related Facilities—Functional and User Requirements and Performance—Tools for Assessment and Comparison*; ISO: Geneva, Switzerland, 2011.
122. ISO 15928-1:2015. *Houses—Description of Performance—Part 1: Structural Safety*; ISO: Geneva, Switzerland, 2015.
123. ISO 4356:1977. *Bases for the Design of Structures—Deformations of Buildings at the Serviceability Limit States*; ISO: Geneva, Switzerland, 1977.
124. ISO 13822:2010. *Bases for Design of Structures—Assessment of Existing Structures*; ISO: Geneva, Switzerland, 2010.
125. ISO 2394:2015. *General Principles on Reliability for Structures*; ISO: Geneva, Switzerland, 2015.
126. ISO 37104:2019. *Sustainable Cities and Communities—Transforming Our Cities—Guidance for Practical Local Implementation of ISO 37101*; ISO: Geneva, Switzerland, 2019.
127. ISO 37105:2019. *Sustainable Cities and Communities—Descriptive Framework for Cities and Communities*; ISO: Geneva, Switzerland, 2019.
128. ISO TS 37107:2019. *Sustainable Cities and Communities—Maturity Model for Smart Sustainable Communities*; ISO: Geneva, Switzerland, 2019.
129. Mufti, A.A. *Guidelines for Structural Health Monitoring*; ISIS Canada: Winnipeg, MB, Canada, 2001.
130. Aktan, A.E.; Catbas, F.N. *Development of a Model Health Monitoring Guide for Major Bridges*; CONTRACT/ORDER NO. DTFH61-01-P-00347; Federal Highway Administration Research and Development: McLean, VA, USA, 2003.
131. *Tunnel Operations, Maintenance, Inspection, and Evaluation Manual*; FHWA-HIF-15-005; Federal Highway Administration: Washington, DC, USA, 2015.
132. Bergmeister, K. *Monitoring and Safety Evaluation of Existing Concrete Structures: State-of-the-Art Report (Fib Task Group 5.1)*; The International Federation for Structural Concrete: Lausanne, Switzerland, 2002.
133. Rucker, W.; Hille, F.; Rohrmann, R. *Guideline for Structural Health Monitoring. Final Report*; SAMCO: Berlin, Germany, 2006.
134. GOST R 53778:2010. *Building and Structures, Technical Inspections and Monitoring Regulations*; RussianGost—Official Regulatory Library: Alief, TX, USA, 2010.

135. Osterreichisches, Ö. *Forschungsgellschaft, RVS 13-03-01: Quality Assurance for Structural Maintenance, Surveillance, Checking and Assessment of Bridges and Tunnels. Monitoring of Bridges and Other Engineering Structures*; Forschungsgellschaft: Eisenstadt, Austria, 2012.
136. Ministry of Housing and Urban-Rural Development of the People's Republic of China. *Technical Code for Monitoring of Buildings and Bridge Structures: GB50982-2014*; China Construction Industry Press: Beijing, China, 2014.

symmetry MDPI

Article

Intelligent Safety Assessment of Prestressed Steel Structures Based on Digital Twins

Zhansheng Liu [1,2,*], Guoliang Shi [1,2], Zedong Jiao [1,2] and Linlin Zhao [1,2]

1 Faculty of Architecture, Civil and Transportation Engineering, Beijing University of Technology, Beijing 100124, China; shiguoliang@emails.bjut.edu.cn (G.S.); jiaozedong@emails.bjut.edu.cn (Z.J.); llzhao@bjut.edu.cn (L.Z.)
2 The Key Laboratory of Urban Security and Disaster Engineering of the Ministry of Education, Beijing University of Technology, Beijing 100124, China
* Correspondence: liuzhansheng@bjut.edu.cn

Abstract: In the development process of intelligent construction, the safety assessment of prestressed steel structures as an important research direction has become more and more attractive in academia. Digital twins (DTs) is the key technology to realize intelligent construction. The virtual and real interaction of the DTs can provide an efficient management and control mechanism for the construction process. This research proposes an intelligent safety assessment method of prestressed steel structures based on DTs. In this research method, the structural safety assessment is divided into two aspects: performance analysis and maintenance. By analyzing the characteristics of the construction safety assessment, a DTs framework for construction safety assessment is built. Driven by the DTs framework, a physical space model and a virtual space model are constructed. On the basis of virtual and actual interaction, multidimensional information fusion of time and space is carried out to realize the analysis of structural safety performance. On this basis, the paper establishes a Bow-tie model for the maintenance modeling of unsafe construction events. Moreover, the theoretical method formed is applied to the construction of a symmetrical structure (wheel–spoke cable truss). The validity of the method is verified by comparing the cable force calculated by the theoretical method and measured on site. The assessment method driven by the DTs ensures the structural safety and improves the intelligence level of safety management and control of the structure construction.

Keywords: digital twin; prestressed steel structure; construction process; safety assessment; intelligent construction

Citation: Liu, Z.; Shi, G.; Jiao, Z.; Zhao, L. Intelligent Safety Assessment of Prestressed Steel Structures Based on Digital Twins. *Symmetry* **2021**, *13*, 1927. https://doi.org/10.3390/sym13101927

Academic Editors: Yang Yang, Ying Lei, Xiaolin Meng and Jun Li

Received: 8 September 2021
Accepted: 9 October 2021
Published: 14 October 2021

1. Introduction

In recent years, the construction industry is gradually developing towards digitalization, informatization, and intelligence [1]. The construction safety of prestressed steel structures is a widespread concern in the engineering field. Under the continuous progress of science and technology, the construction technology level of prestressed steel structures has been significantly improved. Facing the construction process, many new technologies, new processes, and new equipment have emerged. The structural deformation of prestressed steel structures is often the reverse to the deformation under load, so that the structural stiffness is improved. Therefore, it is commonly used in significant buildings, such as large stadiums [2]. Structural construction has the characteristics of complex links and numerous abnormal disturbance factors. Improving the accuracy and intelligence level of the safety assessment of the prestressed steel structure will directly reflect the construction ability of the country. Besides, safety control of the prestressed steel structure construction process has also become a hot topic in the research field of civil engineering. This study uses the mechanical parameters of the structure as an important basis for measuring the safety performance. Through the analysis of mechanical parameters, intelligent control of construction safety is realized.

For the structural construction safety control, researchers have developed various approaches to improve the accuracy of the analysis. Wang et al. [3] focused on the analysis of cable force, which is the most active parameter in the construction process of spatial structure prestressed cables. In the whole process of prestressed cable tension, the safety of the construction process of the structure is ensured by analyzing the cable force. To ensure the stability of the cable dome structure in the construction process, Zhang et al. [4] proposed a kind of square double-strut cable dome structure. This structure effectively improved the safety control precision of the structure tension process. Liu et al. [5] obtained the reliability index of the cable by the response surface method and the Monte Carlo method, and then analyzed the influence of cable relaxation on structural reliability. The research results provide a basis for safety assessment of the wheel–spoke cable truss structure when prestress loss occurs. Bai et al. [6] investigated the seismic behavior of steel beam–column connections with an outer annular stiffener under bi-directional cyclic loadings. The seismic performance is analyzed to ensure the safety performance of the structure, which provides a reference for the normal operation of the structure under complex loads. Liu et al. [7] studied the dynamic response of a typical umbrella membrane structure under heavy rainfall by experimental and numerical methods. They put forward suggestions on the safety and structural stability of the membrane structure. Basta et al. [8] studied the quantitative evaluation of the decomposability of the cable-net structure based on building information modeling (BIM). Alamdari et al. [9] proposed a new damage identification technology for the safety performance evaluation of long-span spatial structures based on the concept of rotation influence line (RIL). This method effectively improves the efficiency of structural safety performance evaluation. Bera et al. [10] studied the active control of the safety performance of long-span cable-stayed bridges. Two finite element (FE) modeling schemes are adopted, namely one-element cables and multi-element cables, with each stay-cable discretized into multiple elements to consider the cable vibration effect. Considering that the long-span cable will vibrate under different weather conditions, D'Auteuil et al. [11] developed a new type of wind- and rain-induced vibration test.

In summary, experts and scholars of civil engineering have performed a lot of research and exploration in structural safety assessment. However, these studies only focus on the performance of the structure in a certain construction stage, ignoring the dynamics and uncertainty of the construction process. Therefore, it is necessary to integrate multidimensional information of the structure to solve the problem of real-time safety analysis during the whole construction process. How to integrate multi-source heterogeneous information in time and space dimensions to improve the intelligence of safety performance assessment is still in the exploratory stage. The above-mentioned issues require real-time analysis of structural mechanical properties and effective maintenance of construction unsafe events. Meantime, how to improve the intelligence of construction safety assessment of prestressed steel structures is a new direction of current research [12]. In order to improve the intelligence of safety assessment, it is a requisite to realize multidimensional information fusion in the construction process. Through the intuitive mapping of the virtual space to the actual construction, the construction of the physical world is guided by the simulation model. By analyzing the virtual model, each step of the construction is ensured to be in a safe state. In order to improve the intelligence of the safety assessment, the dynamic perception of each stage should be realized, and the real-time simulation and analysis of the structure can be carried out from the virtual space. The maintenance measures for unsafe incidents are developed based on the analysis results. The feasibility of decision-making is analyzed in the virtual space and accurately guides the entire construction process. Finally, the intelligent closed-loop control of structural safety is realized. In intelligent manufacturing, digital twins (DTs) are the key enabling technology to achieve multi-source information fusion. The introduction of DTs from manufacturing to the construction industry will provide new ideas for the intelligent construction of structures [13,14].

DT establishes a multidimensional dynamic virtual model within multi-spatial scale, multi-time scale, and multi-physical entities through digitization. The virtual model can simulate the attributes, behaviors, and performances in the real environment [15]. DTs have the characteristics of rapid information exchange and extensive representation of live scenes. DTs are not only widely utilized in the practical construction area but are also applied by researchers in the investigation as one of the necessary parts. Professor Grieves of the University of Michigan formally proposed the concept of DTs in 2003 [16]. The virtual digital model of twins can abstractly map the performance of physical entities by simulating the state and behavior of products. Yu et al. [17] provided new method based on DTs technology and machine learning for accurately and timely predicting pavement performance. The new method solved the problem of pavement performance evaluation. In spite of the difficulty for a single enterprise to achieve large-scale and personalized task requirements, Tao et al. [18] pointed out that by manufacturing service collaboration, driven by DTs technology and industrial Internet platforms, cross-enterprise manufacturing collaboration was realized. Combined with the visual question-answering technology in artificial intelligence, Wang et al. [19] applied the digital twin model to human–computer collaboration tasks. This progress created the possibility for real-time simulation and optimization and formed an on-demand intelligent service system. Ruppert et al. [20] proposed a DTs model by combining the advantages of information fusion and real-time location systems. The DTs model can continuously predict the production status and provide information for production performance monitoring. Gopalakrishnan et al. [21] pointed out that manufacturing requires digital transformation. They created a DT model-based feature information network (MFIN), which realized the digital description of components or systems. Thus, DT is widely used in the manufacturing industry and can effectively improve the intelligence level of manufacturing processes. However, it is relatively less used in the construction industry. In order to improve the intelligence level of the construction industry, it is necessary to introduce DTs. Aiming at the research hotspot about intelligent assessment of structural safety performance, this paper proposes an intelligent safety assessment method of prestressed steel structures based on DTs. In terms of structural safety assessment, previous studies have conducted a detailed analysis of mechanical parameters. In order to improve the intelligence of construction safety assessment, the concept of DTs is introduced, and the whole assessment process is divided into safety performance analysis and unsafe event maintenance. Driven by the DTs, the virtual and real spaces are integrated to realize dynamic perception and intelligent collection of the construction process. At the same time, it is necessary to consider the temporal and spatial evolution of each element of the construction process. Therefore, intelligent algorithms are integrated for real-time analysis of structural safety. The maintenance measures for unsafe events are formulated, and the feasibility of decision-making is analyzed in the twin model. In the end, it accurately guides the construction site. In this study, through the analysis of the characteristics of construction safety assessment, an intelligent assessment framework based on DTs is firstly established. Driven by this framework, three kinds of information from real construction are captured to support the construction of the virtual model. After the completion of the virtual model, the time dimension and space dimension information of the construction process are fused by Markov chain. Thus, the structural safety performance of each construction step is analyzed intelligently. In view of unsafe events, the maintenance model of that in construction is carried out based on the Bow-tie model. The corrective measures are formulated in the maintenance model to ensure the safety of the structure in each construction step. The resulting DTs assessment theory is applied in the construction process of a wheel–spoke cable truss. This research method effectively improves the informatization and intelligence of the safety assessment in the construction process. This research method provides a reference for the health monitoring of the symmetric structure in the operation and maintenance stage. In brief, this work makes the following contributions:

(1) A DTs framework is proposed, and a construction safety assessment method driven by DTs is formed.

(2) Driven by the assessment method, a multidimensional DT model for structural safety analysis is constructed.
(3) In view of the unsafe events in the construction process, the maintenance modeling is carried out to ensure the structure in a safe state in each construction step.
(4) A case study is used to prove the superiority of the proposed method in improving the intelligence of the structural safety assessment.

The remainder of the paper is structured as follows. Section 2 summarizes the characteristics of construction safety assessment, builds a DT framework for construction safety assessment, and forms a DT-driven construction safety assessment method. Section 3 describes the details of the construction of a multidimensional DT model to analyze the structural safety performance. Section 4 establishes the maintenance model of construction unsafe events to ensure the safety of the structure. Section 5 illustrates the case study used to verify the proposed method, with a further discussion in Section 6. Finally, conclusions are drawn in Section 7.

2. Construction Safety Assessment Method Driven by DTs

2.1. Characteristics of Construction Safety Assessment

In the construction process of prestressed steel structures, the reliability of the structure is an important prerequisite to ensure the construction quality [22]. However, in the construction process, the integrity of the structure is not stable enough, the material properties of the components are time-varying, and the structural resistance is not mature. In addition, the environmental effects of the construction process are also uncertain [23]. Therefore, the probability of safety accidents in the construction stage is high. According to the structural safety performance analysis of the construction process, the parameters are nonlinear, linkage, and change with time. Therefore, the construction safety assessment is a multidimensional mechanical problem coupling time and space. The characteristics of the construction process are as follows:

(1) Construction complexity: Prestressed steel structure construction is a complicated systematic project. In the construction process, the materials, equipment, sequence, and technology are interactive. Various types of external factors are superimposed on the structure. In addition, the change of structural mechanical properties is a nonlinear process with noise, which causes difficulty in structural safety assessment.
(2) Time correlation: The external action and mechanical properties of components in the process of structural construction are time-varying. This feature is directly reflected in the data collected by various sensing devices.
(3) Factor linkage: The structural construction process involves the multi-factor conversion of material, energy, and information. This feature has caused problems such as huge construction system modeling and cumbersome multi-element management and control.

2.2. DT Framework for Construction Safety Assessment

According to the characteristics of the construction process, it is necessary to establish a DTs model. This model monitors the structure in real-time and improves the accuracy and intelligence of the safety assessment. DTs visually reproduce the real physical entity by building a virtual space model to simulate the dynamic behavior of the entity in the real environment [24]. Many scholars have applied the framework of DTs in vehicles, ships, power plants, complex electromechanical equipment, satellite/space communication networks, three-dimensional warehouses, medical treatment, aircraft, smart city, and other fields [25]. Combined with the characteristics of construction safety assessment, this study builds a DT framework for the construction safety assessment of prestressed steel structures. The DT framework is shown in Figure 1.

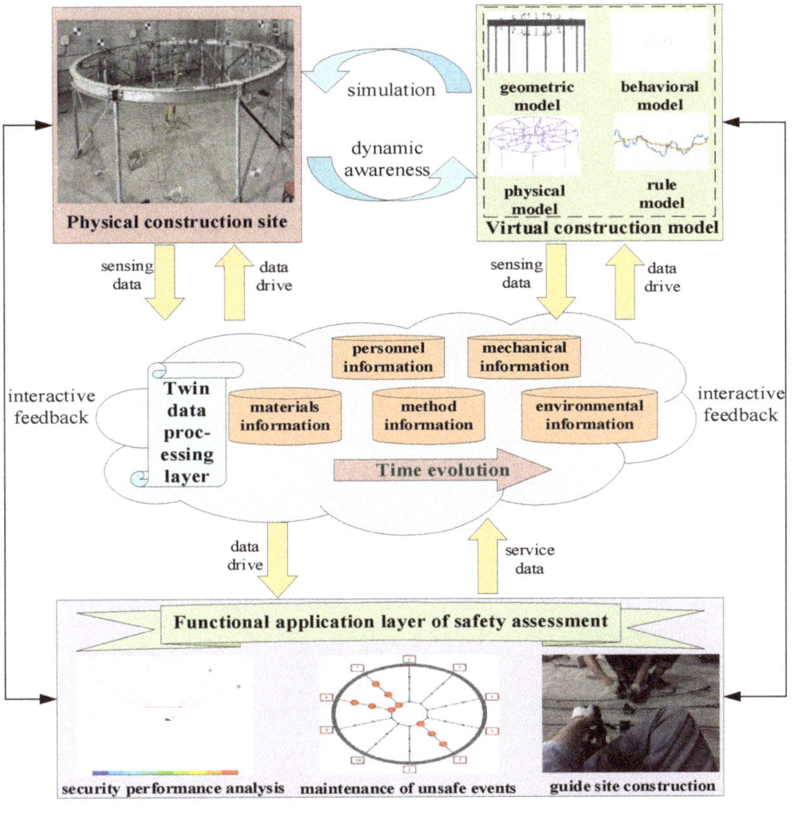

Figure 1. DT framework for construction safety assessment of prestressed steel structures.

The DT framework for the safety assessment of prestressed steel structures' construction is composed of five dimensions, namely, physical construction site, virtual construction model, twin data processing layer, the functional application layer of safety assessment, and the connection layer between components. Based on the physical construction site, the virtual construction model is established. From the perspective of virtual–real interaction, the virtual model includes four levels: geometry, physics, behavior, and rule. The various types of information on the physical construction site are dynamically collected by RFID and other sensing equipment [26]. Information is fed back into the virtual construction model in real-time. In the virtual construction model, the working condition parameters consistent with the actual construction are set. By adjusting the mechanical parameters of the structure, the virtual model simulates the actual construction state of the structure. The twin data of the construction process are formed by the data collected from the physical construction site and the simulation data in the virtual construction model. The twin data contain multiple construction elements, such as personnel, machinery, material, construction method, and environment, in the construction process. In the twin data processing layer, the data are modeled by a machine learning algorithm to realize the high integration of spatial elements and time dimensions in the construction process. Therefore, the structural safety performance of each construction step is analyzed. Based on the analysis of structural safety performance, the maintenance model of construction unsafe events is established to accurately predict and implement the unsafe events. By setting up a construction framework for prestressed steel structures, a construction safety assessment method driven by DTs is formed. Through the dynamic perception of the real construction information, the virtual mode can evaluate the safety performance. The maintenance measures for unsafe events are formulated, and the feasibility of decision-making is analyzed in the twin model. In the end, the measure accurately guides the construction site. The intelligent closed-loop control of the structural safety assessment in the construction process is realized.

This study divides the construction safety assessment method driven by DTs into two levels, as shown in Figure 2. The first level is the construction of the multidimensional model based on DTs. This level includes the information capture of the physical construction site, the construction of the virtual construction model, and the integration of time and space information in the construction process. The second level is the maintenance of construction unsafe events based on the Bow-tie model. This level focuses on the qualitative analysis of unsafe events and the precise maintenance of unsafe events. Logically, the former is the foundation of the latter, and the latter is the expansion and extension of the former.

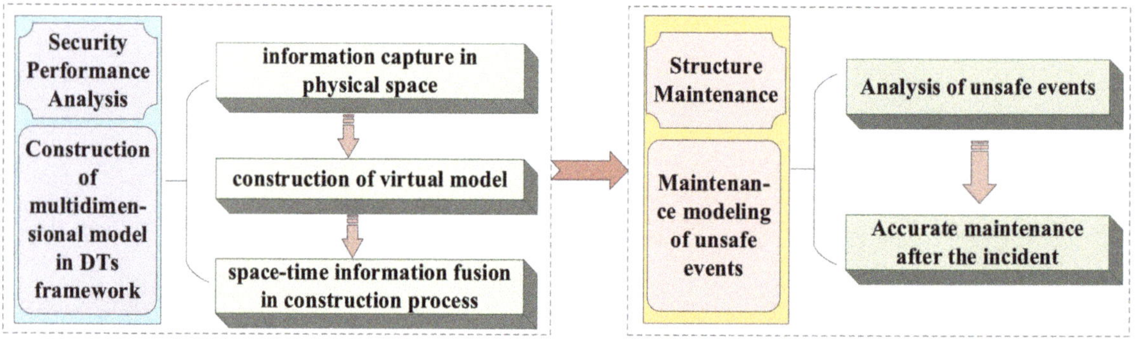

Figure 2. Construction safety assessment method.

3. Construction of Multidimensional DTs Model

Driven by the DT framework, the multidimensional model is constructed. The multidimensional model is composed of the capture of physical space information, the construction of virtual models, and the fusion of spatiotemporal information in the construction process. The research on the three aspects above can realize the integration of multiple factors in the time and space dimensions of the construction process. The fusion of information provides the basis for the safety analysis of each construction step in the construction process. The results of the analysis are the basis for the maintenance of unsafe events. The three components of the DTs model are described as follows.

3.1. Information Capture in Physical Space

In the process of constructing a multidimensional DTs model, the information capture of physical space is the first step to realize the safety assessment. The dynamic information of the construction site can be sensed in real-time through sensing equipment such as RFID. This realizes the one-to-one mapping between the real space and the virtual model [27]. The main analysis objects of the safety assessment of building structures are structural response and environmental loads [28]. Hence, this study divides the information of physical space into two aspects, namely component information (*CI*) and environmental information (*EI*).

(1) Component information capture

There are many kinds of components involved in the construction process of prestressed steel structures, and the construction sequence of each component is complicated. Therefore, the information capture of various components in the construction process is of great significance to the safety performance assessment. In the process of component information capture, the main information is divided into the component symbol (*CS*), basic information (*BI*), construction information (*CI**), and mechanical information

(*MI*). The mathematical language for information capture of components is expressed in Equations (1)–(4):

$$CI_i = (CS_i,\ BI_i,\ CI_i^*\ C_i,\ MI_i) \tag{1}$$

$$BI_i = (CD_i, CL_i, CM_i, Unit_i, Data_i) \tag{2}$$

$$CI_I^* = (CP_i,\ CT_i,\ QR_i,\ T_i) \tag{3}$$

$$MI_i = (E_i, C_{fi}, \varepsilon_i, V_{di}) \tag{4}$$

In the equations, CI_i represents the component information for the *i*th component, CS_i means the unique identifiable ID bound by the *i*th component, and BI_i is the basic attribute set of the *i*th component. BI_i contains the dimension information (CD_i), the locating information (CL_i), the material information (CM_i), the production unit ($Unit_i$), and the date of production ($Data_i$). CI_i^* is the construction information of the *i*th component. CI_i^* contains the construction process (CP_i), construction technology (CT_i), construction quality requirements (QR_i), and the time (T_i) used in this step. MI_i means the mechanical information of the *i*th component in the construction process. MI_i involves the elastic modulus (E_i), cable force (C_{fi}), stress (ε_i), vertical displacement (V_{di}), and other mechanical properties.

In the process of capturing component information, the symbol, basic information, and construction information of the component are filled in the active RFID tag. For structural components, the relevant information is updated through the RFID tag at all times to realize the dynamic perception of the information. For the information change of the construction process component, the editability of the label can be used to modify it through the reader. As a result, the basic information and construction information of the component can be viewed in real-time in the mobile terminal device. For the capture of mechanical information of components, the mechanical parameters of components are collected by sensors for each construction step. Cable force is the key information collected in the study and the basis for evaluating the safety performance of the structure. This type of information is collected in real-time by the column tension-compression sensor in the test. By arranging sensing devices on the components, the efficiency of information collection and update is improved. Therefore, the self-perception, self-decision, and self-execution of the components can be realized to a certain extent. The real-time information is captured by the sensing equipment at each construction step. Structure information of the scene is collected and has interactive feedback with the virtual model. Virtual and real interaction provides twin data for the intelligent assessment of structural safety.

(2) Environmental information capture

In the process of structural construction, there are many reasons causing structural damage. In the structural safety assessment, the effects of various external influence factors are comprehensively analyzed [29]. The component length error (L_e), wind load effect (W_l), and temperature effect (T_e) are captured and perceived emphatically in the process of environmental information capture. The mathematical language expression of environmental information capture in the construction process is shown in Equation (5):

$$EI = (L_e, W_l, T_e) \tag{5}$$

In the process of capturing environmental information, RFID technology is used to collect the size of the component in real-time. The error of the cable length is dynamically updated through the mobile terminal device. The length error of structural components due to processing, installation, and other reasons has an important impact on the safety performance of the structure. In view of the length change of the cable, the key coordinates are picked up by three-dimensional laser scanning in each step of construction. The point cloud data are structured through on-site scanning. According to the point cloud data, the point cloud model is built in real-time. Finally, the length error of the component can be captured. On account of the editable RFID reader, using mobile terminals to identify tags on components can record and update their status. The effect of wind load and temperature is collected by the wind speed sensor and temperature sensor in real-time.

These messages provide the basis for the setting of virtual model conditions and the analysis of structural safety performance. In the process of extracting wind speed and temperature information, a complete set of sensing equipment is applied to perceive the environmental information of the scene. By arranging sensors in the construction process and setting control modules, the collected data are visually presented through the display module. Finally, the environmental information collected on the site is input into the terminal equipment by Wi-Fi transmission.

In this study, the experimental model was built indoors. For the effect of wind load and temperature, the test equipment is used to generate the wind speed and temperature effect, similar to the actual working condition. At the same time, the sensor equipment is used for the setting of virtual model conditions and the evaluation of structural safety performance. The sensing equipment is arranged on the site, and the temperature of the field anemometer is dynamically perceived by the acquisition module. The information data are presented in real-time by the display module. Finally, the information query of each construction step is carried out by the Wi-Fi module in the mobile terminal. Environmental sensing devices and information are shown in Figure 3.

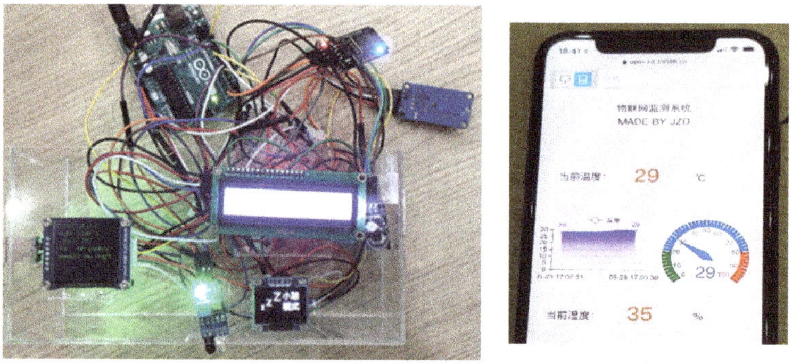

(**a**) (**b**)

Figure 3. Environmental sensing equipment and information view. (**a**) Sensing equipment, (**b**) Environmental information.

Through the collection of component information and environmental information in the construction process, the structural safety performance can be predicted. Information capture in physical space is shown in Figure 4. At the same time, on the basis of collecting construction site information, the virtual model can be built. The safety performance of the structure is simulated by setting the same working condition as the construction site in the virtual model. Accurate simulation achieves the goal of mapping the construction site and guiding the construction process.

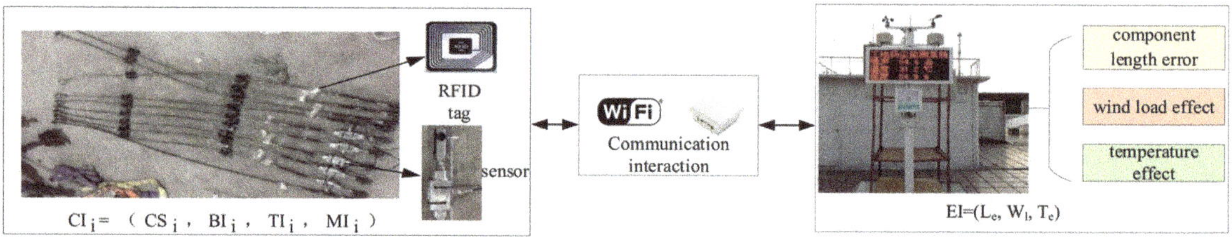

a. Component information capture

b. Environmental information capture

Figure 4. Information capture in physical space.

3.2. Construction of Virtual Model

In the DT framework, the construction of the virtual model is divided into four levels of "geometry–physics–behavior–rule". According to the actual construction process, each dimension model is associated and integrated to realize the deep, multi-angle, and comprehensive simulation.

The first step of virtual model establishment is modeling basic information of appearance, size, and type of components at the geometric level. The geometric model is mainly established by BIM modeling software such as Revit [30]. By establishing a high-fidelity geometric model, the geometric characteristics of the construction process can be truly mapped. Simulation mapping provides strong support for the subsequent analysis of the physical model. In this process, a three-dimensional laser scanner is required to extract the geometric form of the site structure at each construction step. The geometric model is adjusted in real-time from the point cloud data. The scanning accuracy of this method can reach a reliable level of 0.1 mm. This method uses laser scanning technology to scan the three-dimensional information of the component. The generated point cloud data model can be directly converted into CAD or BIM software. The reverse modeling of the component is realized, and the digital model that meets the accuracy requirements is obtained. In this process, the target paper is arranged on the site. The structure is scanned for each construction step to extract the coordinates of key nodes.

During the construction of the test model, the real structure is scanned by a three-dimensional laser scanner to obtain the measured model of the structure. After obtaining the point cloud data, the necessary task is to denoise. This link can remove the outliers deriving from the machine error, human factors, or the external environment. Then, the remaining data are imported into BIM software, while the key points of the cable truss structure are extracted. The coordinate correction of the theoretical BIM model is carried out to obtain the modified BIM model. The coordinates of key nodes are extracted from the modified BIM model to modify the theoretical finite element analysis model. Finally, the modified finite element analysis model considering the time dimension is obtained. At the physical level, the material parameters of construction components are mainly simulated in the physical model by finite element analysis software such as ANSYS. In this process, with the collected data of the sensing equipment, the geometric model and the connection parameters of the components in the model are modified. The calculation of structural mechanical properties during construction is realized by the physical model in the end. As is known, the size of the component will affect the mechanical properties of the structure. In this study, the section area of cable is adjusted by comparing the measured value of cable force with the simulation value. Area adjustment improves the fidelity of the physical model. The geometric model and physical model are established to describe the construction site and provide model support for the safety assessment. At the behavior level, the finite element model can set the working conditions matching the actual construction. The mechanical properties of components and the changes of the parameters of the material itself under the action of working conditions are analyzed. The material parameters and mechanical properties' parameters extracted in this way can be directly used for the assessment of construction safety. According to the analysis of the finite element model, the transition probability of safety state analysis can be obtained. On the basis of the change of working conditions, the safety performance of the structure can be analyzed in advance, as detailed in Section 3.3. Through the analysis of the physical model and the integration of time dimension information, the real-time data collection of the whole construction process can be carried out. The changes of material parameters and mechanical properties in the spatio-temporal evolution process are obtained. In the whole process of building the virtual model, it is necessary to establish a rule model to limit the simulation and ensure the feasibility and scientificity of the analysis. At the rule level, according to the standard specification, the mechanical properties' parameters of the components in the construction process should be quantitatively limited. The rule model is the reference standard for quality control, risk prediction, and decision optimization. The unsafe events of the structure should be avoided by establishing the maintenance model and setting the correction measures. Consequently, the feasibility analysis is carried out in the finite element model to guide the field construction, as shown in Section 4. The internal relations of the four levels of the virtual model are shown in Figure 5.

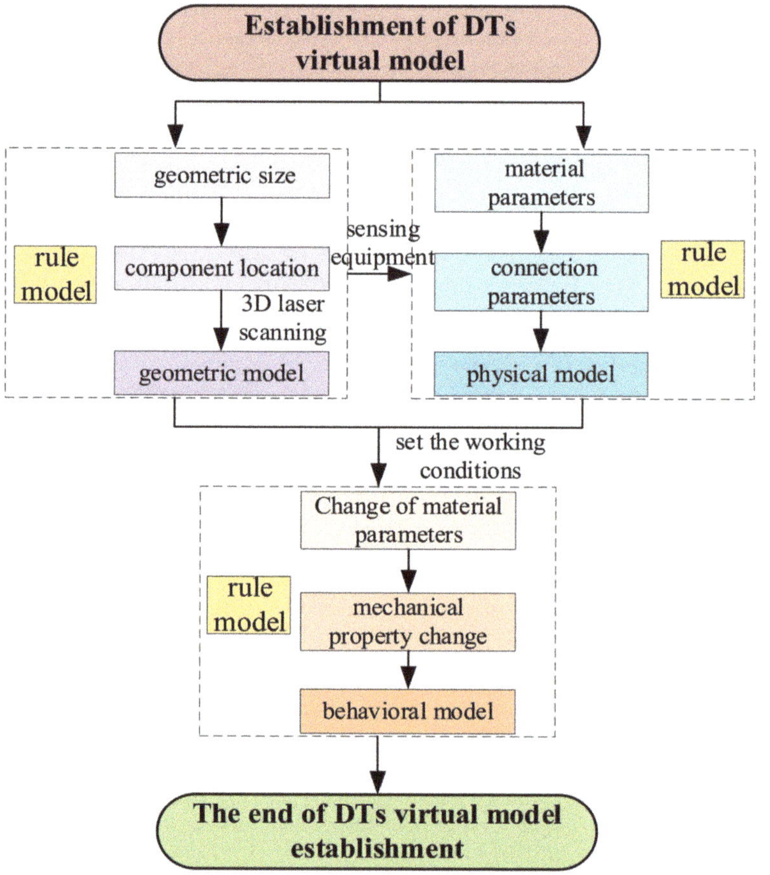

Figure 5. Intrinsic relationship of the four levels in the virtual model.

The mathematical language of DTs' virtual modeling is expressed as Equation (6):

$$VM = (GM_{set} \bowtie PM_{set} \bowtie BM_{set} \bowtie RM_{set}) \tag{6}$$

where VM represents a virtual model in a DT framework for safety assessment of prestressed steel structure construction. The elements in the model set are represented as a geometric model set (GM_{set}), physical model set (PM_{set}), behavior model set (BM_{set}), and rule model set (RM_{set}). All kinds of model sets are connected by a natural connector ($\bowtie$). Model integration realizes the simulation of the full elements, and multidimensional and multi-state of the physical construction site. In the related process of various models, the geometric model and physical model are adjusted by using a three-dimensional laser scanner, sensors, and other equipment. Therefore, the model can map the state of the real structure. At the same time, the information of each construction step is analyzed by Markov chain to realize the information fusion of the time dimension and space dimension in the construction process. Thus, the behavior model is effectively connected with the geometric and physical models to achieve the goal of real-time evaluation of structural safety performance. In Section 3.3, the spatio-temporal information fusion of the construction process based on Markov chain is mainly explored. The rule model runs through the whole process of building the virtual model. The rule model adjusts the geometric, physical, and behavioral models in real-time to ensure that the construction steps are in a safe state.

3.3. Space–Time Information Fusion in Construction Process

During the construction process, the construction elements change with time. The rigidity of the structure is gradually formed, and the structure has undergone large deformations during the construction process. The next construction stage must be based on the previous step and depends on all previous construction steps. It follows that the analysis of structural safety performance belongs to the category of construction mechanics. Based on picking up physical construction information and establishing a virtual model, it is necessary to fuse the information of time and space dimensions and establish a data analysis model for the safety assessment. Markov chain [31] is a tool for the immediate transfer process and an important branch of machine learning. In this mode, the structural state of the next period is only related to the state of this period, while each period before this period is irrelevant. This process is suitable for the safety performance analysis between the adjacent construction steps in the construction process. Combined with the threshold of the main control factors in the construction process, the probability of structural safety and unsafe events is formulated. The structural safety performance of the next construction step is predicted. This requires reference to the safety performance of the current one and the probability of occurrence of risk factors or the degree of structural mechanics parameters' change. The safety risk factors mainly include operational errors and sharp changes in environmental factors. This study takes the cable force of each construction step as the research object. When the cable force is greater than or equal to the design value of the cable force of the construction step, it is denoted that the structural safety performance is at level a. When the cable force is greater than or equal to 93% of the design value of the cable force of the construction step, it is denoted that the structural safety performance is at level b. When the cable force is greater than or equal to 90% of the design value of the cable force of the construction step, it is denoted that the structural safety performance is at level c. When the cable force is less than 85% of the design value of the cable force of the construction step, it is denoted that the structural safety performance is at level d. Structure construction information fusion based on Markov chain is shown in Figure 6.

In the process of Spatio-temporal information fusion, the state of the structure is divided into four categories according to the safety performance level. Assuming that the random variable X_n (n = 1,2,3 ...) represents the structural state of the nth construction step, $X_n = 1$ means that the structural safety performance is at level a, $X_n = 2$ represents that the structural safety performance is at level b, $X_n = 3$ indicates that the structural safety performance is at level c, and $X_n = 4$ means that the structural safety performance is at level d. $a_i(n)$ represents the probability that the nth construction step structure is in state i, namely $a_i(n) = p(X_n = i)$, where i is 1, 2, 3, or 4. p_{ij} represents the probability that the current construction step structure state is i and the next construction step structure state is j, namely $p_{ij} = p(X_{n+1} = j | X_n = i)$, where i, j = 1, 2, 3, or 4. In this study, the conversion probability (p_{ij}) is obtained by comparing the cable force between the construction steps with the condition setting of the finite element model. In this process, the cable force of the structure is obtained in the finite element model, which generates the safety level of the structure. The transition probability can be obtained from the change of safety levels of multiple components of the structure. According to the current safety state and the change of working conditions, the safety state of the structure in the next construction step can be predicted. Thus, the prediction of structural safety performance during construction can be realized, and the prediction formula is expressed as Equation (7):

$$
\begin{aligned}
a_1(n+1) &= a_1(n)p_{11} + a_2(n)p_{21} + a_3(n)p_{31} + a_4(n)p_{41} \\
a_2(n+1) &= a_1(n)p_{12} + a_2(n)p_{22} + a_3(n)p_{32} + a_4(n)p_{42} \\
a_3(n+1) &= a_1(n)p_{13} + a_2(n)p_{23} + a_3(n)p_{33} + a_4(n)p_{43} \\
a_4(n+1) &= a_1(n)p_{14} + a_2(n)p_{24} + a_3(n)p_{34} + a_4(n)p_{44}
\end{aligned} \tag{7}
$$

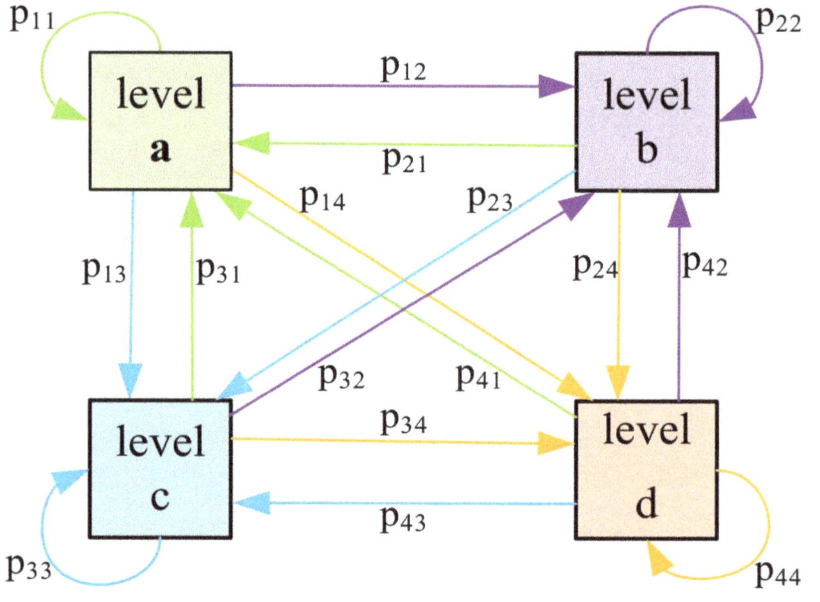

Figure 6. Structural construction information fusion based on Markov chain.

In the construction process of the structure, the mechanical properties of the structure should be analyzed timely and accurately. According to the current structural performance and working conditions, the safety performance of the next construction step is effectively predicted. In this process, the Markov chain is used to connect each construction step. The safety performance of the structure in the next period is analyzed by combining the previous structural state and the changes in construction conditions. In space, the cable force of the structure is collected in real-time, and the mechanical properties of the whole structure are fully considered. The application of Markov chain in the safety assessment of a construction process realizes the integration of construction factors. The whole process fully considers the changes of construction factors in a time dimension. The information fusion of the space–time dimension of the construction process provides a basis for the safety performance analysis of each construction stage. The analysis results provide a reference for the maintenance of unsafe events. The simulation and guidance of real construction can be realized by simulation in virtual space. The data association model of the prestressed steel structure construction process based on Markov chain is shown in Figure 7.

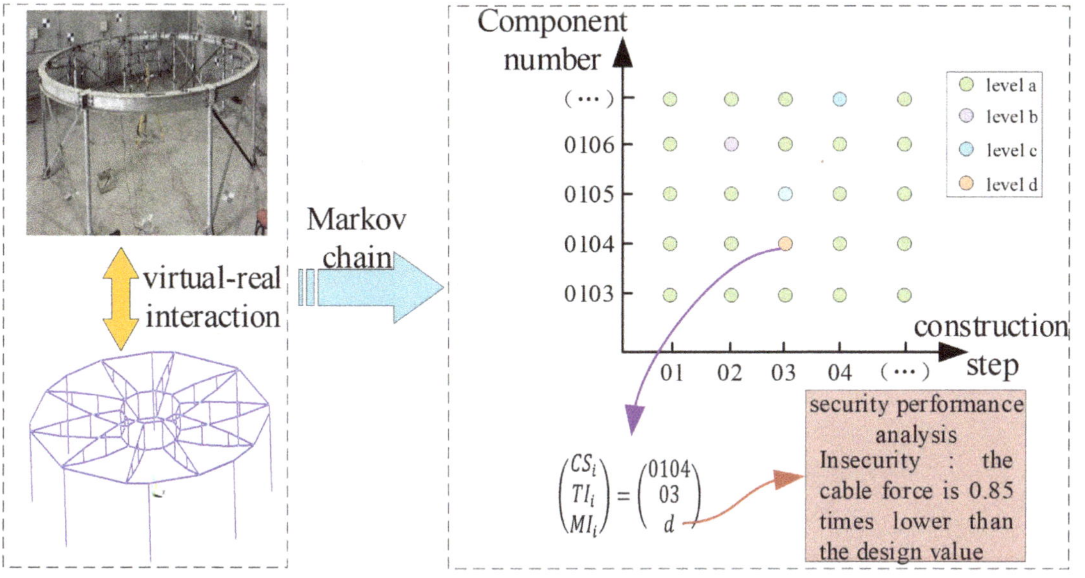

Figure 7. Data association model of the prestressed steel structure construction process.

4. Maintenance Modeling of Construction Insecurity Incidents

In the process of structural construction, the unsafe events should be timely regulated to ensure the quality of structural construction [32]. The premise of analysis and maintenance of structural construction unsafe events is to build a model. This model can describe the occurrence of events qualitatively or quantitatively. The Bow-tie model [33] integrates many factors, such as the cause of the accident, preventive measures, possible consequences, and corresponding control measures. The unsafe event maintenance process is divided into two levels: fault tree (*FT*) and event tree (*ET*). In this study, the maintenance modeling of construction unsafe events is carried out by constructing the Bow-tie model [34].

In the model of this study, the whole maintenance model is divided into two levels: *FT* and *ET*, which are connected by the top event (*TE*). The leftmost side of the Bow-tie model is the risk source of structural unsafe events in the construction process. The risk source mainly includes operational errors and drastic changes in environmental factors during the construction process. The control measures for unsafe events are proposed in the maintenance model. By the above steps, combined with the simulation of the virtual model and the actual construction of the site, the safety performance of the structure is analyzed. If the safety state of the structure is at level a under preventive measures, the next construction step will continue. On the contrary, if the safety performance of the structure cannot reach the standard, the state is judged as an unsafe event, namely *TE* in the Bow-tie model. According to *TE*, the formulated corresponding control measures finally realize the effective maintenance of the structure. The safety performance index of the structure under the construction step is recorded as the result of the Bow-tie model. The principle of the Bow-tie model is shown in Figure 8.

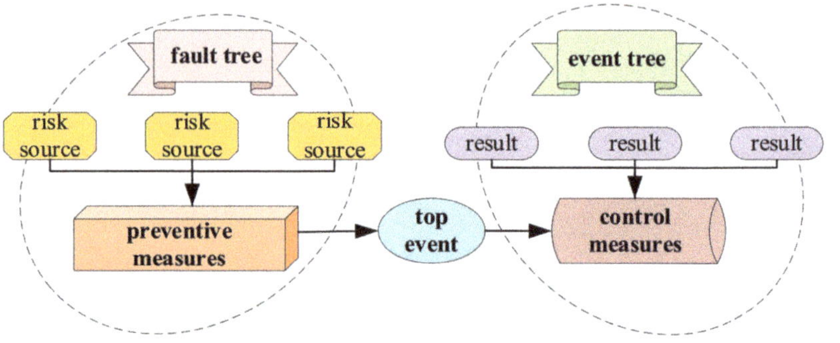

Figure 8. The schematic diagram of the Bow-tie model.

The whole process control from risk source analysis to construction result evaluation is realized by constructing the Bow-tie model. This solves the problems of insufficient model quantification, serious block segmentation, weak intuition, and pertinence in structural safety accident analysis. The maintenance model constitutes a visual summary map to show the causes and consequences of unsafe events.

Based on the qualitative analysis of the causes and consequences of unsafe events, the values of variables in each link involved in the Bow-tie model are determined. Thus, the quantitative analysis of structural maintenance can be carried out according to the sequence of events and their internal relations. In the process of maintaining unsafe events, the whole model is divided into five types of events, namely basic event (*BE*), intermediate event (*IE*), top event (*TE*), control event (*CE*), and result event (*RE*). The formal mathematical language for various events is expressed as Equations (8)–(12):

$$BE = (BE_1, BE_2, \ldots, BE_n) \tag{8}$$

$$IE = (IE_1, IE_2, \ldots, IE_m) \tag{9}$$

$$TE = (TE_1, TE_2) \tag{10}$$

$$CE = (CE_1, CE_2, \ldots, CE_k) \tag{11}$$

$$RE = (RE_1, RE_2, \ldots, RE_s) \tag{12}$$

In the equations, *BE* represents the basic event, corresponding to the risk source in the Bow-tie model, and *IE* means the intermediate event, corresponding to preventive measures in the Bow-tie model. In the analysis, this model assumes that there are n kinds of risk sources and m kinds of preventive measures. *TE* is the top event. The *TE* is divided into two types: security events (structural safety performance at level a) and unsafe events (structural safety performance at levels b, c, and d). Maintenance measures are needed for unsafe events. *CE* means the control event, that is, the maintenance measures taken for unsafe events. *RE* is the result event, which is the state of the structure after the adoption of maintenance measures. In this model, it is assumed that k control measures are adopted for unsafe events, and s result events are finally obtained. The maintenance schematic diagram of construction unsafe events based on the Bow-tie model is shown in Figure 9.

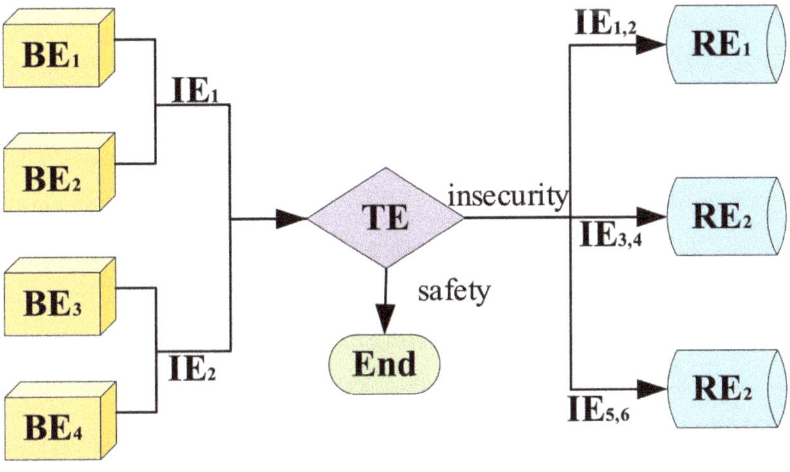

Figure 9. Maintenance schematic diagram of construction unsafe events based on the Bow-tie model.

Assuming that the basic events are independent of each other, the probability of each basic event is p_{BE} in the virtual–real interaction mode. The probability of the intermediate event is p_{IE} through the logical relationship. Combined with Markov chain, the probability of the top event (p_{TE}) can be calculated by spatio-temporal information fusion of the construction process. Calculating the probability of RE occurrence should consider the possibility that the existence of L branches can lead to the occurrence of the ith result event (RE_i). Assuming that the occurrence probability of the control event on the tth (t < l) branch is p_{CEj}, the occurrence probability of the consequent event on the tth branch is expressed as Equation (13):

$$p_{REi} = \sum_{t=1}^{l} p_{TE} \prod_{j=1}^{t} f(p_{CEj}) \tag{13}$$

In the equation, $f(p_{CEj}) = p_{CEj}$ when a link event occurs in a branch. When the link event does not occur, $f(p_{CEj}) = 1 - p_{CEj}$. Therefore, the occurrence probability of the result event (RE_i) can be expressed as a function of the occurrence probability of n basic events and the occurrence probability of q control events, which is specifically expressed as Equation (14):

$$p_{REi} = f(p_{BE1}, p_{BE2}, \cdots, p_{BEn}, p_{CE1}, p_{CE2}, \cdots p_{CEq}) = f(p_{BE}, p_{CE}) \tag{14}$$

The instructions of control events are imported into the finite element model to analyze the feasibility of decision-making. Finally, the control measures are applied to the construction guidance on the site to achieve the closed-loop control of the construction process [35].

In the whole construction safety assessment process, the components and environmental information of the construction site are dynamically perceived first. The information mapping body from four levels in the virtual space is established. Driven by Markov chain, the spatio-temporal integration of virtual and real twin information is realized, and the unsafe events in the construction process are accurately analyzed. In view of the unsafe events, the risk source is analyzed by the Bow-tie model. Then, the corresponding preventive measures and control measures are formulated to ensure the safety of the construction process. The mathematical language of construction process safety assessment is expressed as Equation (15):

$$I_{TR} = \begin{pmatrix} CI_{TS} \\ EI_{TS} \end{pmatrix} \Rightarrow M_m = \begin{pmatrix} R_S \\ M_T \\ R_T \end{pmatrix} \tag{15}$$

In the Equation, I_{TR} denotes construction information fusion of the time dimension and space dimension, CI_{TS} represents component information for spatio-temporal fusion, EI_{TS} is the environmental information of spatio-temporal fusion, M_m indicates the maintenance measures for unsafe events, R_S is the risk source of unsafe events, M_T indicates measures to treat unsafe events, and R_T represents the treatment results of unsafe events. $\Rightarrow$ denotes the transmission of information from the analyzed unsafe events to the maintenance model.

5. Case Study

Driven by the DT framework for the construction safety assessment of prestressed steel structures, physical construction site information collection and virtual model building are carried out. The safety performance of each construction step in the construction process is analyzed through the Markov chain. The Bow-tie model is used to establish a maintenance model for unsafe events. In the structural construction safety assessment modeling, the whole process is divided into two stages. The first stage is the integration of spatio-temporal information to analyze safety performance. On this basis, for unsafe events, the maintenance model outputs the control measures, which ensures the safety of the structure in each construction step. To verify the effectiveness of the assessment method, this study takes the wheel–spoke cable truss as the research object. Compared with the actual project, the scale ratio of the test model was 1:10, the cross-sectional area ratio of the cable was 1:100, and materials were identical. The structure span of the test model was 6 m and consisted of 10 radial cables, ring cables, braces, nodes, outer ring beams, and steel columns. The radial cables include upper and lower radial cables, and the ring cables include upper and lower ring cables. The struts include the outer, middle, and inner struts. The entire structure model is a symmetrical structure. Its geometric shape, supporting conditions, member stiffness, and cross-sectional dimensions are symmetrical about the line where the radial cables are located. The construction plan in the experiment is to tension the upper radial cable. The tension of each cable is synchronized, so the force of the structure is also uniform. The wheel–spoke cable truss test structure is shown in Figure 10.

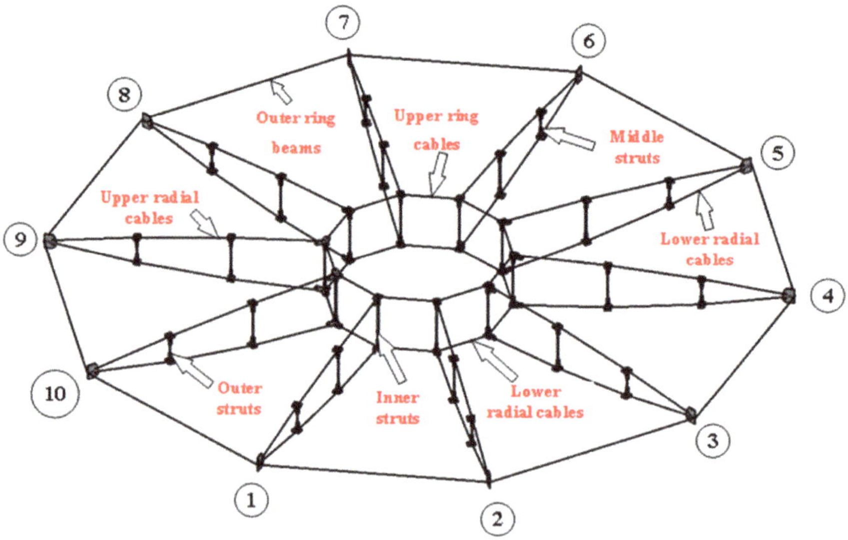

Figure 10. Test structure of the wheel–spoke cable truss.

In this test, the construction scheme of tensioning the upper radial cables is adopted. The construction steps are as follows: (1) Assemble the upper ring cable and the upper radial cable: according to the coordinate, the tenth upper radial cable and the upper ring cable are expanded and paved on the ground and connected to the ring radiation. The clamp connecting the upper radial cable and the upper ring cable is installed, and the radial cable is tensioned by the guide chain tool. (2) The tooling guide chain is used to tension cables to the position where the height from the ground is greater than the length of the internal strut (0.428 m). The inner, middle, and outer struts are installed conveniently in place later. Then, install the lower ring cable and suspend the radial cable. (3) Tensioning the upper radial cable in place: the upper radial cable is tensioned by using the guide chain tooling. The head of the upper radial cable with the ear plate on the outer ring beam is connected by the pin shaft. The upper radial cable is installed in place. (4) Tensioning the lower radial cable in place: the length of the lower radial cable is shortened until the length of the formed state by twisting the sleeve with the wrench. The lower radial cable is installed in place. At this time, the structure is shaped. The whole process is composed of the four main construction steps outlined above. In order to analyze the safety performance of the structure in real-time, the whole construction process is divided into 12 sub-construction steps. In the whole process, the cable force, generated by applying prestress in every stage, is used as the basis to analyze the structural safety performance. The change of cable force in the actual construction process can be expressed as Equation (16) [36]. The main construction process is shown in Figure 11.

$$G_1 = \begin{pmatrix} X_1^1 & X_2^1 & \cdots & X_k^1 \end{pmatrix} \xrightarrow{F^*} G_2 = \begin{pmatrix} X_1^2 & X_2^2 & \cdots & X_k^2 \end{pmatrix} \tag{16}$$

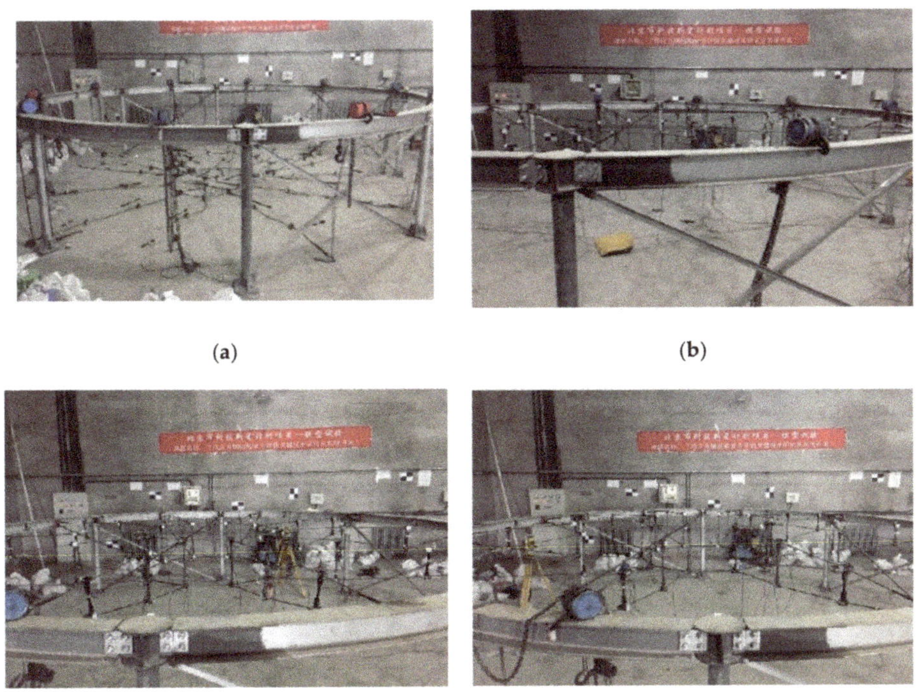

(a) (b) (c) (d)

Figure 11. Main construction processes. (**a**) Step (1), (**b**) Step (2), (**c**) Step (3), (**d**) Step (4).

In the equation, G_i is the form of the structure during the construction phase i. It is assumed that the structure bears k kinds of influencing factors. X_j^i is the jth influencing factor affecting the structure form in the ith construction stage, which can be load, constraint condition, temperature, and prestress. F^* represents the prestress required for the transformation of the two forms.

(1) Acquisition of physical spatial information

The construction scheme adopted in the process of structural tension is to tension the radial cable. The component information and environmental information are collected by sensing equipment in real-time [37]. By using the real-time update and editability of RFID technology, the symbol, basic information, and construction information of the component are extracted and changed in each construction step. Similarly, each construction step is taken as the time interval to collect the mechanical information of the component by the sensor. Data collection provides the field basis for the construction of the virtual space structure model. By the description of the theoretical method in Section 3.1, the basic information and construction information of the component are filled in on the RFID tag. Firstly, the label of the cable and its basic information are input into the RFID electronic label and affixed to the corresponding components. Before the construction, the electronic tags on the components are scanned by the RFID handheld terminal. At the same time, the position and process of the components are clarified to guide the construction. Finally, the mechanical information collected by the cable force sensor is changed in real-time to facilitate the capture of the cable mechanical information. During the test, the cables involved are updated and checked once at the end of each construction step. Due to the indoor test environment, the control of information appears strong. For the change of environmental information, the use of wind load, temperature, and other control equipment is directly generated. Therefore, the extraction of environmental information can be directly extracted by the controller of the device itself, and the sensor equipment in Section 3.1 is used to extract information from the test model. In the test process, the simulated conditions are converted from the actual project. In the test process, the data transmission interval of the mechanical sensing device is 2 s. The component information and environmental information are summarized and analyzed at the completion of each construction step. In the test, for the collection of cable force, the tension-compression sensor, such as the DH3815 acquisition equipment, is utilized by connecting to the cable head and cable body. The upper and lower radial cables are arranged with a measuring point every other cross, with a total of 12 measuring points. The detailed position of sensors and measuring points is shown in Figure 12. Due to the characteristics of cables, the cable force on the whole cable is the same. The column tension-compression sensor is an advanced mechanical parameter acquisition instrument, which can transmit the cable force to the terminal display device in real-time.

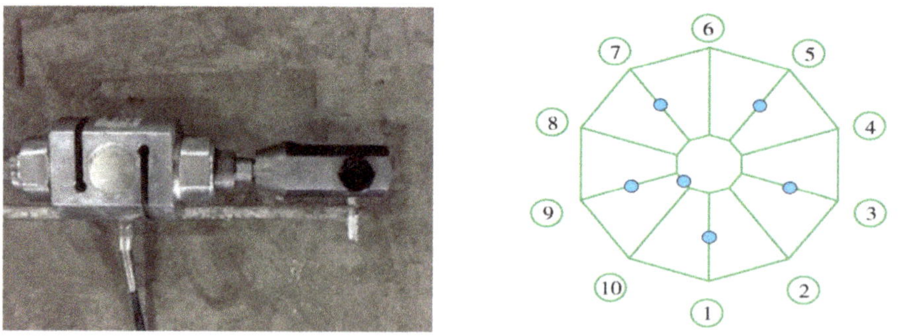

(a) (b)

Figure 12. The positions of the sensors and measuring points. (**a**) column tension-compression sensor, (**b**) measuring points.

In this section, regarding the upper radial cable 9 as the research object, the research shows the physical information capture during the installation of the lower radial cable. According to the information explanation in Section 3.1, the physical information is summarized. Basic information set = (314 mm^2, upper radial cable, Q235 B, A building materials Co., Ltd., 14 August 2019), construction information = (the lower radial cable is installed in place, wrench-fixed connection, ensure firm connection of each node of cable, 11 min), and environmental information = (20 °C, 1.2 m/s, 0 mm). The cable force is captured in terms of structural mechanics information, and the cable force value of this construction step is 5587 N. At the construction site, the physical information is collected in real-time by arranging the sensor equipment, and the various information sets needed for safety performance assessment are integrated. The capture of physical space information is shown in Figure 13.

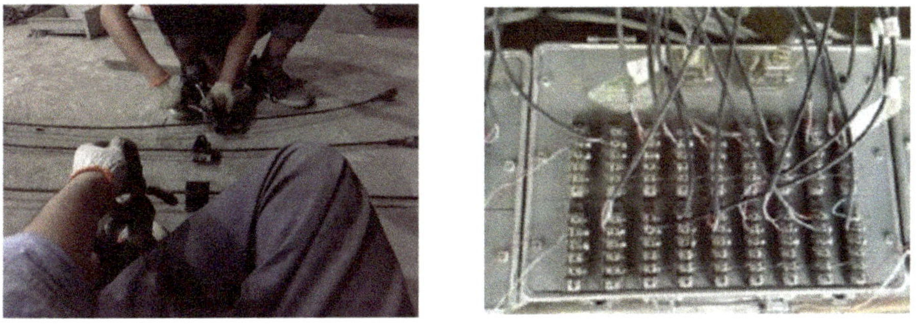

(a) (b)

Figure 13. The capture of physical space information. (**a**) Arrangement of transducers, (**b**) Information collection.

(2) Construction of the virtual model

Based on the data collected by the sensor equipment, building the virtual model is able to realize the mapping of virtual space to reality construction. Revit is used to establish the geometric model of the construction site in all the construction processes. ANSYS is used to establish the physical model. According to the behavior process of structural components and the external effect of the structure, the corresponding working conditions are set in the finite element model. In the whole process, through the analysis of the specification limit constraint model of structural performance, the model fusion of the four levels of 'geometry–physics–behavior–rule' is realized. The visual twin data platform architecture is established.

In the process of building the virtual model, the geometric position of the structure is confirmed by the three-dimensional laser scanner at the end of each construction step. The geometric model established by Revit is corrected in real-time. The model coordinates are rectified through the whole construction process, so as to ensure the high fidelity of the model and improve the robustness of the simulation results. In the correction process of the geometric model, the point cloud model of the experimental model is formed by scanning the structure. Then, the coordinates of the key nodes are picked up from the point cloud model. Finally, the BIM model is corrected to ensure that the geometric model can truly reflect the size of the structure and other information. In this study, a Trimble TX5 3D scanner was used, and the cloud registration of each site was carried out by Realworks 8.0. As for the complex nodes, the displacement meter was also used to extract it. The overall inter-cloud error of registration results is 0.57 mm, the coincidence point reaches 91%, and the reliability reaches 100%. The point cloud data of the test model generated by scanning technology were exported into rcp format files, and the rcp format files were linked in Revit to complete the data conversion. During the work of establishing a revised BIM model based on the point cloud model, the coordinates were compared for each construction step. Finally, a revised finite element analysis model was obtained by updating the key point coordinates in the ANSYS theoretical analysis model. The coordinates of the nodes are obtained by taking the center of the truss as the origin in the model. The node coordinates in the BIM model are called theoretical coordinates, and the node coordinates generated by the point cloud model are called measured coordinates. In the morphological structure, the comparison between the theoretical coordinates and the measured coordinates of the key nodes is shown in Table 1. The correction process of the geometric model is shown in Figure 14.

Table 1. The comparison between the theoretical coordinates and the measured coordinates of the key nodes.

Key Node	Theoretical Coordinates	Measured Coordinates	Coordinate Difference
Upper radial cable head hole 1	(−2110.4, 0, −172.6)	(−2111.3, 0.3, −171.4)	(−0.85, 0.3, 1.25)
Upper radial cable head hole 2	(−2116.4, 0, −179.3)	(−2117.4, 0.3, −178.3)	(−0.92, 0.3, 1.00)
Upper radial cable head hole 3	(−2110.5, 0, −184.5)	(−2111.5, 0.3, −183.7)	(−0.91, 0.3, 0.84)
Upper radial cable head hole 4	(−2104.5, 0, −177.8)	(−2105.5, 0.3, −176.7)	(−0.94, 0.3, 1.19)
Upper radial cable head hole 5	(−2110.4, 20, −172.6)	(−2111.3, 20.3, −171.4)	(−0.85, 0.3, 1.25)
Upper radial cable head hole 6	(−2116.4, 20, −179.3)	(−2117.4, 20.4, −178.5)	(−0.92, 0.4, 0.80)
Upper radial cable head hole 7	(−2110.5, 20, −184.5)	(−2111.5, 20.3, −183.7)	(−0.91, 0.3, 0.84)
Upper radial cable head hole 8	(−2104.5, 20, −177.8)	(−2105.7, 20.3, −176.7)	(−1.14, 0.3, 1.19)

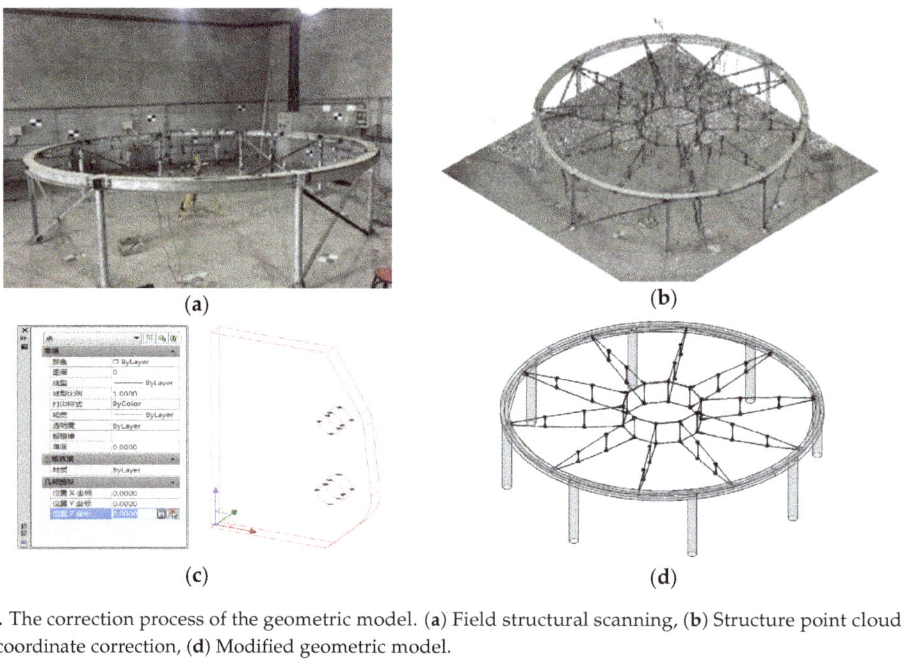

Figure 14. The correction process of the geometric model. (**a**) Field structural scanning, (**b**) Structure point cloud model, (**c**) Node coordinate correction, (**d**) Modified geometric model.

After data collected by cable force sensors were fused by the geometric model, the physical model was established by using APDL language in ANSYS software. The struts use link-8 elements, the cables use link-10 elements, and the ring beams use beam-188 elements. In the finite element model, the model was revised by the coordinates of key nodes. At the same time, under the condition of self-weight, the cross-sectional area of the cable was adjusted to ensure the validity of the physical model. By comparing the simulated value of the cable force with the actual measured value, the basis for the modification was determined. The adjustment of the cable section area is shown in Table 2.

Table 2. Adjustment of cable section area.

Member Type	Location	Cross Sectional Area (mm^2)
Radial cable	Upper level	24.6
	Lower level	33.3
Ring cable	Upper circle	24.6
	Lower circle	49.1
Strut	Outer circle	62.8
	Mid circle	62.8
	Inner circle	62.8
Ring beam	Outer circle	4300

In the finite element model, the command flow of the construction process is compiled. By changing the constraint conditions of components and other parameters, each construction step of the structure is intuitively reflected. At the same time, the temperature sensor, wind speed sensor, and RFID equipment collect the construction environment and the size change of the components in real-time. The collection of information provides the basis for the working condition setting of the physical model. The behavior change of the structure was simulated in the physical model. In this process, the sensor as a data connection device collects the construction information at the end of each construction step. The construction information provides the basis for the simulation of the virtual model and provides data support for the analysis of each stage of the construction process. The tests were conducted indoors, so the temperature and wind speed were relatively uniform. According to the structural specification, the threshold values of cable force, strain, and other parameters were set to ensure the effectiveness of the simulation. The construction of the virtual model can realize the sending and receiving of field data, the visualization of simulation results, the synchronization of structural state, and the acquisition of standard images. In the test process, the establishment of the virtual model for construction safety assessment is shown in Figure 15.

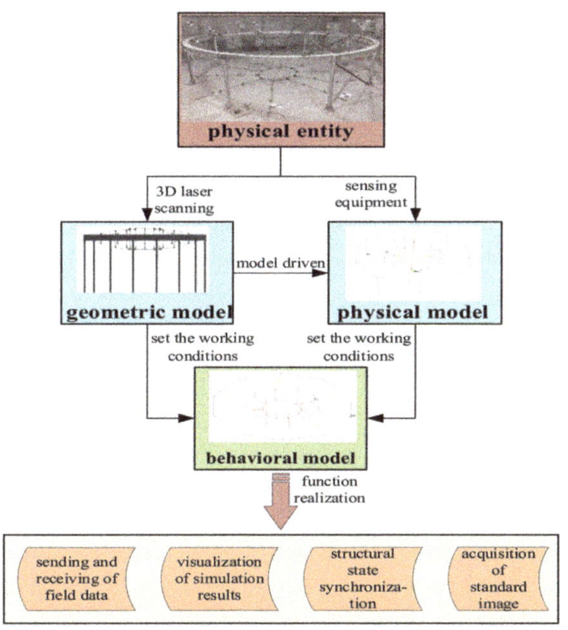

Figure 15. Establishment of the virtual model for construction safety assessment.

(3) Analysis and maintenance of the construction process

After the structural virtual model was completed, the information was fused by the principle of Markov chain. The structural safety performance of the next construction step is predicted according to the probability of the occurrence of risk factors and the degree of changes in structural mechanical parameters. In the process of structural safety analysis, the theoretical method of Section 3.3 was used. In this study, the safety performance of the structure was measured by the cable force. Comparing the calculated cable force with the design value in each construction step, the safety performance was judged. The situation that does not meet the requirements is corrected in time to ensure that each stage of the construction process is safe. The joint analysis of each construction step can highly integrate the heterogeneous information of the time dimension and the space dimension. According to the state of the components, the safety performance of the structure can be analyzed. Then, a data association model for the construction process of the prestressed steel structure was formed. Therefore, the state of each component in each construction step can be intuitively analyzed to provide a basis for the maintenance of construction unsafe events.

The unsafe events were found by Markov chain analysis in the structure, and imported into the Bow-tie model for the maintenance decision. Control measures were brought into the virtual model to verify the feasibility of maintenance, which realizes the goal of field construction guided by the virtual model and ensures the safety of the structure in each construction step. For the unsafe phenomenon of the structure during the construction step, the risk source should be caught in time. After the adjustment measures were provided, the instructions were input into the finite element model for feasibility analysis. The revised finite element model can accurately reflect the state of the structure. After analyzing the effectiveness of the adjustment measures, the on-site maintenance was guided to ensure the safety of the structure. For example, due to the length error of the component, it was analyzed that the connection node of the cable was loose. The length error affects the safety performance of the structure and needs to be maintained. Through the analysis of the Bow-tie model, it is necessary to perform a reinforcement treatment and import the instructions into the finite element model for the final feasibility analysis. The twin model analyzes and guides site maintenance, as shown in Figure 16. The looseness of joints was analyzed, which affects the safety performance of the structure. Therefore, it needs to be maintained, and the construction site was timely processed by the analysis of the DT modeling method to ensure the normal construction.

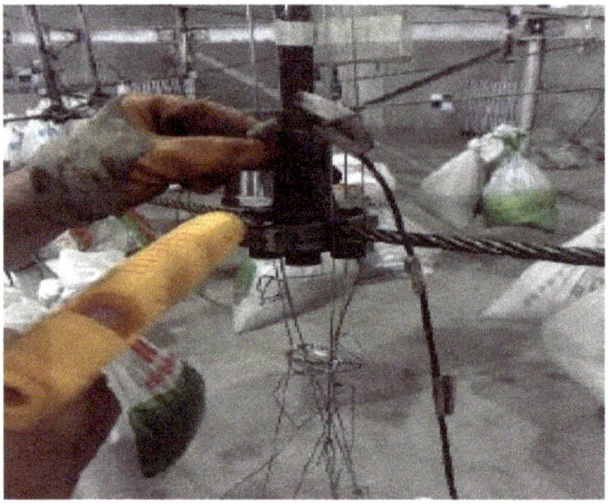

Figure 16. Maintenance of unsafe events in site construction.

Driven by the DT modeling, the safety performance assessment of the construction process structure was completed. The whole process was divided into 12 sub-construction steps. Starting from the first construction step, the cable force was collected on the site and simulated in the virtual model. The cable force of the next construction step was predicted by the environmental change of the site and the Markov chain. The assessment of the cable force was predicted by setting the working conditions in the virtual model. So far, the integration of spatio-temporal information is realized, and the safety performance of the structure is analyzed. The failure of the cable force to ensure the safety of the structure underwent maintenance in a timely manner. Through the evaluation of the whole construction process, the theoretical value and measured value of the cable force in each construction step were formed. According to the cable force value, the safety level was guaranteed to be at level a. During the tension of the lower radial cable, the upper radial cable was in a state of relaxation with zero internal force for a long time. The DTs model of construction process safety assessment is shown in Figure 17. The test model is a symmetrical structure, and the construction plan adopted is that each cable is tensioned synchronously. Therefore, the cable force of each upper and lower radial cable were the same. In the figure, the cable force of one of the cables represents the overall cable force change. The data in the figure is the cable force at the end of each construction step.

(4) Analysis of the modeling method

According to the radial cable force development process image in Figure 17, the cable force values of each construction step were extracted, as shown in Table 3. After the assessment of the intelligent method, the structural safety state of each construction step is finally guaranteed to be at level a.

Table 3. Cable force of each construction step of the structure.

	Upper Radial Cable		**Lower Radial Cable**	
	Theoretical Value (N)	**Measured Value (N)**	**Theoretical Value (N)**	**Measured Value (N)**
Construction step 1	0	0	0	0
Construction step 2	285	372	0	0
Construction step 3	914	1062	0	0
Construction step 4	0	0	356	482
Construction step 5	0	0	365	363
Construction step 6	0	0	387	442
Construction step 7	0	0	427	502
Construction step 8	603	574	498	518
Construction step 9	1985	2079	1024	1482
Construction step 10	2387	2549	1621	1612
Construction step 11	3964	3912	2992	3010
Construction step 12	5503	5587	4374	4383

The validity of the modeling method can be judged by the fitting degree of data. In this study, the fitting degree was obtained from the theoretical value analyzed by the modeling method and the measured value of the cable force in the construction process. The calculation formula of the fitting degree is expressed as Equation (17):

$$R^2 = 1 - \frac{\sum_1^{12}(y_{MVi} - y_{TVi})^2}{\sum_1^{12}(y_{MVi} - \overline{y_{MV}})^2} \tag{17}$$

In the equation, R^2 represents the fitting degree, y_{MVi} represents the measured value of the cable force of each construction step in the construction process, and $\overline{y_{MV}}$ means the average value of the measured value of the cable force of each construction step in the construction process. y_{TVi} is the theoretical value of the cable force in the virtual model simulation. By analyzing the changes of upper and lower radial cable forces in each construction step, the fitting degree was above 95%. On the one hand, the comparative analysis of the theoretical value and the measured value showed that the cable force analyzed by the research method can effectively reflect the real situation of the structure. On the other hand, the research modeling method can intelligently evaluate the construction process of the structure. The cable force can meet the requirements of the specification in each construction step to ensure that the structure is always in the safety level a. The effectiveness of the intelligent assessment method for the construction safety of prestressed steel structures based on the DTs was verified. The DT modeling method can effectively guide the construction site and realize the intelligent assessment of structural safety performance.

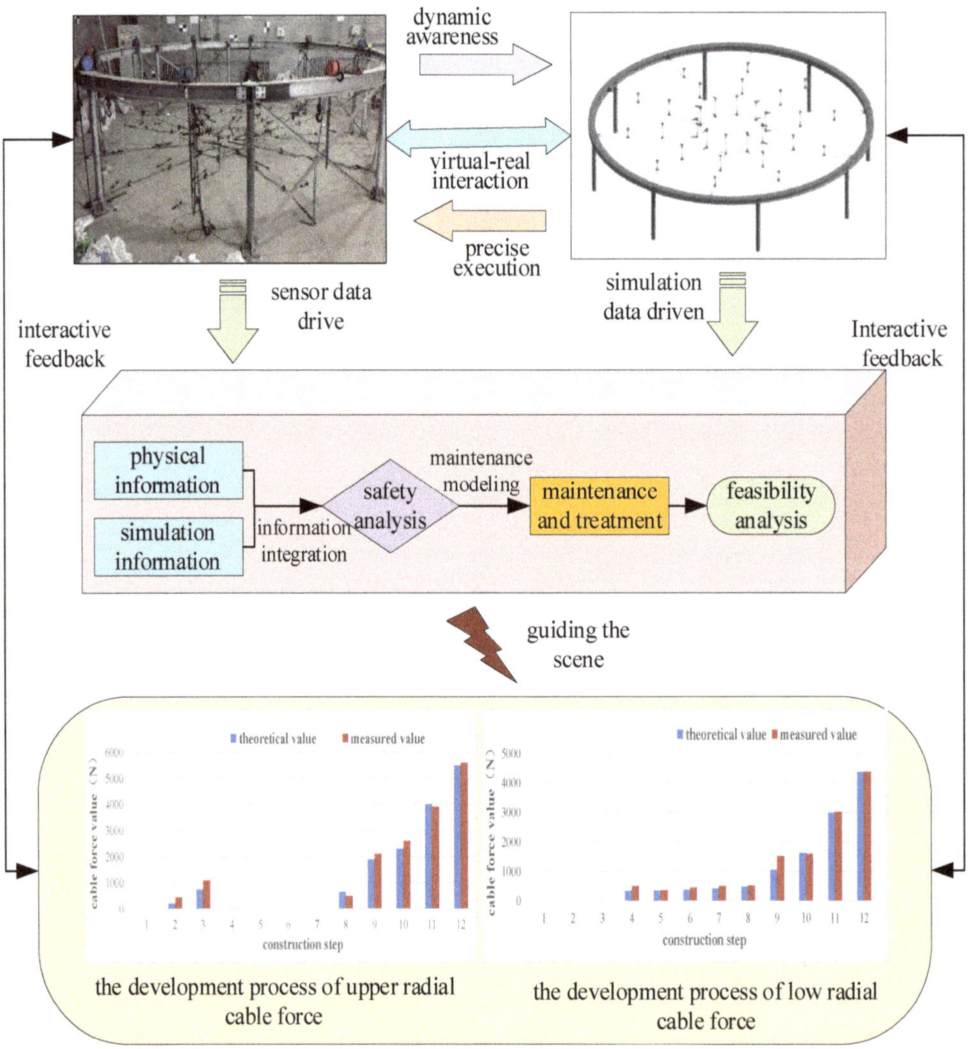

Figure 17. DTs model for safety assessment of the construction process.

6. Discussion

In order to improve the intelligence level of safety assessment in the process of structural construction, this study proposed a digital twin modeling method. In the whole research process, the components and environmental information were captured based on the sensing equipment in the physical space. In the virtual space, three-dimensional laser scanning was carried out at each construction step to form the point cloud model of the actual construction process. The geometric model of the structure was corrected by extracting the key node coordinates in the point cloud model. The physical model with high fidelity was finally established to realize the interactive mapping between the virtual space and physical space. The mechanical properties of the structure can be accurately analyzed by the physical model. The conversion probability of the structural state before and after the construction stage of the structure can be obtained. According to the transition probability, the safety state of the structure in the next stage can be predicted by the state of the current structure. In view of the unsafe state, the reasons were traced by the maintenance modeling and the corrective measures were proposed. The feasibility of the measures was analyzed in the physical model. Finally, the construction process of the site was guided to ensure that the structure is in a safe state at all construction stages. Based on the research of the theoretical method, it was applied in the construction process of the wheel–spoke cable truss. The effectiveness of the proposed method was verified by the experimental results, and it improved the intelligence level of construction. At the same time, the safety intelligent assessment methods also provide new ideas and methods for the development of intelligent monitoring of symmetrical structures.

7. Conclusions

Based on the characteristics of the construction process of prestressed steel structures, this study built a DT framework for the safety assessment of prestressed steel structure construction. Driven by the DT framework, an intelligent method for structural safety assessment was proposed. In this research, the interactive mapping between the physical space and digital space was realized by capturing three kinds of information of physical construction sites and establishing a four-level virtual model. In the construction process, the safety assessment of the structure was divided into two stages, namely the analysis of structural safety performance and the maintenance of unsafe events. From the field information capture to the final maintenance decision output, the assessment theory realizes the intelligent closed-loop control of the construction process. In the research process of the intelligent safety assessment method, the following main findings were obtained:

(1) Capturing the information of components and the environment in the construction process is the basis of analyzing the safety performance of the structure. It is also the basis of establishing the virtual model mapping the construction site, so as to provide twin data for the safety performance analysis of the structure.
(2) In the virtual model, the performance of the structure was simulated under the same working condition as the site. The heterogeneous information of the time dimension and the space dimension of the construction process were fused by Markov chain. The safety performance of each construction step can be predicted.
(3) Based on the Bow-tie model, the construction unsafe events were maintained. The maintenance decision was imported into the virtual model to analyze the feasibility of control measures. Each construction step was taken as the control object to ensure the reliability of the structure.

The intelligent assessment method driven by DTs was applied to the tension process of the wheel–spoke cable truss. The validity of the assessment method can be analyzed by the fitting degree between the theoretical value and the measured value. The guidance of the virtual space to the field construction was realized. The method proposed in this paper ensures that the structure is in a safe state in each construction step, improves the intelligence level of structural construction safety assessment, and provides new ideas for health monitoring of symmetrical structures. In this study, the application of DTs in the safety assessment of the structural construction process is the first step to realize the intelligent analysis of the whole construction process. Therefore, based on the experience and lessons of this study, future research is needed to improve the intelligence level of the construction process by considering more realistic conditions. Driven by DTs, the fusion of sensing technology and artificial intelligence technology to analyze the safety performance change of the whole lifecycle of the structure will be the next research focus.

Author Contributions: Conceptualization, Z.L.; methodology, Z.L.; software, Z.L.; validation, Z.L., G.S., Z.J. and L.Z.; writing—original draft preparation, G.S.; writing—review and editing, Z.L.; project administration, Z.L.; funding acquisition, Z.L. All authors have read and agreed to the published version of the manuscript.

Funding: 1. The research was funded by the National Key R&D Program for the 13th-Five-Year Plan of China (grant number 2018YFF0300300). 2. The research was funded by the Natural Science Foundation of Beijing, Beijing, China (grant number 8202001).

Institutional Review Board Statement: Not applicable.

Informed Consent Statement: Not applicable.

Data Availability Statement: The data presented in this study are available upon request from the corresponding author. The data are not publicly available due to the confidentiality.

Acknowledgments: The authors would like to thank Beijing University of Technology, Beijing, China, for their support throughout the research project.

Conflicts of Interest: The authors declare no conflict of interest. The funders had no role in the study's design; in the collection, analysis, or interpretation of data; in the writing of the manuscript, or in the decision to publish the results.

References

1. Zhao, L.; Liu, Z.; Mbachu, J. Development of Intelligent Prefabs Using IoT Technology to Improve the Performance of Prefabricated Construction Projects. *Sensors* **2019**, *19*, 4131. [CrossRef] [PubMed]
2. Dong, S.L.; Tu, Y. Structural system innovation of cable dome structures. *Jianzhu Jiegou Xuebao/J. Build. Struct.* **2018**, *39*, 85–92.
3. Wang, Y.; Guo, Z.; Luo, B.; Shi, K. Study on the determination method for the equivalent pre-tension in cables of spatial prestressed steel structure. *China Civ. Eng. J.* **2013**, *46*, 53–61.
4. Zhang, A.L.; Sun, C.; Jiang, Z.Q. Calculation method of prestress distribution for levy cable dome with double struts considering self-weight. *Eng. Mech.* **2017**, *34*, 211–218.
5. Liu, Z.S.; Han, Z.B.; He, J.; Wang, Z.Q. Sensitive Test on Relaxation of Cable and Reliability Assessment of Spoke Cable-truss Structure. *J. Tongji Univ. (Nat. Sci.)* **2019**, *47*, 946–956.
6. Bai, Y.; Wang, S.; Mou, B.; Wang, Y.; Skalomenos, K.A. Bi-directional seismic behavior of steel beam-column connections with outer annular stiffener. *Eng. Struct.* **2021**, *227*, 111443. [CrossRef]
7. Liu, C.; Wang, F.; He, L.; Deng, X.; Liu, J.; Wu, Y. Experimental and numerical investigation on dynamic responses of the umbrella membrane structure excited by heavy rainfall. *J. Vib. Control.* **2020**, *27*, 107754632093269. [CrossRef]
8. Basta, A.; Serror, M.H.; Marzouk, M. A BIM-based framework for quantitative assessment of steel structure deconstructability. *Autom. Constr.* **2020**, *111*, 103064. [CrossRef]
9. Alamdari, M.M.; Kildashti, K.; Samali, B.; Goudarzi, H.V. Damage diagnosis in bridge structures using rotation influence line: Validation on a cable-stayed bridge. *Eng. Struct.* **2019**, *185*, 1–14. [CrossRef]
10. Bera, K.K.; Chandiramani, N.K. Controlling flutter of a cable-stayed bridge with output feedback driven winglets. *J. Wind Eng. Ind. Aerodyn.* **2020**, *206*, 104372. [CrossRef]
11. D'Auteuil, A.; Mctavish, S.; Raeesi, A. A new large-scale dynamic rig to evaluate rain-wind induced vibrations on stay cables: Design and commissioning-ScienceDirect. *J. Wind Eng. Ind. Aerodyn.* **2020**, *206*, 104334. [CrossRef]
12. Liu, Z.; Shi, G.; Zhang, A.; Huang, C. Intelligent Tensioning Method for Prestressed Cables Based on Digital Twins and Artificial Intelligence. *Sensors* **2020**, *20*, 7006. [CrossRef]
13. Boje, C.; Guerriero, A.; Kubicki, S.; Rezgui, Y. Towards a semantic Construction Digital Twin: Directions for future research. *Autom. Constr.* **2020**, *114*, 103179. [CrossRef]
14. Tahmasebinia, F.; Fogerty, D.; Wu, L.O.; Li, Z.; Sepasgozar, S.M.E.; Zhang, K.; Sepasgozar, S.; Marroquin, F.A. Numerical Analysis of the Creep and Shrinkage Experienced in the Sydney Opera House and the Rise of Digital Twin as Future Monitoring Technology. *Buildings* **2019**, *9*, 137. [CrossRef]
15. Teng, S.Y.; Touš, M.; Leong, W.D.; How, B.S.; Lam, H.L.; Máša, V. Recent advances on industrial data-driven energy savings: Digital twins and infrastructures. *Renew. Sustain. Energy Rev.* **2021**, *135*, 110208. [CrossRef]
16. Grieves, M. *Virtually Perfect: Driving Innovative and Lean Products through Product Lifecycle Management*; Space Coast Press: Cocoa Beach, FL, USA, 2011.
17. Yu, G.; Zhang, S.; Hu, M.; Wang, Y.K. Prediction of Highway Tunnel Pavement Performance Based on Digital Twin and Multiple Time Series Stacking. *Adv. Civ. Eng.* **2020**, *2020*, 1–21.
18. Tao, F.; Zhang, Y.; Cheng, Y.; Ren, J.; Wang, D.; Qi, Q.; Li, P. Digital twin and blockchain enhanced smart manufacturing service collaboration and management-ScienceDirect. *J. Manuf. Syst.* **2020**, *2020*. [CrossRef]
19. Wang, T.; Li, J.; Kong, Z.; Liu, X.; Snoussi, H.; Lv, H. Digital twin improved via visual question answering for vision-language interactive mode in human–machine collaboration. *J. Manuf. Syst.* **2020**, *2020*. [CrossRef]

20. Ruppert, T.; Abonyi, J. Integration of real-time locating systems into digital twins. *J. Ind. Inf. Integr.* **2020**, *20*, 100174. [CrossRef]
21. Gopalakrishnan, S.; Hartman, N.W.; Sangid, M.D. Model-Based Feature Information Network (MFIN): A Digital Twin Framework to Integrate Location-Specific Material Behavior within Component Design. *Manuf. Perform. Anal.* **2020**, *9*, 394–409.
22. Erdélyi, J.; Kopáik, A.; Kyrinovi, P. Spatial Data Analysis for Deformation Monitoring of Bridge Structures. *Appl. Sci.* **2020**, *10*, 8731. [CrossRef]
23. Guo, Y.; Zhang, X. Influences of temperature changes and cable length errors on tension structures using un-adjustable cable length design. *Tumu Gongcheng Xuebao/China Civ. Eng. J.* **2017**, *50*, 11–22, 61.
24. Pei, W.A.; Ming, L.B. A digital twin-based big data virtual and real fusion learning reference framework supported by industrial internet towards smart manufacturing-ScienceDirect. *J. Manuf. Syst.* **2021**, *58*, 16–32.
25. Tao, F.; Liu, W.; Zhang, M.; Hu, T.; Qi, Q.; Zhang, H.; Sui, F.; Wang, T.; Xu, H.; Huang, Z. Five-dimension digital twin model and its ten applications. *Jisuanji Jicheng Zhizao Xitong/Comput. Integr. Manuf. Syst. CIMS* **2019**, *25*, 1–18.
26. Fedtschenko, T.; Utz, A.; Stanitzki, A.; Hennig, A.; Lüdecke, A.; Haas, N.; Kokozinski, R. A New Configurable Wireless Sensor System for Biomedical Applications with ISO 18000-3 Interface in 0.35 µm CMOS. *Sensors* **2019**, *19*, 4110. [CrossRef] [PubMed]
27. D'Emilia, G.; Gaspari, A.; Galar, D.P. Improvement of Measurement Contribution for Asset Characterization in Complex Engineering Systems by an Iterative Methodology. *Int. J. Serv. Sci. Manag. Eng. Technol.* **2018**, *9*, 85–103. [CrossRef]
28. Zhang, Y.; Lei, Y. Data Anomaly Detection of Bridge Structures Using Convolutional Neural Network Based on Structural Vibration Signals. *Symmetry* **2021**, *13*, 1186. [CrossRef]
29. Chen, W.T.; Tsai, I.; Merrett, H.C.; Lu, S.T.; Lee, Y.I.; You, J.K.; Mortis, L. Construction Safety Success Factors: A Taiwanese Case Study. *Sustainability* **2020**, *12*, 6326. [CrossRef]
30. Hu, Z.Z.; Tian, P.L.; Li, S.W.; Zhang, J.P. BIM-based integrated delivery technologies for intelligent MEP management in the operation and maintenance phase. *Adv. Eng. Softw.* **2017**, *115*, 1–16. [CrossRef]
31. Sonnenberg, A.H.; Herrmann, J.; Grinstaff, M.W.; Suki, B. A Markov chain model of particle deposition in the lung. *Sci. Rep.* **2020**, *10*, 13573. [CrossRef]
32. Lobato, T.C.; Hauser-Davis, R.A.; Oliveira, T.F.; Silveira, A.M.; Silva, H.A.; Tavares, M.R.; Saraiva, A.C. Construction of a novel water quality index and quality indicator for reservoir water quality evaluation: A case study in the Amazon region. *J. Hydrol.* **2015**, *522*, 674–683. [CrossRef]
33. Rafiei, V.; Sharifi, G.; Karamzadeh, S.; Kartal, M. Beam-steering high-gain array antenna with FP Bow-tie slot antenna element for pattern stabilisation. *IET Microw. Antennas Propag.* **2020**, *14*, 1185–1189. [CrossRef]
34. Cui, L.J.; Chen, H.R.; Ren, B.; Zhang, J.K. Quantitative analysis method of aviation unsafe events under mixed uncertain conditions. *J. Natl. Univ. Def. Technol.* **2020**, *42*, 92–97.
35. Francis, S.; Daniel, C.; Derek, C.C.; Tong, Y. Editorial: Smart buildings. *Intell. Build. Int.* **2021**, *13*, 1–3.
36. Lou, R.; Luo, Y.Z.; Shen, Y.B. Research on the cable tension control of spatial steel sturucures considering temperature effect. *Eng. Mech.* **2010**, *27*, 164–168.
37. Wang, M.; Wang, C.C.; Sepasgozar, S.; Zlatanova, S. A Systematic Review of Digital Technology Adoption in Off-Site Construction: Current Status and Future Direction towards Industry 4.0. *Buildings* **2020**, *10*, 204. [CrossRef]

symmetry MDPI

Article

Substructure Shake Table Testing of Frame Structure–Damper System Using Model-Based Integration Algorithms and Finite Element Method: Numerical Study

Bo Fu [1,2], Huanjun Jiang [3] and Jin Chen [1,4,*]

1 School of Civil Engineering, Chang'an University, Xi'an 710061, China; 90_bofu@chd.edu.cn
2 State Key Laboratory of Green Building in Western China, Xi'an University of Architecture and Technology, Xi'an 710055, China
3 College of Civil Engineering, Tongji University, Shanghai 200092, China; jhj73@tongji.edu.cn
4 State Key Laboratory of Mechanical Behavior and System Safety of Traffic Engineering Structures, Shijiazhuang Tiedao University, Shijiazhuang 050043, China
* Correspondence: chenjin5310@126.com

Abstract: Substructure shake table testing (SSTT) is an advanced experimental technique that is suitable for investigating the vibration control of secondary structure-type dampers such as tuned mass dampers (TMDs). The primary structure and damper are considered as analytical and experimental substructures, respectively. The analytical substructures of existing SSTTs have mostly been simplified as SDOF structures or shear-type structures, which is not realistic. A common trend is to simulate the analytical substructure via the finite element (FE) method. In this study, the control effects of four dampers, i.e., TMD, tuned liquid damper (TLD), particle damper (PD) and particle-tuned mass damper (PTMD), on a frame were examined by conducting virtual SSTTs. The frame was modeled through stiffness-based beam-column elements with fiber sections and was solved by a family of model-based integration algorithms. The influences of the auxiliary mass ratio, integration parameters, time step, and time delay on SSTT were investigated. The results indicate that the TLD had the best performance. In addition, SSTT using model-based integration algorithms can provide satisfactory results, even when the time step is relatively large. The effects of integration parameters and time delay are not significant.

Keywords: substructure shake table testing; integration algorithm; finite element method; damper

Citation: Fu, B.; Jiang, H.; Chen, J. Substructure Shake Table Testing of Frame Structure–Damper System Using Model-Based Integration Algorithms and Finite Element Method: Numerical Study. *Symmetry* **2021**, *13*, 1739. https://doi.org/10.3390/sym13091739

Academic Editor: Yang Yang

Received: 4 August 2021
Accepted: 13 September 2021
Published: 18 September 2021

1. Introduction

Substructure shake table testing (SSTT) is one of the most advanced experimental techniques in structural and earthquake engineering [1]. It combines the advantages of the real-time dynamic loading of a shake table and the substructure technique from hybrid simulation (HS) and real-time hybrid simulation (RTHS). The best use of SSTT is for investigating the vibration control effects of dampers, such as the classical TMDs and TLDs, and emerging dampers, such as particle dampers (PDs) [2] and particle-tuned mass dampers (PTMDs) [3]. These dampers can be regarded as secondary structures with respect to the primary structure. To conduct SSTT, the damper and the primary structure are taken as the experimental and analytical substructures, which are experimentally tested on the shake table and numerically simulated on a computer, respectively.

The shake table tests are generally used for obtaining the dynamic responses and dynamic characteristics of structures [4,5]. The conventional experimental method of investigating the control effects of dampers is to carry out shake table tests for complete structure–damper systems. For instance, Kang et al. [6] conducted a series of 1:30 scaled model shake table tests and numerical simulations for a coal-fired power plant equipped with a large mass ratio multiple-tuned mass damper (LMTMD), and found that the LMTMD is effective and robust in reducing structural dynamic responses. Wang et al. [7] evaluated the performance of a pendulum pounding-tuned mass damper (PPTMD) by carrying out a series of shake table tests. They reported that the inherent damping of the primary structure decreases the control efficiency of the PPTMD. Zhao et al. [8] conducted shake table tests on a 1:8 scaled transmission tower equipped with and without TMDs, and found that the TMD's control performance is related to earthquake type and excitation intensity. Vafaei et al. [9] proposed a modified-tuned liquid damper (MTLD) to attenuate multiple mode vibration. The effectiveness of the MTLD was demonstrated by several shake table tests on a scaled three-story structure with and without an MTLD. Lu et al. [3] explored the damping performance of a PTMD by comparing the shake table results of a scaled five-story steel frame with and without a PTMD. Based on the shake table tests, they conducted comprehensive parametric analyses on the reduction effects of PTMD, including the auxiliary mass ratio, gap clearance, and the mass ratio of particles to the total auxiliary mass. Shen et al. [10] investigated the influence of a double-layer-tuned particle damper (DTPD) on the seismic performance of super high-rise structures by conducting a series of 1:20 scaled model shake table tests on high-rise structures with and without DTPDs. They concluded that the effectiveness of the DTPD is closely related to the ground motion.

Compared with the conventional shake table tests for a complete structural system, one of the notable advantages of SSTTs is that any size effect of the specimen can be reduced. SSTT has also been applied to structure–damper systems. For instance, numerous researchers [11–14] investigated the performance of TLDs in controlling the seismic response of structures using a SSTT. Fu et al. [15,16] recently conducted the first SSTT of a single degree-of-freedom (SDOF)-PD system and compared the vibration control effects of TLDs and PDs based on the experimental results of a series of SSTTs. It should be noted, however, that the analytical substructures in most existing studies were optimized to be SDOF structures or shear-type structures with few DOFs. This is because these structures do not require considerable computational time to solve their equations of motion (EOMs). However, oversimplification of the analytical substructure may hinder the application of SSTT in real engineering practice. Therefore, it is essential that more refined models of the analytical substructure are applied and examined for SSTT.

It is well known that the finite element (FE) method is an accurate and reliable approach for simulating structures [17–22]. However, it requires significantly longer computational time compared with the simplified SDOF or shear-type model. Therefore, if the FE method is applied to simulate the analytical substructure in SSTT, an integration algorithm with high computational efficiency must be used. In the past two decades, a new class of integration algorithms, called model-based integration algorithms [23], have been developed for the application of HS and RTHS. Model-based integration algorithms are computationally competitive because they combine the advantages of explicit displacement formulation and unconditional stability, which is not possible for traditional integration algorithms such as Newmark algorithms. Model-based integration algorithms have been successfully applied in actuator-type RTHS with the FE model of the analytical substructure [24–26]. However, there are few studies on SSTT with the FE method using model-based integration algorithms. In this study, we numerically investigate the seismic response reduction effects of four types of dampers, i.e., TMD, TLD, PD, and PTMD, on a four-story steel frame by conducting a series of virtual SSTTs. The steel frame is simulated by stiffness-based beam-column elements with fiber sections and bar elements. The EOM of the frame structure is solved using a family of model-based integration algorithms.

The formulation and basic features of model-based integration algorithms are summarized in Section 2. Section 3 explains the FE modeling of the four-story steel frame. The analytical models of the four types of dampers are given in Section 4. Section 5 provides the procedure of SSTT and numerical results of a series of SSTTs. The influences of the auxiliary mass ratio, integration parameters, time step, and time delay on the SSTT are also investigated in Section 5. Some important conclusions are drawn in Section 6.

2. Model-Based Integration Algorithms

To numerically obtain the structural responses induced by earthquakes or other dynamic actions, the following equation of motion (EOM) should be solved using a direct integration algorithm:

$$\mathbf{M}\ddot{\mathbf{X}}_{i+1} + \mathbf{C}\dot{\mathbf{X}}_{i+1} + \mathbf{R}_{i+1} = \mathbf{F}_{i+1} \tag{1}$$

where $\mathbf{M}$ and $\mathbf{C}$ are the mass and damping matrices, respectively; $\dot{\mathbf{X}}$ and $\ddot{\mathbf{X}}$ are the velocity and acceleration vectors, respectively; the subscripts $i+1$ and i represent the next and current time steps, respectively; $\mathbf{R}$ and $\mathbf{F}$ are the restoring force and external force vectors, respectively. The restoring force $\mathbf{R}_{i+1}$ is generally displacement-dependent and can be degraded into $\mathbf{K}\mathbf{X}_{i+1}$ if the structure is linear elastic, where $\mathbf{K}$ and $\mathbf{X}$ are the stiffness matrix and displacement vector, respectively.

The main task for an integration algorithm is to predict the velocity and displacement at the next step. For instance, the difference equations of the classical Newmark family of algorithms are expressed as:

$$\dot{\mathbf{X}}_{i+1} = \dot{\mathbf{X}}_i + \Delta t\left[(1-\gamma)\ddot{\mathbf{X}}_i + \gamma\ddot{\mathbf{X}}_{i+1}\right], \mathbf{X}_{i+1} = \mathbf{X}_i + \Delta t\dot{\mathbf{X}}_i + \Delta t^2\left[(1/2-\beta)\ddot{\mathbf{X}}_i + \beta\ddot{\mathbf{X}}_{i+1}\right] \tag{2}$$

where Δt is the time step; γ and β are two integration parameters. The Newmark algorithms are explicit only if $\beta = 0$. Furthermore, the Newmark explicit algorithms are conditionally stable and still implicit for velocity. The Newmark algorithms are unconditionally stable when $2\beta \geq \gamma \geq 1/2$. Therefore, it is impossible for the Newmark algorithms to achieve unconditional stability within the framework of an explicit displacement formulation. As γ and β are parameters independent of the structural model, the Newmark algorithms are called model-independent algorithms.

In contrast to model-independent algorithms, model-based integration algorithms are unconditionally stable and have an explicit displacement formulation. That is, model-based integration algorithms predict displacement based on equilibrium at the current time step and calculate the acceleration by satisfying equilibrium at the next time step. They can be used to solve general dynamic problems, i.e., the structural responses of linear and nonlinear SDOF and MDOF structures subjected to dynamic loads. Compared with implicit algorithms, explicit algorithms are computationally more efficient at solving nonlinear dynamic problems without iterations to obtain tangent stiffness. To solve the EOM of an MDOF system with a large number of DOFs, a very small time step is required for a conditionally stable algorithm to ensure stability, because the time step limit is inversely proportional to the highest natural frequency of the structural system. Therefore, the unconditionally stable and explicit model-based integration algorithms have very promising computational efficiency, particularly for the nonlinear MDOF dynamic problems.

Model-based integration algorithms are explicit for displacement, whereas they are not always explicit for velocity. Therefore, model-based integration algorithms can be classified as dual-explicit and semi-explicit according to their formulation of velocity. Specifically, dual-explicit means that the algorithm is explicit for both displacement and velocity, whereas semi-explicit indicates that the algorithm is explicit only for displacement, and implicit for velocity. For instance, the first model-based integration algorithm, which was developed by Chang [27], is semi-explicit, whereas the well-known CR algorithm [28] is dual-explicit. Dual-explicit algorithms are clearly more suitable for SSTT, particularly when the experimental substructure is velocity-dependent. In this study, a family of dual-explicit model-based integration algorithms [29], called GCR algorithms, is adopted. The formulation and characteristics of GCR algorithms are briefly summarized as follows.

Chen and Ricles [28] proposed the first dual-explicit model-based integration algorithm using the discrete control theory. The difference equations of the CR algorithm for the MDOF system are expressed as:

$$\dot{\mathbf{X}}_{i+1} = \dot{\mathbf{X}}_i + \Delta t\boldsymbol{\alpha}_1\ddot{\mathbf{X}}_i, \mathbf{X}_{i+1} = \mathbf{X}_i + \Delta t\dot{\mathbf{X}}_i + \Delta t^2\boldsymbol{\alpha}_2\ddot{\mathbf{X}}_i \tag{3}$$

where $\boldsymbol{\alpha}_1$ and $\boldsymbol{\alpha}_2$ are model-based integration parameter matrices, and are formulated as:

$$\boldsymbol{\alpha}_1 = \boldsymbol{\alpha}_2 = 4\left(4\mathbf{M} + 2\Delta t\mathbf{C} + \Delta t^2\mathbf{K}\right)^{-1}\mathbf{M} \quad (4)$$

The CR algorithm has been demonstrated to be unconditionally stable for both linear elastic and nonlinear softening systems. In addition, the second-order accurate CR algorithm has no numerical damping. To obtain a family of model-based integration algorithms with more general and versatile numerical features, Fu et al. [29] recently developed GCR algorithms, which stands for generalized CR algorithms. The difference equations of the GCR algorithm are inherited from the CR algorithm, and the integration parameter matrices are updated by introducing two additional coefficients, κ_1 and κ_2, which are shown in Equation (5):

$$\boldsymbol{\alpha}_1 = \left(\mathbf{M} + \kappa_1\Delta t\mathbf{C} + \kappa_2\Delta t^2\mathbf{K}\right)^{-1}\mathbf{M}, \boldsymbol{\alpha}_2 = (1/2 + \kappa_1)\boldsymbol{\alpha}_1 \quad (5)$$

It was found that GCR algorithms with κ_1 and κ_2 possess numerical properties identical to the classical Newmark algorithms with γ and β. The mapping relation is $[\kappa_1,\ \kappa_2] = [\gamma,\ \beta]$. The CR algorithm is a special type of GCR algorithm, with $\kappa_1 = 1/2$, $\kappa_2 = 1/4$; its counterpart includes Newmark algorithms with $\gamma = 1/2, \beta = 1/4$, which is the well-known constant average acceleration (CAA) algorithm. The members of the subfamily of GCR algorithms with $\kappa_1 = 1/2$ have no numerical damping and are second-order accurate. GCR algorithms with $2\kappa_2 \geq \kappa_1 \geq 1/2$ are unconditionally stable for linear elastic systems. GCR algorithms with $\kappa_1 > 1/2$, $\kappa_2 \geq (\kappa_1 + 1/2)^2/4$ have numerical damping. The period elongation (PE) and equivalent damping ratios are two widely used indices of evaluating the accuracy of the integration algorithm [30]. The two accuracy indices of five GCR algorithms, i.e., $[\kappa_1,\ \kappa_2] = [1/2,\ 1/4], [1/2,\ 1/2], [1/2,\ 1], [1,\ 1/2], [1,\ 1]$, are depicted in Figure 1. The abscissa in Figure 1 is $\Omega = \omega\Delta t$, where ω is the circular frequency. Figure 1 shows that for a certain κ_1, the PE increases with the increase in κ_2. GCR algorithms with $[\kappa_1,\ \kappa_2] = [1,\ 1/2]$ have PE close to that of $[1/2,\ 1/4]$ (CR algorithm) and are minimum. Regarding the equivalent damping ratio, GCR algorithms with $\kappa_1 = 1/2$ are zero, whereas GCR algorithms with $\kappa_1 = 1$ are positive. Overall, the original CR algorithm has excellent accuracy and can be used as a reference for other GCR algorithms.

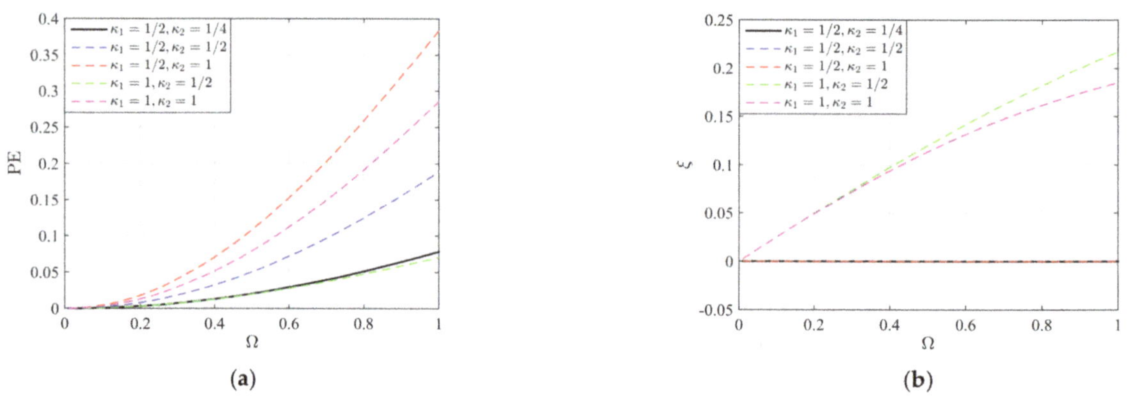

Figure 1. Numerical properties of GCR algorithms. (**a**) Period elongation; (**b**) equivalent damping ratio.

Figure 2 provides a general flowchart for solving the EOM for MDOF systems using GCR algorithms. The first step is to select appropriate integration coefficients κ_1 and κ_2 and the time step Δt. Then, the structural model matrices **M**, **K**, and **C**, and the external force **F** are calculated. The model-based integration parameter matrices $\boldsymbol{\alpha}_1$ and $\boldsymbol{\alpha}_2$ are then obtained using Equation (5). Then, the initial conditions are calculated as follows:

$$\ddot{\mathbf{X}}_0 = (\mathbf{M})^{-1}\left(\mathbf{F}_0 - \mathbf{C}\dot{\mathbf{X}}_0 - \mathbf{K}\mathbf{X}_0\right) \tag{6}$$

where $\dot{\mathbf{X}}_0$, $\mathbf{X}_0$, and $\ddot{\mathbf{X}}_0$ are the initial velocity, displacement, and acceleration, respectively.

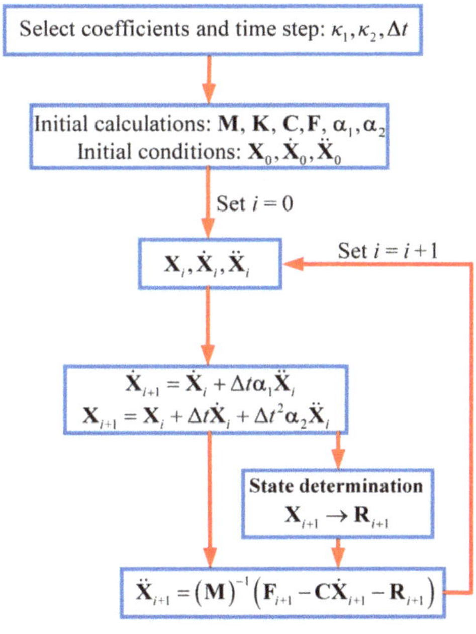

Figure 2. General flowchart of solving EOM for MDOF systems using GCR algorithms.

Then, the difference equations of the GCR algorithms, i.e., Equation (3), are adopted to predict the velocity $\dot{\mathbf{X}}_{i+1}$ and displacement $\mathbf{X}_{i+1}$ at the next time step. The displacement $\mathbf{X}_{i+1}$ can be used to determine the restoring force $\mathbf{R}_{i+1}$, which is called state determination. If the structural system is linear elastic, the restoring force $\mathbf{R}_{i+1}$ can be directly calculated as $\mathbf{K}\mathbf{X}_{i+1}$; otherwise, it can be obtained using the FE method, e.g., the FE modeling in Section 3. The final step is to calculate the acceleration by rewriting the EOM, i.e., Equation (1):

$$\ddot{\mathbf{X}}_{i+1} = (\mathbf{M})^{-1}\left(\mathbf{F}_{i+1} - \mathbf{C}\dot{\mathbf{X}}_{i+1} - \mathbf{R}_{i+1}\right) \tag{7}$$

3. Finite Element Modeling of Frame Structure

In this study, stiffness-based beam-column elements with fiber sections along with P-Δ effects are adopted to simulate the frame structure. The stiffness-based beam-column elements are used to acquire the first-order restoring force, and the P-Δ effects are considered to obtain the second-order restoring force. The following content summarizes the basic principles and procedures.

3.1. *Stiffness-Based Beam-Column Element with Fiber Sections*

3.1.1. Initialization

The model-based integration parameter matrices, i.e., $\boldsymbol{\alpha}_1$ and $\boldsymbol{\alpha}_2$, should be predetermined before using model-based integration algorithms. As shown in Equation (5), model-based integration parameter matrices are functions of model matrices of structure, i.e., **M**, **C**, and **K**, and two controlling coefficients κ_1 and κ_2. Therefore, the first step is to find an appropriate combination of κ_1 and κ_2 based on the numerical properties of the GCR algorithms. Then, the model matrices of the structures should be determined. Among the three model matrices of structure, the Rayleigh damping assumption is widely adopted for establishing the damping matrix **C**:

$$\mathbf{C} = a_m\mathbf{M} + a_k\mathbf{K} = \frac{2\xi\omega_i\omega_j}{\omega_i + \omega_j}\mathbf{M} + \frac{2\xi}{\omega_i + \omega_j}\mathbf{K} \tag{8}$$

where a_m and a_k are the combination coefficients of **M** and **K**, respectively; ξ is the damping ratio of structure; ω_i and ω_j are the ith and jth circular frequencies, respectively, which can be obtained by conducting modal analysis of the structure. Therefore, determination of the three model matrices is reduced to two matrices: **M** and **K**.

The formulation of the mass matrix can be classified into two types: lumped mass and consistent mass. The detailed construction process can be found in [31].

As a stiffness-based beam-column element with fiber sections is used, the initial stiffness matrix **K** of the structure should be sequentially established from different levels. There are four levels from bottom to top: fiber, section, element, and structure. Regarding the section with fibers, the stiffness matrix $\boldsymbol{k}_s(x)$ is:

$$\boldsymbol{k}_s(x) = \begin{bmatrix} \sum_{j=1}^{N_f} E_jA_j & -\sum_{j=1}^{N_f} E_jA_jy_j \\ -\sum_{j=1}^{N_f} E_jA_jy_j & \sum_{j=1}^{N_f} E_jA_j{y_j}^2 \end{bmatrix} \tag{9}$$

where E_j, A_j, and y_j are the elastic modulus, area, and centroid coordinate of the jth fiber, as shown in Figure 3; N_f represents the total number of fibers for a section.

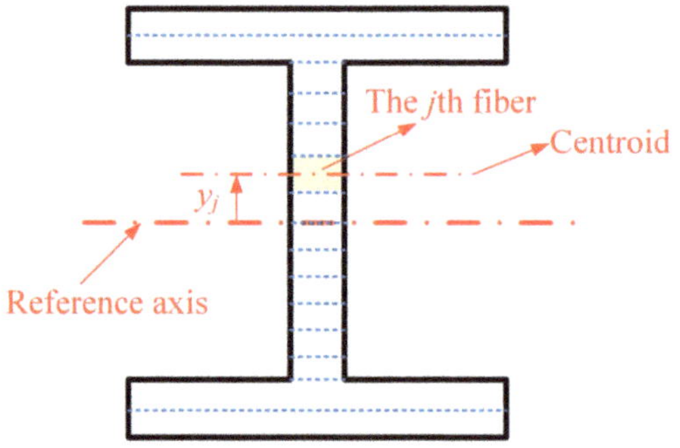

Figure 3. Fiber section.

Then, the element stiffness matrix in the local coordinate system can be calculated as:

$$\overline{\boldsymbol{k}}_e = \int_0^{L_e} B_d(x)^T\boldsymbol{k}_s(x)B_d(x)dx \tag{10}$$

where L_e is the element length; $B_d(x)$ is the differential of the displacement interpolation function with the expression of:

$$B_d(x) = \begin{bmatrix} -\frac{1}{L_e} & 0 & 0 & \frac{1}{L_e} & 0 & 0 \\ 0 & \frac{12x}{{L_e}^3} - \frac{6}{{L_e}^2} & \frac{6x}{{L_e}^2} - \frac{4}{L_e} & 0 & \frac{6}{{L_e}^2} - \frac{12x}{{L_e}^3} & \frac{6x}{{L_e}^2} - \frac{2}{L_e} \end{bmatrix} \tag{11}$$

It is impractical to integrate Equation (10) directly, so the Gauss–Legendre integration method is typically used to indirectly solve Equation (10). In this study, the Gauss–Legendre integration method with five integration points is used; then, Equation (10) can be rewritten as:

$$\overline{k}_e = \sum_{k=1}^{5} B_d(x_k)^T k_s(x_k) B_d(x_k) \omega_k L_e \quad (12)$$

where x_k and ω_k are the coordinate and weight at the kth integration point. The coordinates and weights of five integration points are listed in Table 1.

Table 1. Coordinates and weights of five integration points of the Gauss–Legendre integration method.

k	1	2	3	4	5
x_k	0.046910 L_e	0.230765 L_e	0.500000 L_e	0.769234 L_e	0.953090 L_e
ω_k	0.118463	0.239314	0.284444	0.239314	0.118463

As the initial stiffness is elastic, direct use of the element stiffness matrix of the elastic beam-column element without fiber sections is also applicable and simpler (Equation (13)). The fiber section is adopted to formulate the element stiffness in order to make it consistent with the state determination in Section 3.1.2. It should be noted that the element stiffness matrix shown in Equation (13) is a symmetric matrix:

$$\overline{k}_e = \begin{bmatrix} EA/L_e & 0 & 0 & -EA/L_e & 0 & 0 \\ 0 & 12EI/{L_e}^3 & 6EI/{L_e}^2 & 0 & -12EI/{L_e}^3 & 6EI/{L_e}^2 \\ 0 & 6EI/{L_e}^2 & 4EI/L_e & 0 & -6EI/{L_e}^2 & 2EI/L_e \\ -EA/L_e & 0 & 0 & EA/L_e & 0 & 0 \\ 0 & -12EI/{L_e}^3 & -6EI/{L_e}^2 & 0 & 12EI/{L_e}^3 & -6EI/{L_e}^2 \\ 0 & 6EI/{L_e}^2 & 2EI/L_e & 0 & -6EI/{L_e}^2 & 4EI/L_e \end{bmatrix} \quad (13)$$

where E is the elastic modulus of the material; A and I are the area and inertia moment of the cross-section, respectively. It should be noted that only flexural deformation (without shear deformation) is considered in Equation (13).

Then, the element stiffness matrix $\boldsymbol{k}_e$ in the global coordinate system can be transformed from the element stiffness matrix $\overline{\boldsymbol{k}}_e$ in the local coordinate system:

$$\boldsymbol{k}_e = \boldsymbol{T}_e^T \overline{\boldsymbol{k}}_e \boldsymbol{T}_e \quad (14)$$

where $\boldsymbol{T}_e$ is the coordinate transformation matrix. After calculating $\boldsymbol{k}_e$ for all elements, the initial stiffness $\mathbf{K}$ of the structure can finally be assembled by mapping the degrees of freedom (DOFs).

3.1.2. State Determination

State determination for using the stiffness-based beam-column element with fiber sections includes the following steps:

1. Structure level: Predict the displacement $\mathbf{X}_{i+1}$ at the (i + 1) time step using the GCR algorithms and obtain the incremental displacement $\Delta\mathbf{X}_{i+1} = \mathbf{X}_{i+1} - \mathbf{X}_i$.
2. Element level: Calculate the incremental element displacement $\Delta\mathbf{d}_{e,i+1}$ in the global coordinate system by mapping the DOFs and transform it into the incremental element displacement $\Delta\overline{\mathbf{d}}_{e,i+1} = \boldsymbol{T}_e \Delta\mathbf{d}_{e,i+1}$ in the local coordinate system.
3. Section level: The incremental section deformation $\Delta\mathbf{v}_{S,i}(x_k)$ at the integration points can be obtained as $\Delta\mathbf{v}_{S,i+1}(x_k) = B_d(x_k)\Delta\overline{\mathbf{d}}_{e,i+1} = [\Delta\varepsilon(x_k), \Delta\phi(x_k)]^T$, where $\Delta\varepsilon(x_k)$ and $\Delta\phi(x_k)$ are the incremental axial strain and curvature, respectively.
4. Fiber level: The incremental strain of the jth fiber is $\Delta\varepsilon_{j,i+1} = -\Delta\varepsilon(x_k) + \Delta\phi(x_k)y_j$. Then, the linear or nonlinear constitutive relationships of the material can be applied to obtain the stress $\sigma_{j,i+1}$ of the jth fiber.

5. Section level: The section internal force $\mathbf{S}_{i+1}(x_k)$ at the integration points can be assembled as $\mathbf{S}_{i+1}(x_k) = [N_{i+1}(x_k), M_{i+1}(x_k)]^T = \left[\sum_{j=1}^{N_f} \sigma_{j,i+1}A_j, \sum_{j=1}^{N_f} \sigma_{j,i+1}A_j y_j\right]^T$.
6. Element level: The element force $\bar{\mathbf{f}}_{e,i+1}$ in the local coordinate system can be integrated by the section forces at the integration points as $\bar{\mathbf{f}}_{e,i+1} = \sum_{k=1}^{5} B_d(x_k)^T \mathbf{S}_{i+1}(x_k)\omega_k L_e$. Then, the element force $\mathbf{f}_{e,i+1} = \boldsymbol{T}_e^T \bar{\mathbf{f}}_{e,i+1}$ in the global coordinate system can be obtained.
7. Structure level: The element forces of all elements can be assembled as the first-order restoring force $\mathbf{R}_{i+1}$.

3.2. P-Δ Effects

In this study, the P-Δ effects are taken into account with the lean-on column, which is subjected to the gravity of each floor. The lean-on column is simulated by several bar elements and connected to the moment-resisting frame by a rigid diaphragm at each floor. The discretization of the cross-section, i.e., fiber section, is not used to simulate the bar elements of the lean-on column, which is assumed to behave elastically. The second-order restoring force is the product of the structural geometric stiffness $\mathbf{K}_g$ and the structural displacement $\mathbf{X}_{i+1}$. The structural geometric stiffness $\mathbf{K}_g$ is the assembly of the geometric stiffness $\boldsymbol{k}_g$ of the bar elements, which is expressed as:

$$\boldsymbol{k}_g = \frac{P}{L_e}\begin{bmatrix} 1 & 0 & -1 & 0 \\ 0 & 1 & 0 & -1 \\ -1 & 0 & 1 & 0 \\ 0 & -1 & 0 & 1 \end{bmatrix} \tag{15}$$

where P is the gravity subjected to the lean-on column. It should be noted that the geometric stiffness is a symmetric matrix.

3.3. Time History Analysis of a Four-Story Frame Subjected to Earthquake

The primary structure is a four-story steel frame, as shown in Figure 4. The stiffness-based beam-column elements with fiber sections are used to simulate the beams and columns of the moment resisting frame, which is completely symmetric, and the bar elements are adopted to model the lean-on column. There are 24 fibers for each section of the beam-column elements. The elastic modulus and yield strength of the steel are 200 GPa and 345 MPa, respectively. The elastic–perfectly plastic constitutive relationship is adopted for steel. A consistent mass is used to build the mass matrix. The formulation of the initial stiffness matrix follows the procedure in Section 3.1. Shear deformation is not considered in the finite element model. According to the modal analysis, the first and second natural periods of the frame are 1.02 and 0.32 s, respectively. The Rayleigh damping assumption with a 2% damping ratio for the first and second order modes is applied to generate the damping matrix.

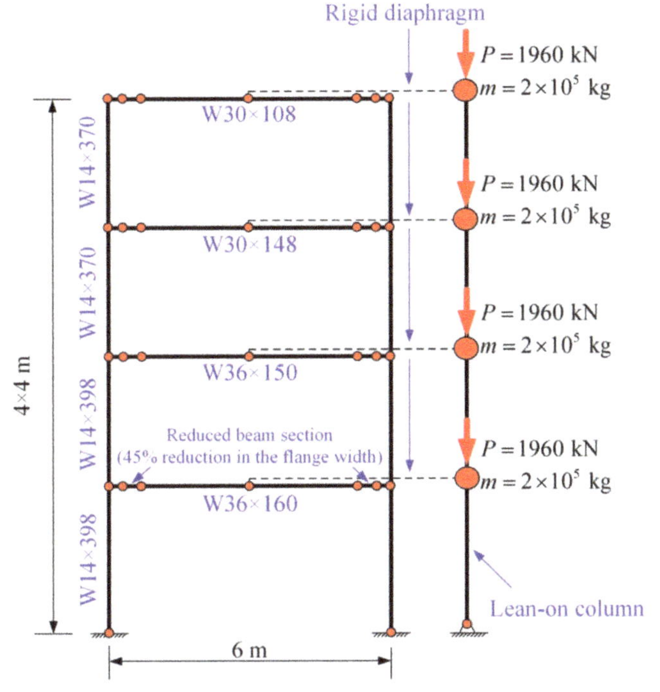

Figure 4. Four-story steel frame.

The structure is subjected to the unscaled 1940 El Centro NS ground motion. GCR algorithms with $\kappa_1 = 1/2, \kappa_2 = 1/4$ and two time steps of 0.01 and 0.001 s are used to conduct the time history analysis. The Newmark CAA algorithm with a time step of 0.001 s is taken for comparison. It should be noted that the GCR algorithms are explicit, whereas the CAA is implicit, so the computing efficiency of the GCR algorithms far exceeds that of the CAA algorithm with the same time step. If a larger time step is adopted, the efficiency of the GCR algorithm can be further improved. Figure 5 compares the lateral displacements of different stories using variant algorithm schemes.

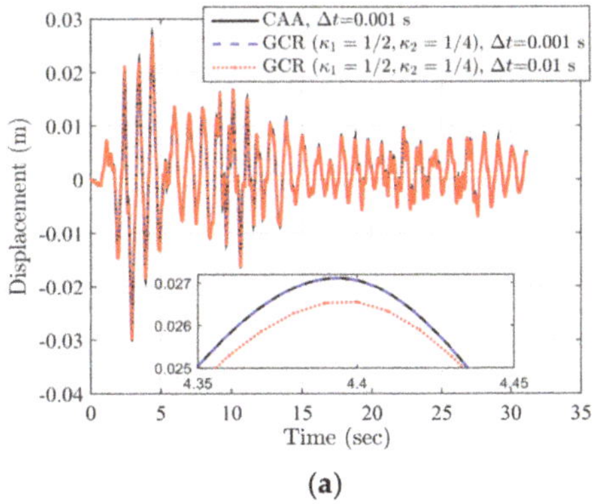

(a)

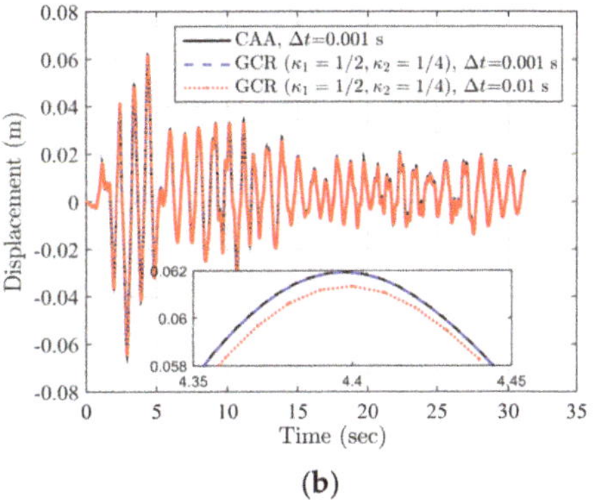

(b)

Figure 5. *Cont.*

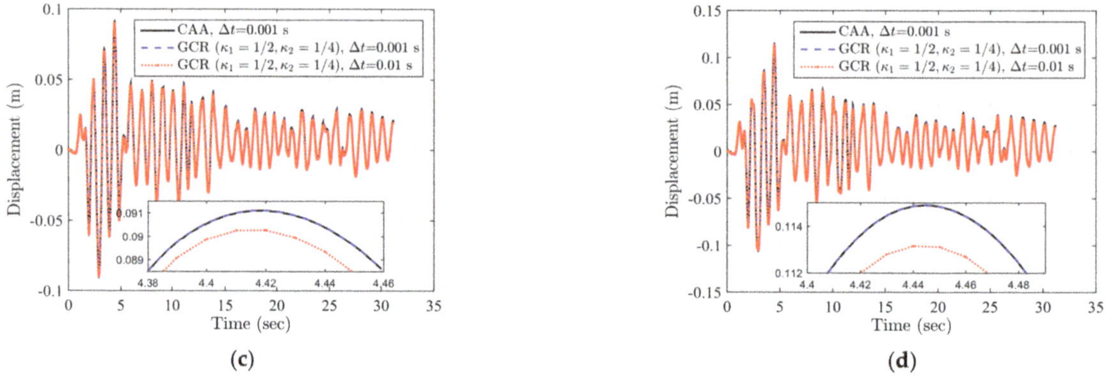

Figure 5. Time history curves of lateral displacements. (**a**) First story; (**b**) second story; (**c**) third story; (**d**) fourth story.

Figure 5 shows that the numerical results of the GCR algorithms match well with those of the CAA algorithm. Furthermore, two error indices are adopted to measure the errors between the reference model (CAA) and computing models (GCR):

$$\text{NEE} = \left| \frac{\sum_{i=1}^{N} x_{RM,i}^2 - \sum_{i=1}^{N} x_{CM,i}^2}{\sum_{i=1}^{N} x_{CM,i}^2} \right| \tag{16}$$

$$\text{NRMSE} = \frac{\sqrt{\sum_{i=1}^{N} \frac{(x_{RM,i} - x_{CM,i})^2}{N}}}{\max(x_{CM}) - \min(x_{CM})} \tag{17}$$

where x_{RM} and x_{CM} are the structural responses of the reference model and computing model, respectively; N is the sampling number. NEE and NRMSE are sensitive to the amplitude and frequency errors, respectively. Table 2 provides the corresponding error indices for Figure 5. It can be concluded from Table 2 that the differences between the GCR algorithms and the CAA algorithm with the same time step of 0.001 s are extremely small; even for the GCR algorithm with a larger time step of 0.01 s, the maximum NEE and NRMSE are less than 4% and 1%, respectively, which are acceptable in engineering practice. This indicates that GCR algorithms are viable for solving nonlinear dynamic problems with superior computational efficiency and accuracy.

Table 2. Error indices of lateral displacements using different integration algorithms (unit: %).

Integration Algorithm Scheme	Story	NEE	NRMSE
GCR ($\kappa_1 = 1/2, \kappa_2 = 1/4$), Δt = 0.001 s	1	1.09×10^{-2}	2.75×10^{-3}
	2	1.49×10^{-2}	3.26×10^{-3}
	3	1.66×10^{-2}	3.73×10^{-3}
	4	1.62×10^{-2}	3.95×10^{-3}
GCR ($\kappa_1 = 1/2, \kappa_2 = 1/4$), Δt = 0.01 s	1	2.33	0.64
	2	3.06	0.65
	3	3.54	0.73
	4	3.61	0.82

4. Analytical Models of Dampers

Four types of dampers, i.e., tuned mass dampers (TMDs), tuned liquid dampers (TLDs), particle dampers (PDs), and particle-tuned mass dampers (PTMDs), are selected to mitigate the seismic responses of the four-story steel frame. All four types of damper are installed on the top story and can be regarded as secondary structures to the primary structure, so they are ideal experimental substructures for SSTT. The four structure–damper systems are illustrated in Figure 6. The TMD can absorb the vibration energy at the tuned frequency and, therefore, reduce the structural response of the primary structure. The TLD is a tank containing liquid and can dissipate the vibration energy by liquid boundary layer friction, free surface contamination, and wave breaking. Similarly, the PD is a container equipped with particles that dissipate the energy by particle-to-particle and particle-to-container wall collision and friction. The PTMD is a combination of a TMD and PD, so it integrates the energy dissipation mechanisms of the two dampers.

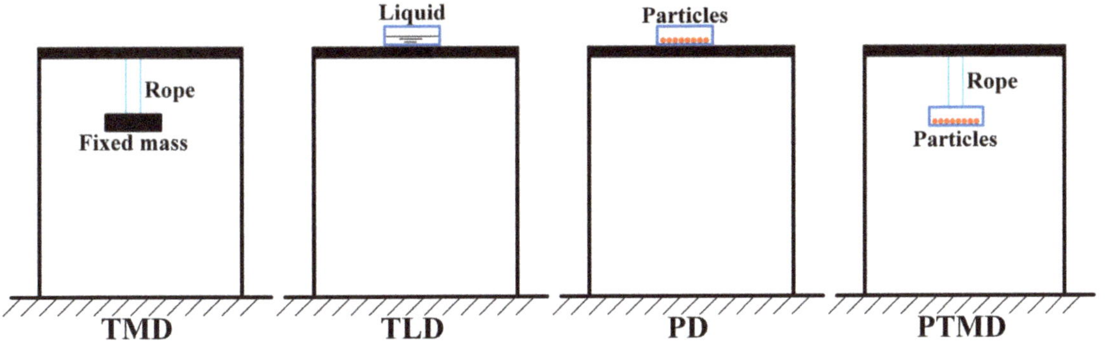

Figure 6. Primary structure–secondary structure (damper) systems.

4.1. *Tuned Mass Damper (TMD)*

The tuned mass damper (TMD) can be idealized as an SDOF system with mass, stiffness, and damping coefficients. To design a TMD, the mass ratio γ_M is the first parameter that should be determined:

$$\gamma_M = \frac{M_{TMD}}{M_s} \tag{18}$$

where M_{TMD} and M_s are the mass of the TMD and the primary structure, respectively. According to Section 3, the total mass of the primary structure is 8×10^5 kg. Four TMDs with various auxiliary mass ratios, i.e., 1%, 2%, 3%, and 4%, are adopted. The corresponding mass of the TMD can easily be calculated using Equation (18).

According to the classical optimizing parameters for TMD proposed by Den Hartog [32], the optimized frequency ratio γ_F and the damping ratio ξ_{TMD} can be expressed as the function of γ_M:

$$\gamma_F = \frac{f_{TMD}}{f_s} = \frac{1}{1+\gamma_M} \tag{19}$$

$$\xi_{TMD} = \sqrt{\frac{3\gamma_M}{8(1+\gamma_M)}} \tag{20}$$

where f_{TMD} and f_s are the frequency of the TMD and the primary structure, respectively. The first-order frequency of the primary structure ($f_1 = 0.98$ Hz) can be assigned to f_s. Other parameters of the TMD, such as stiffness $K_{TMD} = M_{TMD}(2\pi f_{TMD})^2$ and damping coefficient $C_{TMD} = 2M_{TMD}\xi_{TMD}(2\pi f_{TMD})$, can be determined and are listed in Table 3.

Table 3. Parameters of dampers with different mass ratios (mass unit: kg; stiffness unit: N/m; damping coefficient unit: N.s/m).

Auxiliary Mass Ratio	TMD	TLD	PD	PTMD
1%	$M_{TMD} = 8 \times 10^3$ $K_{TMD} = 2.95 \times 10^5$ $C_{TMD} = 5.92 \times 10^3$	$M_{TLD} = 8 \times 10^3$ $K_{TLD} = 2.05 \times 10^5\ \kappa$ $C_{TLD} = 8.11 \times 10^4$ $\xi_{TLD}\sqrt{\kappa}$	$M_{PD} = 8 \times 10^3$ $K_{PD} = 1.26 \times 10^8$ $C_{PD} = 4.02 \times 10^6$	$M_{TMD} = 1.6 \times 10^3, M_{PD} = 6.4 \times 10^3$ $K_{TMD} = 5.91 \times 10^4, K_{PD} = 1.01 \times 10^8$ $C_{TMD} = 1.18 \times 10^3, C_{PD} = 3.22 \times 10^5$
2%	$M_{TMD} = 1.6 \times 10^4$ $K_{TMD} = 5.79 \times 10^5$ $C_{TMD} = 1.65 \times 10^4$	$M_{TLD} = 1.6 \times 10^4$ $K_{TLD} = 4.11 \times 10^5\ \kappa$ $C_{TLD} = 1.62 \times 10^5$ $\xi_{TLD}\sqrt{\kappa}$	$M_{PD} = 1.6 \times 10^4$ $K_{PD} = 2.53 \times 10^8$ $C_{PD} = 8.04 \times 10^6$	$M_{TMD} = 3.2 \times 10^3, M_{PD} = 1.28 \times 10^4$ $K_{TMD} = 1.16 \times 10^5, K_{PD} = 2.02 \times 10^8$ $C_{TMD} = 3.30 \times 10^3, C_{PD} = 6.43 \times 10^5$
3%	$M_{TMD} = 2.4 \times 10^4$ $K_{TMD} = 8.52 \times 10^5$ $C_{TMD} = 2.99 \times 10^4$	$M_{TLD} = 2.4 \times 10^4$ $K_{TLD} = 6.16 \times 10^5\ \kappa$ $C_{TLD} = 2.43 \times 10^5$ $\xi_{TLD}\sqrt{\kappa}$	$M_{PD} = 2.4 \times 10^4$ $K_{PD} = 3.79 \times 10^8$ $C_{PD} = 1.21 \times 10^6$	$M_{TMD} = 4.8 \times 10^3, M_{PD} = 1.92 \times 10^4$ $K_{TMD} = 1.70 \times 10^5, K_{PD} = 3.03 \times 10^8$ $C_{TMD} = 5.98 \times 10^3, C_{PD} = 9.65 \times 10^5$
4%	$M_{TMD} = 3.2 \times 10^4$ $K_{TMD} = 1.11 \times 10^6$ $C_{TMD} = 4.54 \times 10^4$	$M_{TLD} = 3.2 \times 10^4$ $K_{TLD} = 8.21 \times 10^5\ \kappa$ $C_{TLD} = 3.24 \times 10^5$ $\xi_{TLD}\sqrt{\kappa}$	$M_{PD} = 3.2 \times 10^4$ $K_{PD} = 5.05 \times 10^8$ $C_{PD} = 1.61 \times 10^6$	$M_{TMD} = 6.4 \times 10^3, M_{PD} = 2.56 \times 10^3$ $K_{TMD} = 2.23 \times 10^5, K_{PD} = 4.04 \times 10^8$ $C_{TMD} = 9.07 \times 10^3, C_{PD} = 1.29 \times 10^6$

4.2. Tuned Liquid Damper (TLD)

There are several existing analytical or numerical models for TLD. According to [33], the TLD model developed by Yu et al. [34] has good predictions in both weak and strong wave breaking and in a broad range of frequency ratios. Therefore, Yu et al.'s model is adopted in this study.

Yu et al.'s model is an equivalent nonlinear-stiffness–damping (NSD) model of the TLD. The structural parameters of the TLD model are summarized as follows.

1. Mass: Assume the mass M_{TLD} of TLD equals the mass of liquid; that is, the mass of the container is neglected. Water is typically used as the liquid, so M_{TLD} can be calculated as $M_{TLD} = M_w = \rho_w BLh$, where ρ_w is the density of water; B and L are the width and length (the excitation direction) of the rectangular tank, respectively; h is the water depth.
2. Stiffness: The initial linear stiffness of TLD is $K_w = M_{TLD}(2\pi f_w)^2$, where f_w is the linear fundamental natural frequency for water and expressed as:

$$f_w = \frac{\sqrt{\frac{\pi g}{L}\tanh\left(\frac{\pi h}{L}\right)}}{2\pi} \tag{21}$$

where g is the gravitational constant. According to Yu et al. [34], the nonlinear stiffness K_{TLD} of TLD can be determined by the stiffness hardening ratio of κ and K_w:

$$\kappa = \frac{K_{TLD}}{K_w} = \left(\frac{f_{TLD}}{f_w}\right)^2 = \begin{cases} 1.075\Lambda^{0.007}, \text{for } \Lambda \leq 0.03 \text{ weak wave breaking} \\ 2.52\Lambda^{0.25}, \text{ for } \Lambda > 0.03 \text{ strong wave breaking} \end{cases} \tag{22}$$

$$K_{TLD} = M_{TLD}\kappa\frac{\pi g}{L}\tanh\left(\frac{\pi h}{L}\right) \tag{23}$$

where $\Lambda = A/L$ is the non-dimensional displacement amplitude, and A is the displacement amplitude of excitation.

3. Damping: The damping coefficient of the TLD can be obtained as $C_{TLD} = 2M_{TLD}\xi_{TLD}(2\pi f_{TLD})$, where the damping ratio ξ_{TLD} is also a function of Λ:

$$\xi_{TLD} = 0.5\Lambda^{0.35} \tag{24}$$

$$C_{TLD} = 2M_{TLD}\xi_{TLD}\sqrt{\kappa\frac{\pi g}{L}\tanh\left(\frac{\pi h}{L}\right)} \tag{25}$$

To design a TLD, the first parameter is also the mass ratio γ_M, which is the ratio of M_{TLD} to M_s. In the same manner as for the TMDs, four TLDs with various auxiliary mass ratios, i.e., 1%, 2%, 3%, and 4%, are adopted. The tank length L is assigned as a constant: 1 m. To achieve the maximum effectiveness of the TLD, the nonlinear natural frequency f_{TLD} of the TLD should be equal to the natural frequency f_s of the primary structure. Then, the water depth h is determined as:

$$h = \frac{L}{\pi}\tanh^{-1}\left(\frac{4\pi L f_s^2}{g\kappa}\right) \tag{26}$$

Based on the numerical results presented in Section 3, the displacement amplitude A of the fourth story is 0.1149 m; the non-dimensional displacement amplitude Λ and the corresponding stiffness hardening ratio κ can be calculated. By substituting the values of L, κ, and f_s into Equation (26), the water depth h is determined to be 0.3821 m. Therefore, the mass for a certain TLD is a constant, whereas the stiffness and damping are displacement amplitude-related variables and should be updated during time-history analysis. The parameters of the four TLDs with different mass ratios are also listed in Table 3.

4.3. Particle Damper (PD)

The conventional computing model of the PD uses the discrete element method (DEM), which is very time consuming and complicated. Based on correlational studies by Papalou and Masri [35], Lu et al. [3] proposed a simplified analytical method for the PD. The simplified model has been verified to have computational efficiency and a satisfactory degree of accuracy in practical applications. Therefore, Lu et al.'s model is adopted in this study.

The essence of the model is to transform multiple particles into an equivalent single particle, as shown in Figure 7. The mass of the single particle equals the total mass of the multiple particles, assuming that the collisions between the particles can be neglected; then, the damping forces of the PD mainly originate from the collisions between the particles and the container wall.

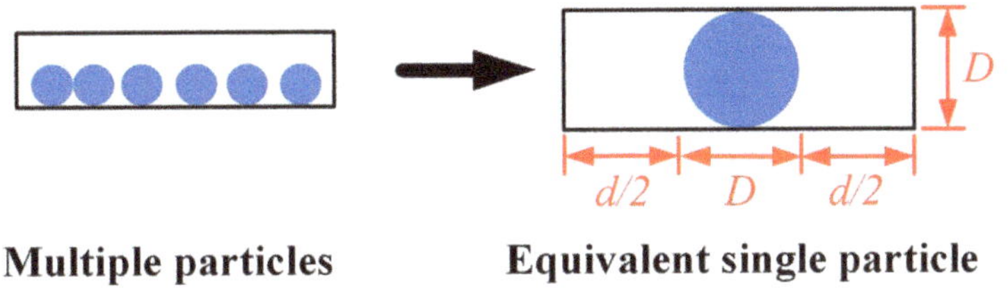

Figure 7. Equivalent single particle of multiple particles (adapted from [3]).

In Figure 7, the clearance d of the single particle is a critical parameter governing the damping force and can be determined by the following equation:

$$\left(\frac{1}{\rho_{PD}} - 1\right)\frac{M_{PD}}{\rho} = \frac{M_{PD}}{2\rho} + \frac{\pi}{4}\left(\frac{6M_{PD}}{\pi\rho}\right)^{\frac{2}{3}} d \tag{27}$$

where M_{PD} is the mass of the PD; ρ is the density of the particle material; ρ_{PD} is the packing density of the multiple particles. Hales et al. [36] suggested that ρ_{PD} should not exceed 0.74; a value of 0.6 is assigned to ρ_{PD}.

The mass M_{PD} of PD is the product of the mass ratio γ_M and the mass M_s of the primary structure. Four mass ratios ranging from 1% to 4% are selected. The stiffness K_{PD} of the equivalent single PD is determined as $K_{PD} = M_{PD}(2\pi f_{PD})^2$. Masri and Ibrahim [37] recommended that $f_{PD} \geq 20f_1$, so the frequency of PD is set as $20f_1 \approx 20$ Hz. The damping coefficient C_{PD} of the equivalent single PD can be written as $C_{PD} = 2M_{PD}\xi_{PD}(2\pi f_{PD})$, where the damping ratio ξ_{PD} is related to the coefficient of restitution e [38]. In this study, a steel particle with $e = 0.5$ is adopted; thus, ξ_{PD} is determined to be 0.2. The damping force of the single PD can be expressed as:

$$f_{PD} = C_{PD}H\left(x_{PD}^r, \dot{x}_{PD}^r\right) + K_{PD}G(x_{PD}^r) \tag{28}$$

where x_{PD}^r and $\dot{x}_{PD}^r$ are the relative displacement and velocity of the PD with respect to the primary structure. $H\left(x_{PD}^r, \dot{x}_{PD}^r\right)$ and $G(x_{PD}^r)$ are two nonlinear functions with the expressions of:

$$H\left(x_{PD}^r, \dot{x}_{PD}^r\right) = \begin{cases} \dot{x}_{PD}^r, for\ x_{PD}^r \leq -d/2\ and\ x_{PD}^r \geq -d/2 \\ 0, for - d/2 < x_{PD}^r < d/2 \end{cases} \tag{29}$$

$$G(x_{PD}^r) = \begin{cases} x_{PD}^r + d/2, \text{for } x_{PD}^r \leq -d/2 \\ 0, \text{for } - d/2 < x_{PD}^r < d/2 \\ x_{PD}^r - d/2, \text{for } x_{PD}^r \geq d/2 \end{cases} \tag{30}$$

4.4. Particle-Tuned Mass Damper (PTMD)

The particle-tuned mass damper (PTMD) is the combination of TMD and PD. The simplified analytical model proposed by Lu et al. [3] is also adopted. The PTMD can be idealized as a 2DOF system, which includes an SDOF TMD connected to the primary structure and an SDOF PD attached to the TMD. For the PTMD, the mass of the container cannot be neglected because it constitutes the TMD. Therefore, the total auxiliary mass of the PTMD is divided into two parts: PD and TMD. Lu et al. [3] investigated the influence of the ratio of the particle mass (M_{PD}) to the total auxiliary mass (M_{PTMD}) on the structural control effects of PTMD and found that the vibration attenuation of the PTMD can be improved to a certain extent by increasing the mass proportion of the PD. Therefore, an 80% mass ratio of M_{PD} to M_{PTMD} is used. Four PTMDs with varying auxiliary mass ratios from 1% to 4% are selected. The parameters of the TMD and PD are determined following the procedures presented in Sections 4.1 and 4.3, respectively. The corresponding parameters of PTMDs can be found in Table 3.

5. Modeling of Substructure Shake Table Testing of Frame Structure–Damper System

5.1. Procedure of Substructure Shake Table Testing of Frame Structure–Damper System

To carry out SSTT of the frame structure–damper system, the primary structure (frame) and the secondary structure (damper) are assigned as the analytical and experimental substructures, respectively. The analytical substructure is numerically simulated, and the damper is mounted on and excited by the shake table [15,16]. A schematic diagram of SSTT of the frame structure–damper system is illustrated in Figure 8.

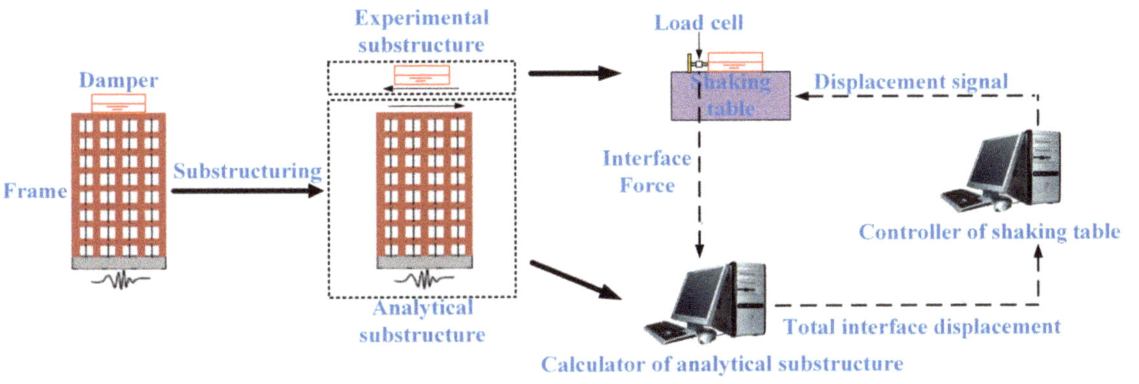

Figure 8. Schematic diagram of shake table testing of the frame structure–damper system.

Figure 9 shows a flowchart of SSTT of the frame structure–damper system using GCR algorithms and the finite element method.

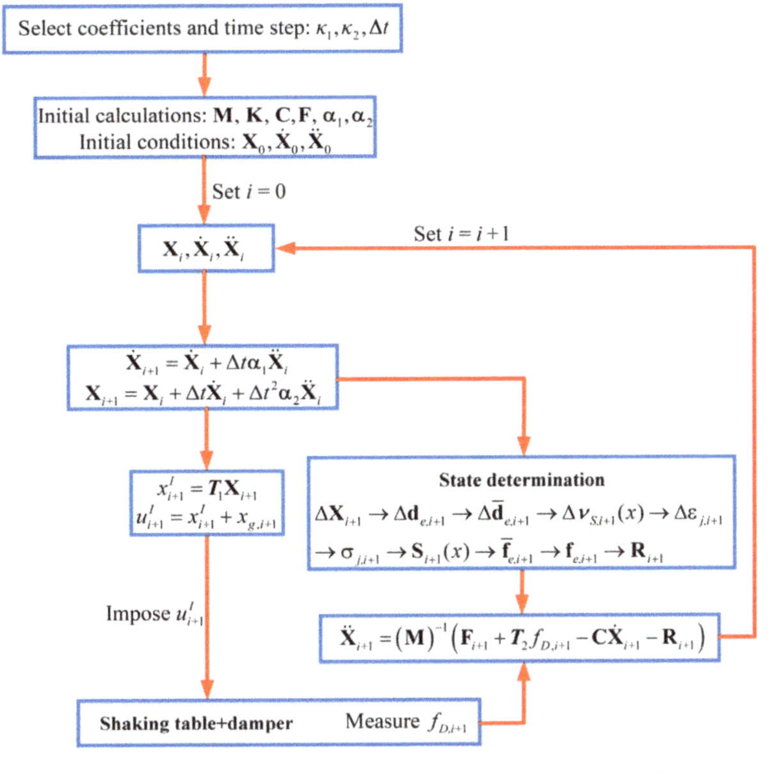

Figure 9. Flowchart of substructure shake table testing of the frame structure–damper system using GCR algorithms and the finite element method.

In Figures 8 and 9, it should be noted that the excitation signal of the shake table is the total displacement at the interface instead of the relative displacement. Therefore, two additional calculation steps are required:

$$x_{i+1}^I = \boldsymbol{T}_1 \mathbf{X}_{i+1} \tag{31}$$

$$u_{i+1}^I = x_{i+1}^I + x_{g,i+1} \tag{32}$$

where x_{i+1}^I and u_{i+1}^I are the relative and total displacements at the interface, respectively; $x_{g,i+1}$ is the ground displacement; $\boldsymbol{T}_1$ is the matrix transforming the DOFs of the structure to the interface DOF.

In addition, the EOM of the primary structure should be modified due to introduction of the damper force (interface force):

$$\mathbf{M}\ddot{\mathbf{X}}_{i+1} + \mathbf{C}\dot{\mathbf{X}}_{i+1} + \mathbf{R}_{i+1} = \mathbf{F}_{i+1} + \boldsymbol{T}_2 f_{D,i+1} \tag{33}$$

where $f_{D,i+1}$ is the damper force; $\boldsymbol{T}_2$ is the matrix converting the interface DOF to the DOFs of the structure. This study aimed to conduct a series of virtual SSTTs, so the damper forces of the four types of dampers are numerically simulated instead of experimentally measured in a real SSTT. All numerical simulations were performed using MATLAB software and the Simulink toolbox.

5.2. *Numerical Results and Discussions*

It can be seen from Figures 8 and 9 that the SSTT system is composed of four parts: the experimental substructure (damper), the analytical substructure (frame), the integration algorithm, and the shake table. Therefore, all four components influence the results of the SSTT system. The four-story frame in Section 3.3 is taken as the analytical substructure and remains unchanged. The four types of damper in Section 4 with different auxiliary mass ratios are taken as the experimental substructures. Therefore, the effects of the auxiliary mass ratio are investigated first. As discussed in Section 2, the integration parameters κ_1 and κ_2 greatly influence the numerical properties of the GCR algorithms. Thus, the effects of different sets of integration parameters are considered. Similarly, the time step is a critical factor determining the accuracy of the integration algorithm, and thus, is also studied. Finally, the time delay, which is a critical property of the shake table dynamics, is also taken into account.

5.2.1. Effects of the Auxiliary Mass Ratio

GCR algorithms with $\kappa_1 = 1/2, \kappa_2 = 1/4$ and a time step of 0.001 s are adopted to solve the virtual SSTT of the four-story steel frame (Section 3) with four types of damper (Section 4). Figure 10 shows the lateral displacements at the fourth story for the steel frame, in addition to four dampers with four auxiliary mass ratios. The seismic responses of the uncontrolled structure without dampers are used for comparison. The corresponding damper forces are provided in Figure 11. Two widely used reduction factors based on the maximum and root-mean-square (RMS) structural responses are used to quantify the reduction effects of dampers:

$$R_{\max} = 1 - \frac{\max|x_{controlled}|}{\max|x_{uncontrolled}|} \tag{34}$$

$$R_{\mathrm{RMS}} = 1 - \frac{\mathrm{RMS}(x_{controlled})}{\mathrm{RMS}(x_{uncontrolled})} \tag{35}$$

where $x_{controlled}$ and $x_{uncontrolled}$ are the structural responses of the controlled structure with dampers and the uncontrolled structure without dampers, respectively. The two reduction effects of the different dampers with varying mass ratios are compared in Figure 12.

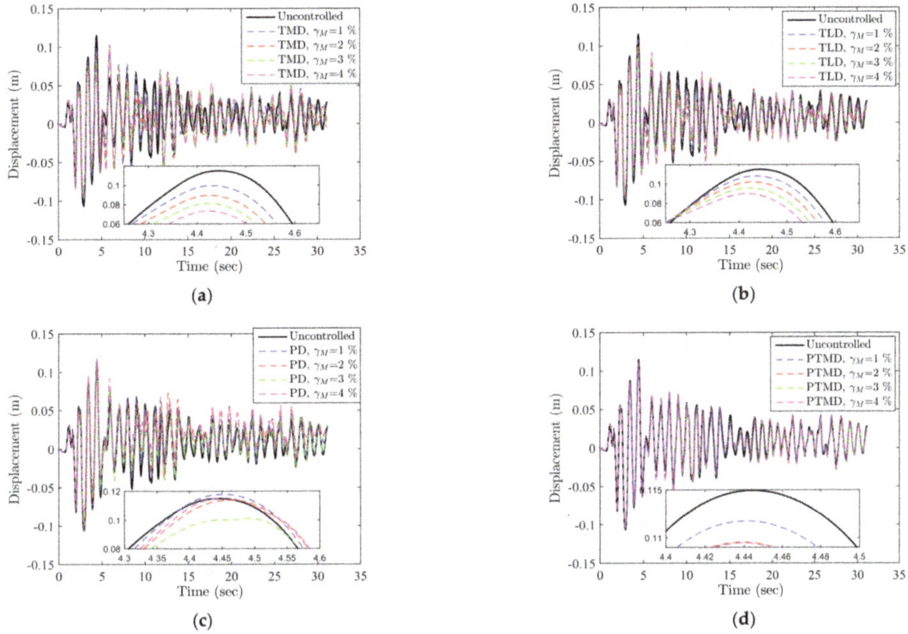

Figure 10. Comparisons of lateral displacements with different auxiliary mass ratios. (**a**) TMD; (**b**) TLD; (**c**) PD; (**d**) PTMD.

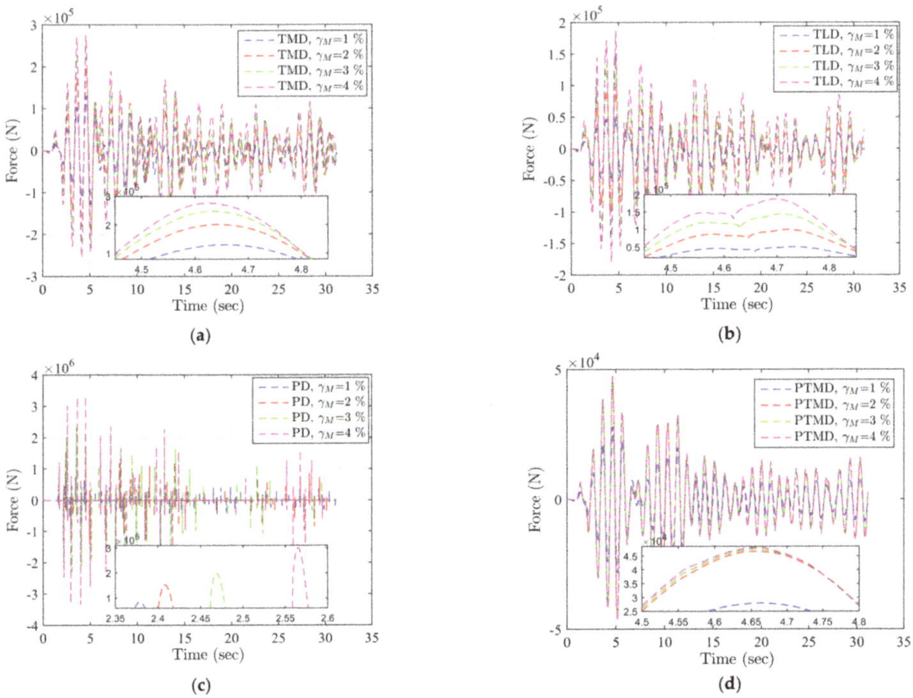

Figure 11. Time–history curves of damper forces. (**a**) TMD; (**b**) TLD; (**c**) PD; (**d**) PTMD.

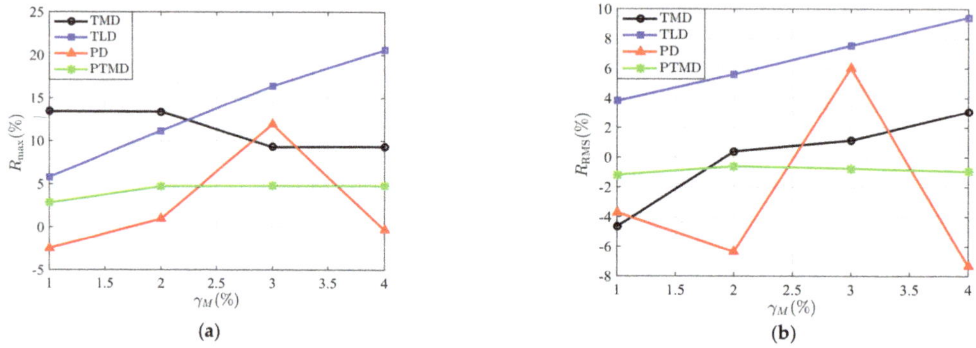

Figure 12. Reduction factors with respect to auxiliary mass ratios for four types of dampers. (**a**) R_{max}; (**b**) R_{RMS}.

Figures 10 and 12 show that the TLD has the best performance in vibration control among the four dampers. Both the reduction factors of the TLD linearly increase with respect to the mass ratio. This result is consistent with the shake table results of structure–damper systems, such as [3,16]. Most TMDs possess positive vibration reduction effects, with the exception of R_{RMS} of the TMD, with a mass ratio of 1%. The R_{RMS} of the TMD increases with the increase in mass ratio. According to [6], the mass ratio of the TMD can only reach 0.1%–5% due to installation difficulties and economic costs. These mass ratios are much smaller than the optimal mass ratio, so the limited tuned mass cannot effectively reduce the structural vibrations. This explains the fair performance of the TMDs. A PD with a mass ratio of 3% has a satisfactory performance in terms of both R_{max} and R_{RMS}, whereas it does not provide a positive control effect for other mass ratios. The mass ratio has negligible influence on the reduction effects of the PTMD. The R_{max} of the PTMDs is positive, whereas the R_{RMS} is negative. Figure 11 shows that the damper forces have a positive correlation with the mass ratios. In addition, the damper forces of the PDs have a magnitude of 10^6 N, which is much larger than those of the counterpart TLDs. This indicates that a larger damper force cannot ensure better control performance. In sum, for the particular structure and ground motion in this study, these dampers are not always effective at controlling the seismic responses of the four-story steel frame. According to previous studies [3,6–10], the control performance of dampers is influenced by many factors, such as the dynamic characteristics of the primary structure, the frequency characteristics and intensities of the seismic inputs, and the damper parameters. Therefore, the conclusions drawn from Figures 10–12 are specific and not generalizable.

5.2.2. Effects of the Integration Parameters of GCR Algorithms

As mentioned in Section 2, the integration parameters, i.e., κ_1 and κ_2, of the GCR algorithms may greatly influence the numerical properties of the algorithms. Therefore, the influences of the two integration coefficients on SSTT are investigated. GCR algorithms with four sets of $[\kappa_1, \kappa_2] = [1/2, 1/2], [1/2, 1], [1, 1/2], [1, 1]$ are selected, whereas GCR algorithms with $\kappa_1 = 1/2, \kappa_2 = 1/4$ are used as the reference model. A time step of 0.001 s is adopted for the integration algorithms. The structural responses of the steel frame with four dampers with a mass ratio of 1% are considered. Figure 13 depicts the lateral displacement at the fourth story using GCR algorithms with different sets of integration coefficients. Table 4 further provides the error indices of the structural responses.

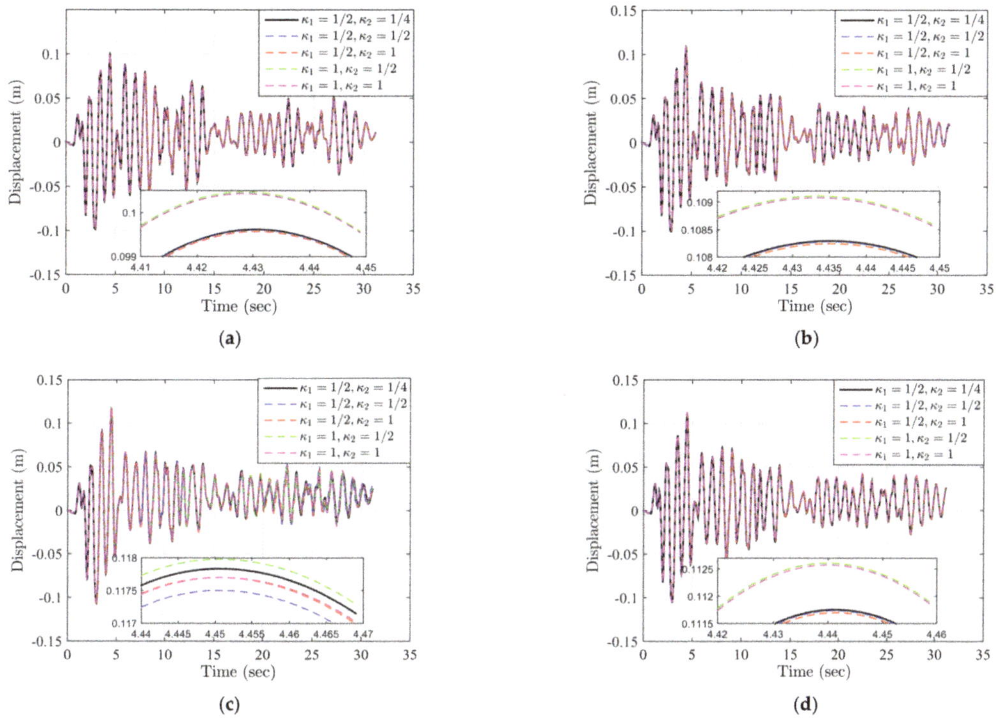

Figure 13. Effects of the integration parameters on structural responses. (**a**) TMD; (**b**) TLD; (**c**) PD; (**d**) PTMD.

Table 4. Error indices of top lateral displacements with different integration parameters (unit: %).

Integration Parameters	Damper	NEE	NRMSE
$\kappa_1 = 1/2, \kappa_2 = 1/2$	TMD	2.68×10^{-2}	6.80×10^{-3}
	TLD	3.86×10^{-2}	9.19×10^{-3}
	PD	1.48	1.77
	PTMD	1.05×10^{-2}	1.66×10^{-2}
$\kappa_1 = 1/2, \kappa_2 = 1$	TMD	7.98×10^{-2}	2.03×10^{-2}
	TLD	0.12	2.76×10^{-2}
	PD	2.49	1.31
	PTMD	1.50×10^{-2}	4.92×10^{-2}
$\kappa_1 = 1, \kappa_2 = 1/2$	TMD	0.76	0.35
	TLD	0.30	0.37
	PD	1.26	0.83
	PTMD	0.45	0.52
$\kappa_1 = 1, \kappa_2 = 1$	TMD	0.80	0.35
	TLD	0.37	0.36
	PD	0.43	0.81
	PTMD	0.46	0.53

Figure 13 and Table 4 show that the error indices of the PDs are generally larger than their counterpart TMDs, TLDs, and PTMDs, particularly when $\kappa_1 = 1/2, \kappa_2 = 1/2$ and $\kappa_1 = 1/2, \kappa_2 = 1$. In addition, the error indices of $\kappa_1 = 1$ normally exceed those of $\kappa_1 = 1/2$ because there exists numerical damping for GCR algorithms with $\kappa_1 = 1$, as shown in Figure 1. However, all the error indices are relatively small (the maximum

NEE and NRMSE are less than 3%). This can be explained as follows: the fundamental frequency of the frame structure f_1 is 0.98 Hz, thus the corresponding $\Omega = 2\pi f_1 \Delta t = 0.006$. As the value of Ω is small, the PE and equivalent damping ratio, as indicated in Figure 1, of the integration algorithm is very small. For instance, when $\kappa_1 = 1, \kappa_2 = 1$, the PE and equivalent damping ratios of the GCR algorithms are only 1.2×10^{-5} and 0.0015, respectively. The relatively low error of the integration algorithm leads to small errors in the SSTT results. Therefore, the influences of the integration parameters of the GCR algorithms on SSTT of the frame structure with dampers can be neglected. It should be noted, however, that if a larger time step and a stiffer structure with a larger frequency are adopted, the influences of the integration parameters may be significant.

5.2.3. Effects of the Time Step

The time step is a crucial factor of the integration algorithms and SSTT. A larger time step means saving more computing time while reducing accuracy. It is well known that SSTT requires high computational efficiency for integration algorithms. If an integration algorithm maintains a relatively high accuracy for a larger time step, it is definitely a promising choice for the application of SSTT. Therefore, we studied the influence of the time steps on SSTT. GCR algorithms with $\kappa_1 = 1/2, \kappa_2 = 1/4$ and four time steps ($\Delta t = 0.002$ s, 0.005 s, 0.01s, 0.02 s) are adopted. The results for GCR algorithms with $\kappa_1 = 1/2, \kappa_2 = 1/4$ and a time step of $\Delta t = 0.001$ s are used for comparison. The lateral displacements at the top story and the steel frame attached to the four types of dampers with a mass ratio of 1% are provided in Figure 14. The corresponding error indices are tabulated in Table 5.

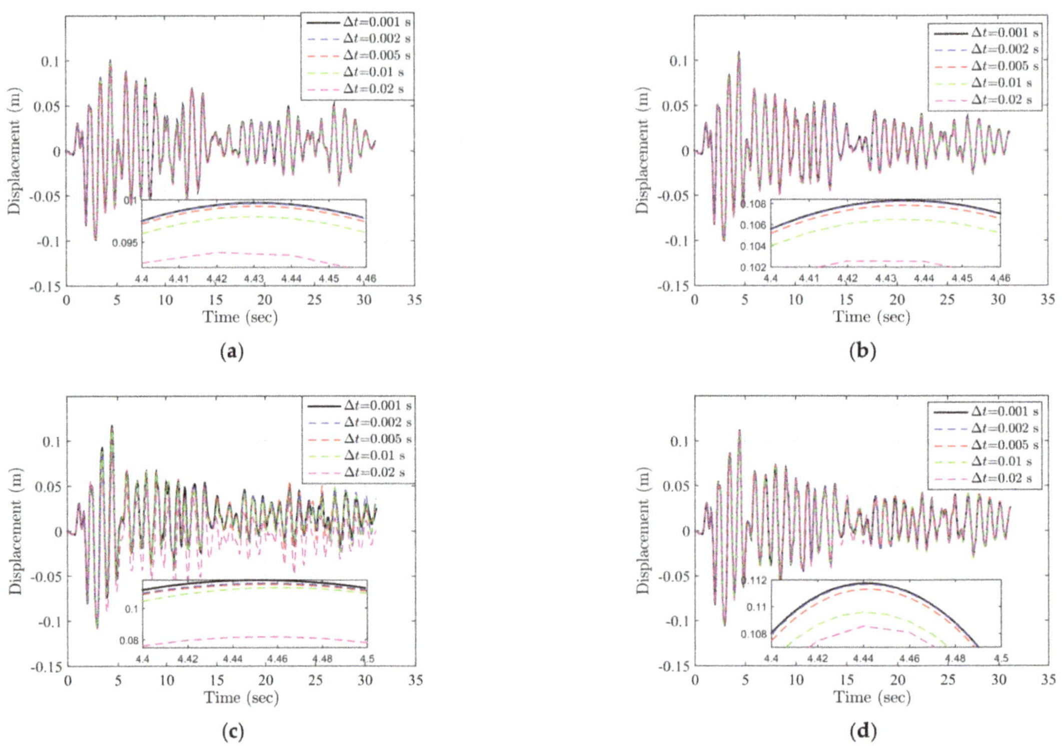

Figure 14. Effects of the time step on structural responses. (**a**) TMD; (**b**) TLD; (**c**) PD; (**d**) PTMD.

Table 5. Error indices of top lateral displacements with different time steps (unit: %).

Time Step	Damper	NEE	NRMSE
$\Delta t = 0.002$ s	TMD	7.14×10^{-2}	1.74×10^{-2}
	TLD	9.07×10^{-2}	2.20×10^{-2}
	PD	0.50	1.48
	PTMD	0.29	0.15
$\Delta t = 0.005$ s	TMD	0.55	0.13
	TLD	0.70	0.17
	PD	3.46	2.02
	PTMD	0.81	0.67
$\Delta t = 0.01$ s	TMD	1.82	0.43
	TLD	2.36	0.60
	PD	9.16	2.07
	PTMD	5.22	1.55
$\Delta t = 0.02$ s	TMD	4.89	1.25
	TLD	3.68	1.09
	PD	28.62	13.16
	PTMD	10.10	2.10

Figure 14 and Table 5 indicate that with the increase in the time step, the error indices for all damper cases increase. Regarding the TMDs and TLDs, even when the time step is very large, i.e., $\Delta t = 0.02$ s, the error indices are less than 5%. However, for the PDs and PTMDs, the error indices for $\Delta t = 0.02$ s are relatively large and almost reach 30%. The TMD and PD cases have the smallest and largest errors, respectively. The error indices of the PTMD cases are between those of the TMD cases and the PD cases, because the PTMD is a combination of the TMD and PD. A possible reason for the large errors of the PD cases is that the impulsive force induced by the PD has a negative impact on the integration algorithms.

5.2.4. Effects of the Time Delay

The time delay induced by the dynamics of the shake table is a significant factor when conducting SSTT. According to previous studies [39], the time delay can be regarded as having a negative damping effect because it introduces additional energy into the SSTT system. GCR algorithms with $\kappa_1 = 1/2, \kappa_2 = 1/4$ and a time step of $\Delta t = 0.01$ s are used to solve the EOM. Four time delays ($\tau = 0.01$ s, 0.02 s, 0.05s, 0.1 s) are considered. Figure 15 presents the lateral displacements of the controlled structure with a mass ratio of 1% for different levels of time delay. The errors are calculated and listed in Table 6.

Figure 15 and Table 6 show that the time delay has a negative influence on the structural responses. Essentially, the errors increase with the increase in time delay. For $\tau = 0.01$ s, 0.02 s, 0.05 s, the PD cases have the largest errors. The TMD case has relatively large errors close to 20% when $\tau = 0.1$ s. The errors of the TLD and PTMD cases are less than 7% for all time delays. This means that the influence of the time delay on SSTT of the frame structure with the TLD and PTMD is less significant. However, if the time delay continues to increase, the negative impact will undoubtedly be enhanced and delay compensation techniques [40–42] will be required to eliminate any adverse effects.

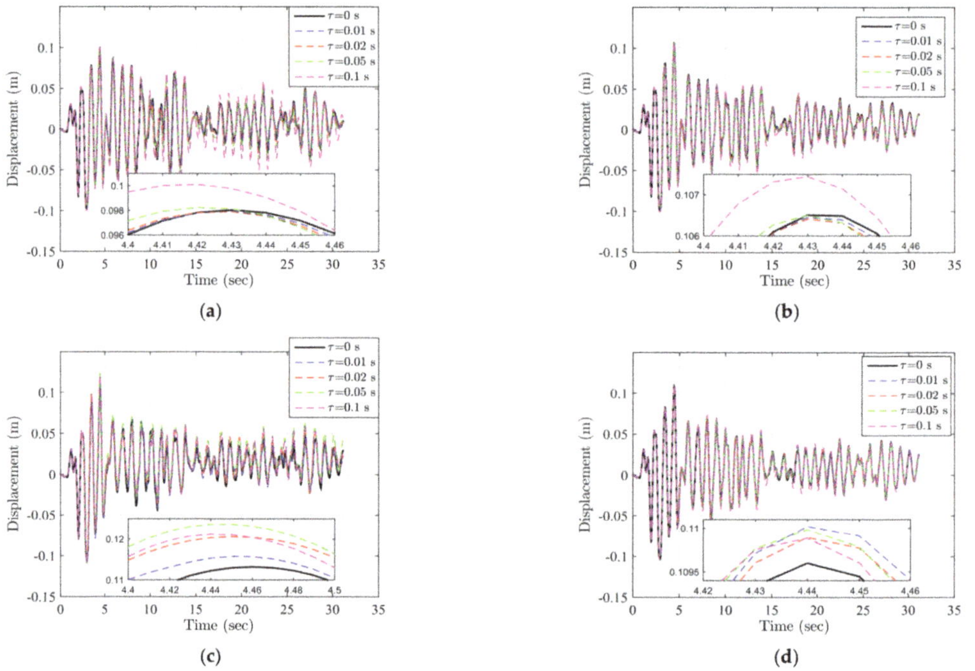

Figure 15. Effects of the time delay on structural responses. (**a**) TMD; (**b**) TLD; (**c**) PD; (**d**) PTMD.

Table 6. Error indices of top lateral displacements with different time delays (unit: %).

Time Delay	Dampers	NEE	NRMSE
$\tau = 0.01$ s	TMD	0.28	0.51
	TLD	5.55×10^{-2}	0.26
	PD	3.88	1.93
	PTMD	1.70	0.65
$\tau = 0.02$ s	TMD	0.94	1.04
	TLD	0.23	0.51
	PD	10.98	2.48
	PTMD	0.32	0.40
$\tau = 0.05$ s	TMD	5.08	2.71
	TLD	1.51	1.25
	PD	24.42	3.96
	PTMD	1.83	0.62
$\tau = 0.1$ s	TMD	19.68	6.26
	TLD	5.32	2.41
	PD	12.63	3.48
	PTMD	6.62	2.00

6. Conclusions

In this study, a series of virtual SSTTs on frame–damper systems were conducted using GCR algorithms and stiffness-based beam-column elements with fiber sections. Four types of secondary structure-type dampers were adopted to attenuate the seismic vibration of the primary structure. The effects of the mass ratio, integration parameters of the GCR algorithms, the time step, and time delay on the SSTT results were studied. Some important conclusions are summarized as follows:

1. The GCR algorithms can provide accurate numerical results, even when the time step is relatively large. Compared with the traditional implicit CAA algorithm, no iteration is required for the GCR algorithms to determine the restoring force, which can save a considerable amount of computational time.
2. For the specific structure and ground motion in this study, the TLD has the best performance in structural control and its control effects are enhanced by increasing the mass ratio, whereas for the other three types of damper, they are not always effective at controlling the vibration induced by an earthquake. However, the above conclusions are not generalizable and may not be correct for other structures with different dynamic characteristics and ground motions exhibiting different energy contents.
3. When the time step is 0.001 s, the GCR algorithms with four typical sets of integration parameters can provide satisfactory SSTT results, because the PE and equivalent damping ratio of the integration algorithm are very small. Therefore, the integration parameters of the GCR algorithms have negligible effects on the SSTT results. It should be noted, however, that if a larger time step and a stiffer structure with larger frequency are adopted, the influences of the integration parameters may be significant.
4. The influences of the time step on the SSTT results are insignificant for the TMD and TLD cases. However, for the PD cases, a large time step of 0.02 s may lead to relatively large errors.
5. The time delay has a negative impact on the SSTT results. However, if the time delay is within a certain level, any adverse effects can be ignored. If the time delay is very large, delay compensation should be used to offset its negative influence.

Author Contributions: Conceptualization, B.F. and J.C.; methodology, B.F. and H.J.; validation, B.F. and J.C.; investigation, B.F. and J.C.; data curation, B.F.; writing—original draft preparation, B.F. and H.J.; writing—review and editing, J.C.; funding acquisition, B.F. and J.C. All authors have read and agreed to the published version of the manuscript.

Funding: This research was funded by the National Natural Science Foundation of China (Grant Nos. 51908048, 52008074), the Natural Science Foundation of Shaanxi Province (Grant No. 2020JQ-382), the Young Talent Fund of University Association for Science and Technology in Shaanxi, China (Grant No. 20200412), the Fundamental Research Funds for the Central Universities, CHD (Grant No. 300102280107), the Open Project of State Key Laboratory of Green Building in Western China (Grant No. LSKF202007), and the Open Project of State Key Laboratory of Mechanical Behavior and System Safety of Traffic Engineering Structures (Grant No. KF2021-03).

Informed Consent Statement: Not applicable.

Data Availability Statement: The data presented in this study are available on request from the corresponding author.

Conflicts of Interest: The authors declare no conflict of interest.

References

1. Fu, B.; Kolay, C.; Ricles, J.; Jiang, H.; Wu, T. Stability analysis of substructure shake table testing using two families of model-based integration algorithms. *Soil Dyn. Earthq. Eng.* **2019**, *126*, 105777. [CrossRef]
2. Lu, Z.; Wang, Z.; Masri, S.F.; Lu, X. Particle impact dampers: Past, present, and future. *Struct. Health Monit.* **2018**, *25*, e2058. [CrossRef]
3. Lu, Z.; Chen, X.; Zhang, D.; Dai, K. Experimental and analytical study on the performance of particle tuned mass dampers under seismic excitation. *Earthq. Eng. Struct. Dyn.* **2016**, *46*, 697–714. [CrossRef]
4. Yang, Y.; Li, J.L.; Zhou, C.H.; Law, S.S.; Lv, L. Damage detection of structures with parametric uncertainties based on fusion of statistical moments. *J. Sound Vib.* **2019**, *442*, 200–219. [CrossRef]
5. Yang, Y.; Li, C.; Ling, Y.; Tan, X.; Luo, K. Research on new damage detection method of frame structures based on generalized pattern search algorithm. *China J. Sci. Instrum.* **2021**, *42*, 123–131.
6. Kang, Y.J.; Peng, L.Y.; Pan, P.; Xiao, G.Q.; Wang, H.S. Shaking table test and numerical analysis of a coal-fired power plant equipped with large mass ratio multiple tuned mass damper (LMTMD). *J. Build. Eng.* **2021**, *43*, 102852. [CrossRef]

7. Wang, W.; Yang, Z.; Hua, X.; Chen, Z.; Wang, X.; Song, G. Evaluation of a pendulum pounding tuned mass damper for seismic control of structures. *Eng. Struct.* **2021**, *228*, 111554. [CrossRef]
8. Zhao, B.; Wu, D.; Lu, Z. Shaking table test and numerical simulation of the vibration control performance of a tuned mass damper on a transmission tower. *Struct. Infrastruct. Eng.* **2021**, *17*, 1110–1124. [CrossRef]
9. Vafaei, M.; Pabarja, A.; Alih, S.C. An innovative tuned liquid damper for vibration mitigation of structures. *Int. J. Civ. Eng.* **2021**, *19*, 1071–1090. [CrossRef]
10. Shen, B.; Xu, W.; Wang, J.; Chen, Y.; Yan, W.; Huang, J.; Tang, Z. Seismic control of super high-rise structures with double-layer tuned particle damper. *Earthq. Eng. Struct. Dyn.* **2020**, *50*, 791–810. [CrossRef]
11. Ashasi-Sorkhabi, A.; Malekghasemi, H.; Mercan, O. Implementation and verification of real-time hybrid simulation (RTHS) using a shake table for research and education. *J. Vib. Control* **2015**, *21*, 1459–1472. [CrossRef]
12. Wang, J.; Gui, Y.; Zhu, F.; Jin, F.; Zhou, M.X. Real-time hybrid simulation of multi-story structures installed with tuned liquid damper. *Struct. Health Monit.* **2016**, *23*, 1015–1031. [CrossRef]
13. Zhu, F.; Wang, J.; Jin, F.; Lu, L.; Gui, Y.; Zhou, M.X. Real-time hybrid simulation of the size effect of tuned liquid dampers. *Struct. Health Monit.* **2017**, *24*, e1962. [CrossRef]
14. Ashasi-Sorkhabi, A.; Malekghasemi, H.; Mercan, O.; Ghaemmaghami, G. Experimental investigations of tuned liquid damper-structure interactions in resonance considering multiple parameters. *J. Sound Vib.* **2017**, *388*, 141–153. [CrossRef]
15. Fu, B.; Jiang, H.; Wu, T. Experimental study of seismic response reduction effects of particle damper using substructure shake table testing method. *Struct. Health Monit.* **2019**, *26*, e2295. [CrossRef]
16. Fu, B.; Jiang, H.; Wu, T. Comparative studies of vibration control effects between structures with particle dampers and tuned liquid dampers using substructure shake table testing methods. *Soil Dyn. Earthq. Eng.* **2019**, *121*, 421–435. [CrossRef]
17. Quesada, A.; Gauchia, A.; Álvarez-Caldas, C.; Román, J. Material characterization for FEM simulation of sheet metal stamping processes. *Adv. Mech. Eng.* **2014**, *2014*, 167147. [CrossRef]
18. Gohari, S.; Sharifi, S.; Burvill, C.; Mouloodi, S.; Izadifar, M.; Thissen, P. Localized failure analysis of internally pressurized laminated ellipsoidal woven GFRP composite domes: Analytical, numerical, and experimental studies. *Arch. Civ. Mech. Eng.* **2019**, *19*, 1235–1250. [CrossRef]
19. Yang, Y.; Xiang, C.; Jiang, M.; Li, W.; Kuang, Y. Bridge damage identification method considering road surface roughness by using indirect measurement technique. *China J. Highw. Transp.* **2019**, *32*, 99–106.
20. Yang, Y.; Cheng, Q.; Zhu, Y.; Wang, L.; Jin, R. Feasibility study of tractor-test vehicle technique for practical structural condition assessment of beam-like bridge deck. *Remote Sens.* **2020**, *12*, 114. [CrossRef]
21. Gohari, S.; Mozafari, F.; Moslemi, N.; Mouloodi, S.; Sharifi, S.; Rahmanpanah, H.; Burvill, C. Analytical solution of the elec-tro-mechanical flexural coupling between piezoelectric actuators and flexible-spring boundary structure in smart composite plates. *Arch. Civ. Mech. Eng.* **2021**, *21*, 31. [CrossRef]
22. Yang, Y.; Liang, J.; Yuan, A.; Lu, H.; Luo, K.; Shen, X.; Wan, Q. Bridge element bending stiffness damage identification based on new indirect measurement method. *China J. Highw. Transp.* **2021**, *34*, 188–198.
23. Kolay, C.; Ricles, J.M. Assessment of explicit and semi-explicit classes of model-based algorithms for direct integration in structural dynamics. *Int. J. Numer. Methods Eng.* **2016**, *107*, 49–73. [CrossRef]
24. Kolay, C.; Ricles, J.M.; Marullo, T.M.; Mahvashmohammadi, A.; Sause, R. Implementation and application of the unconditionally stable explicit parametrically dissipative KR-α method for real-time hybrid simulation. *Earthq. Eng. Struct. Dyn.* **2015**, *44*, 735–755. [CrossRef]
25. Kolay, C.; Ricles, J.M. Force-based frame element implementation for real-time hybrid simulation using explicit direct integration algorithms. *J. Struct. Eng.* **2018**, *144*, 04017191. [CrossRef]
26. Kolay, C.; Ricles, J.M. Improved explicit integration algorithms for structural dynamic analysis with unconditional stability and controllable numerical dissipation. *J. Earthq. Eng.* **2019**, *23*, 771–792. [CrossRef]
27. Chang, S.Y. Explicit pseudodynamic algorithm with unconditional stability. *J. Eng. Mech.* **2002**, *128*, 935–947. [CrossRef]
28. Chen, C.; Ricles, J.M. Development of direct integration algorithms for structural dynamics using discrete control theory. *J. Eng. Mech.* **2008**, *134*, 676–683. [CrossRef]
29. Fu, B.; Feng, D.C.; Jiang, H. A new family of explicit model-based integration algorithms for structural dynamic analysis. *Int. J. Struct. Stab. Dyn.* **2019**, *19*, 1950053. [CrossRef]
30. Chopra, A.K. *Dynamics of Structures*, 4th ed.; Prentice Hall: Upper Saddle River, NJ, USA, 2011; p. 07458.
31. Kang, D.H. *An Optimized Computational Environment for Real-Time Hybrid Simulation*; University of Colorado: Boulder, CO, USA, 2010.
32. Den Hartog, J.P. *Mechanical Vibrations*; Dover Publications: New York, NY, USA, 1985.
33. Malekghasemi, H.; Ashasi-Sorkhabi, A.; Ghaemmaghami, A.R.; Mercan, O. Experimental and numerical investigations of the dynamic interaction of tuned liquid damper-structure systems. *J. Vib. Control* **2015**, *21*, 2707–2720. [CrossRef]
34. Yu, J.K.; Wakahara, T.; Reed, D.A. A non-linear numerical model of the tuned liquid damper. *Earthq. Eng. Struct. Dyn.* **1999**, *28*, 671–686. [CrossRef]
35. Papalou, A.; Masri, S.F. Response of impact dampers with granular materials under random excitation. *Earthq. Eng. Struct. Dyn.* **1996**, *25*, 253–267. [CrossRef]
36. Hales, T.C. The sphere packing problem. *J. Comput. Appl. Math.* **1992**, *44*, 41–76. [CrossRef]

37. Masri, S.F.; Ibrahim, A.M. Response of the impact damper to stationary random excitation. *J. Acoust. Soc. Am.* **1973**, *53*, 200–211. [CrossRef]
38. Lu, Z.; Lu, X.; Lu, W.; Masri, S.F. Experimental studies of the effects of buffered particle dampers attached to a multi-degree-of-freedom system under dynamic loads. *J. Sound Vib.* **2012**, *331*, 2007–2022. [CrossRef]
39. Xu, W.J.; Chen, C.; Guo, T.; Chen, M.H. Evaluation of frequency evaluation index based compensation for benchmark study in real-time hybrid simulation. *Mech. Syst. Signal Pr.* **2019**, *130*, 649–663. [CrossRef]
40. Ouyang, Y.; Shi, W.; Shan, J.; Spencer, B.F. Backstepping adaptive control for real-time hybrid simulation including servo-hydraulic dynamics. *Mech. Syst. Signal Pr.* **2019**, *130*, 732–754. [CrossRef]
41. Ning, X.Z.; Wang, Z.; Wang, C.P.; Wu, B. Adaptive feedforward and feedback compensation method for real-time hybrid simulation based on a discrete physical testing system model. *J. Earthq. Eng.* **2020**. (Early Access). [CrossRef]
42. O'Brien, C.; Mazurek, L.; Christenson, R. Effective compensation of nonlinear actuator dynamics using a proposed linear time-varying compensation. *J. Eng. Mech.* **2021**, *147*, 04021048. [CrossRef]

symmetry

MDPI

Article

A Real-Time Detection Method for Concrete Surface Cracks Based on Improved YOLOv4

Gang Yao [1,2], Yujia Sun [1,2], Mingpu Wong [1,2] and Xiaoning Lv [3,*]

1 School of Civil Engineering, Chongqing University, Chongqing 400044, China; yaogang@cqu.edu.cn (G.Y.); sunyujia@cqu.edu.cn (Y.S.); wongmingpu@cqu.edu.cn (M.W.)
2 Key Laboratory of New Technology for Construction of Cities in Mountain Area, Ministry of Education, Chongqing 400044, China
3 Science & Technology on Integrated Information System Laboratory, Institute of Software Chinese Academy of Sciences, Beijing 100190, China
* Correspondence: xiaoning@iscas.ac.cn

Abstract: AbstractMany structures in civil engineering are symmetrical. Crack detection is a critical task in the monitoring and inspection of civil engineering structures. This study implements a lightweight neural network based on the YOLOv4 algorithm to detect concrete surface cracks. In the extraction of backbone and the design of neck and head, the symmetry concept is adopted. The model modules are improved to reduce the depth and complexity of the overall network structure. Meanwhile, the separable convolution is used to realize spatial convolution, and the SPP and PANet modules are improved to reduce the model parameters. The convolutional layer and batch normalization layer are merged to improve the model inference speed. In addition, using the focal loss function for reference, the loss function of object detection network is improved to balance the proportion of the cracks and the background samples. To comprehensively evaluate the performance of the improved method, 10,000 images (256 × 256 pixels in size) of cracks on concrete surfaces are collected to build the database. The improved YOLOv4 model achieves an mAP of 94.09% with 8.04 M and 0.64 GMacs. The results show that the improved model is satisfactory in mAP, and the model size and calculation amount are greatly reduced. This performs better in terms of real-time detection on concrete surface cracks.

Keywords: structural health monitoring; deep learning; crack detection; improved YOLOv4; concrete surface

Citation: Yao, G.; Sun, Y.; Wong, M.; Lv, X. A Real-Time Detection Method for Concrete Surface Cracks Based on Improved YOLOv4. *Symmetry* **2021**, *13*, 1716. https://doi.org/10.3390/sym13091716

Academic Editors: Yang Yang, Ying Lei, Xiaolin Meng and Jun Li

Received: 1 September 2021
Accepted: 14 September 2021
Published: 16 September 2021

1. Introduction

As one of the most common materials in civil engineering, concrete is widely used in dams, buildings, tunnels, bridges, and other infrastructure. Owing to the influence of internal and external factors (such as temperature change, foundation deformation, shrinkage, etc.), cracks often appear on concrete surfaces. As a common defect in civil engineering, cracks not only affect the health of structures, but also lead to other problems [1]. Therefore, crack detection is an essential part of structural health monitoring.

The traditional detection method relies on human vision, which has high cost but low detection efficiency, and the detection results depend on subjective human judgment. To solve these problems, researchers have proposed many methods of automatically detecting concrete surface defects [2,3]. However, these methods usually have heavy workloads and low precision, and thus adequately cannot meet the demand. With the continuous development of computer vision technology, image processing technology (IPT) has been widely used to detect defects on various structural surfaces [4–6]. Yeum et al. [7] used IPT combined with sliding window technology to detect cracks, which clearly shows the potential of IPT. Due to the existence of background noise, the differences between the database, the limited number of features, and the diversity of application scenarios, traditional image processing methods may produce undesirable results. The adaptability of the model is also limited [8]. IPTs are usually used to help inspectors detect defects, but the final results are still obtained relying on manual judgment [9].

At present, with the improvement of computing capabilities and image acquisition equipment, many machine learning algorithms (such as deep learning) are being used to recognize objects with acceptable results [10–14]. Deep learning can automatically extract high-level semantic information from original images, which provides a new method for the automatic crack detection on concrete surfaces. The convolutional neural network (CNN), which has a strong capacity to learn depth features directly from training data, greatly improves the efficiency and accuracy of detecting concrete surface defects. In addition, it has been emphasized in image classification and object detection [15]. Researchers have developed many methods based on deep learning to detect pavement cracks [16–18], concrete cracks [19–22], concrete bug holes [23–25], and other defects [26–31]. CNN-based crack detection methods generally have problems, such as excessive training parameters and complex network structures. To overcome these problems, the object detection algorithms are further explored.

There are two categories in object detection models: one-stage models and two-stage models. The SSD [32] and YOLO series represent one-stage models. They regard object detection as a regression problem. The Faster-RCNN [15], SPP-NET [12], etc. represent two-stage models. In the two-stage model training process, the network for object regions detection is trained after the region proposal network (RPN) is trained. Consequently, the two-stage model has high precision but slow speed. For the purpose of completing the whole detection process without using RPN and realizing the end-to-end object detection, the initial anchors are used in the one-stage model to predict the category and locate the object area. Correspondingly, the one-stage model has fast speed but low precision. The accuracy and reasoning speed of object detection algorithms are the critical problems in object detection. Balancing the efficiency and accuracy of the detection is a crucial technical problem. YOLOv4 has good processing speed and performance, which just meets the requirement. However, it is difficult to use in embedded devices, which cannot meet the needs of accurate real-time detection.

In this study, we choose YOLOv4 [33] (a one-stage model) as the basic model, and improvements are made to achieve accurate real-time crack detection on concrete surfaces. We make the following major contributions. (1) The model framework is improved. The SwishBlock bottleneck module is established and replaces the ResBlock_body as the main framework of the YOLOv4 model. Meanwhile, the number of original YOLOv4 model modules is changed, and the number of layers in the entire network is reduced, in order to compress the YOLOv4 model. (2) The SPP structure and PANet module are improved. With the aim of maintaining channel separation, the common convolution is replaced by separable convolution, which reduces model parameters and further compresses the network model. (3) Model reasoning speed is improved. In the SwishBlock bottleneck module, the parameters in the batch normalization layer and the convolutional layer are merged to improve the forward reasoning speed of the model. (4) The loss function is improved. Using the focal loss function of the RetinaNet network for reference, a modulation coefficient α is added to the cross-entropy loss function of the object detection to balance the proportion of foreground and background data samples. (5) The trained model is deployed to the Jetson Xavier NX embedded platform for testing to verify that it meets the requirements for accurate real-time detection of concrete surface cracks, which can provide support for the development of mobile monitoring systems.

2. Lightweight Model for Concrete Crack Detection

2.1. The Principles of YOLOv4

The backbone network CSPDarknet53 of YOLOv4 is the core of the algorithm and is used to extract the target features. CSPNet can maintain accuracy and reduce computing bottlenecks and memory costs while being simplified. Drawing from the experience of

CSPNet, YOLOv4 adds CSP to each large residual block of Darknet53. It divides the feature mapping of the base layer into two parts, and then merges them through a cross-stage hierarchical structure to reduce the amount of calculations while ensuring accuracy. The base layer of CSPDarknet53 uses the Mish function as the activation function, and the feature extraction layer network uses the Leaky_relu function. Experiments have shown that the above activation function setting makes the object detection more accurate. Unlike the YOLOv3 algorithm, which uses FPN for upsampling, YOLOv4 draws on the idea of information circulation in the PANet network. The semantic information of the layer features is propagated to the low-level network by upsampling and is then fused with the high-resolution information of the underlying features to improve the small target detection effect. Next, the information transmission path from the bottom to the top is increased, and the feature pyramid is enhanced through downsampling. Finally, the feature maps of different layers are fused to make predictions. The specific network framework is shown in Figure 1. The ResBlock_body is the residual block of CSPDarknet53, which can extract the target features of the image and reduce the computational bottleneck and memory cost, as shown in Figure 2.

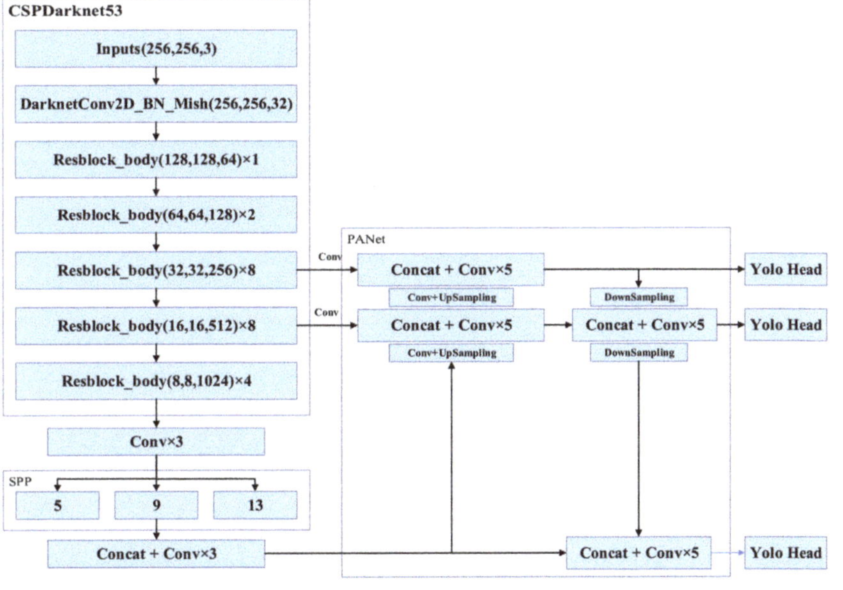

Figure 1. YOLOv4 network architecture [33].

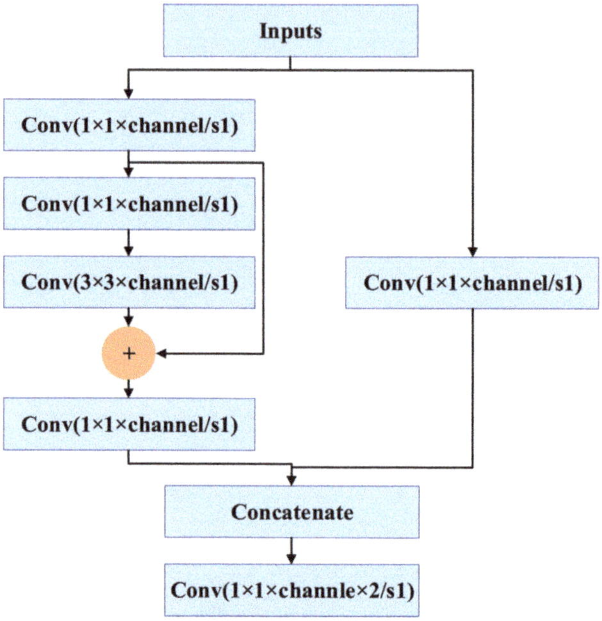

Figure 2. ResBlock module structure.

2.2. Improvements to Model Framework

Three criteria need to be met during the construction of the neural network model [34]. First, the residual neural network is used to increase the depth of the network, and the feature extraction is achieved by the deeper neural network. Second, the number of feature layers extracted from each layer is changed to achieve more feature extraction layers, get more features, and increase the width. Third, by increasing the resolution of the input picture, the features of network learning and expression can be enriched, which is conducive to improving the accuracy. The above criteria are followed in the YOLOv4 model compression. The SwishBlock bottleneck module is established based on the depthwise separable convolution and the construction concept of reverse residual structure. The characteristics of the network are expanded from three aspects at the same time, and the ResBlock_body is replaced as the overall design concept of the main YOLOv4 framework. Meanwhile, the SENet Channel Attention idea is used for reference into the network structure, and different weights are assigned to the extracted feature maps to extract more critical feature information without increasing the model calculation and storage costs.

The actual architecture of the SwishBlock bottleneck module is an inverted residual structure. In a residual structure, there are fewer feature map channels in the middle and more feature map channels on both sides, while in reverse residual structure, there are more feature map channels in the middle and fewer feature map channels on both sides. When processing deep separable convolution, the method of first raising and then reducing the dimension of the feature map greatly improves the feature extraction ability of the network, and also speeds up the calculation. The Shortcut connection used in the module ensures that the gradient will not be affected during the propagation of the deep convolutional neural network, and the reverse residual structure has been shown to improve the memory utilization efficiency. Before the 3 $\times$ 3 separable convolution structure, the 1 $\times$ 1 convolution is used to increase the dimension to improve the feature extraction of the image. At the same time, a channel attention mechanism structure is added after the 3 $\times$ 3 network structure. First, the global pooling is performed, and the neural network is used to train the weight value of each channel to extract more important feature information. Then, a large residual edge is added after a 1 × 1 convolution dimensionality reduction to avoid the gradient disappearance of the network. The SwishBlock module structure is shown in Figure 3.

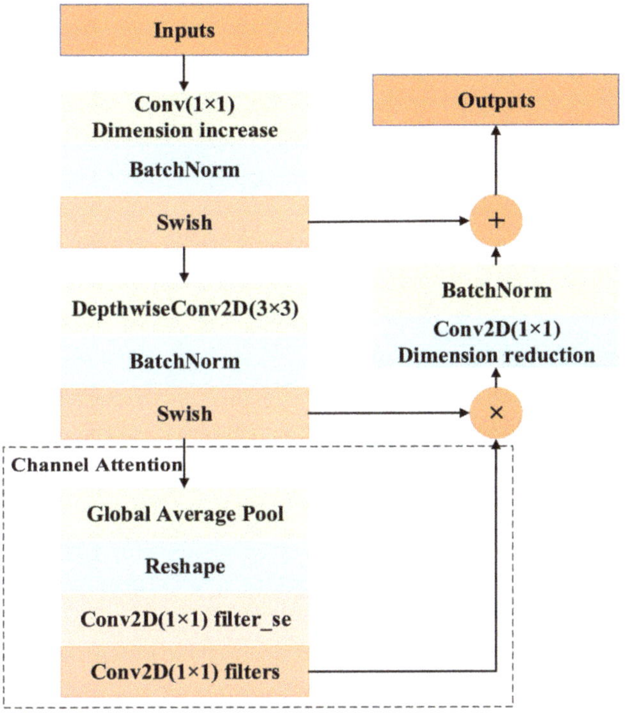

Figure 3. SwishBlock module structure.

In this study, using the advantages of the basic SPP and PANet architecture in the YOLOv4 module, the SwishBlock bottleneck is used to replace the ResBlock_body structure in the main YOLOv4 framework. The neural network model is constructed using the SwishBlock bottleneck structure. Meanwhile, the number of original YOLOv4 network model modules is changed and the number of layers in the whole network is reduced, such that the YOLOv4 network model can be compressed. The compressed network architecture of the YOLOv4 network model is shown in Figure 4.

The improved YOLOv4 prediction network still predicts three feature maps with different sizes to generate the location and target category of the detection box. Considering the imbalance between the cracks and background pixels in the image, the input image size of the model is 256 × 256, and the dimensions of the last three layers are 8 × 8, 16 × 16 and 32 × 32. Among them, the 8 × 8 feature map constructed by combining the two downsamplings of the shallow network with the deep network is mainly used to detect large objects, the once-upsampled 16 × 16 feature map spliced with the middle layer feature map of the backbone network is mainly used to detect medium objects, and the twice-upsampled 32 × 32 feature map spliced with the middle layer feature map of the backbone network is mainly used to detect small targets. The feature fusion and multi-scale detection method can make full use of different scales to extract different layers of semantic information, enhance the feature expression ability of the network, and improve the accuracy of object detection.

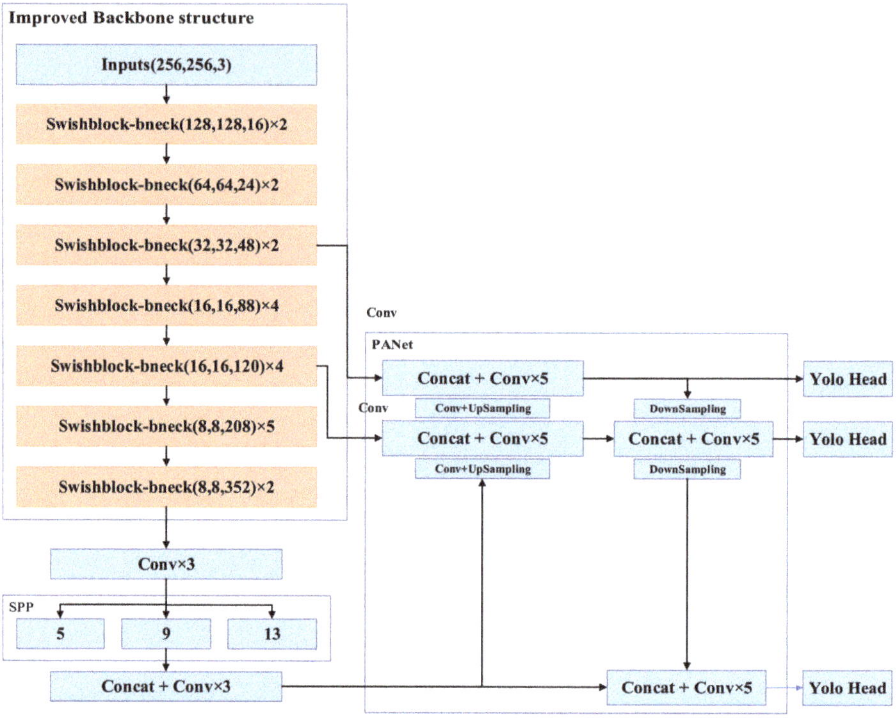

Figure 4. Compressed network architecture of the YOLOv4 network model.

2.3. Improvements to SPP Structure and PANet Module

The SPP structure and PANet module in the YOLOv4 network model play a role in enhancing the network feature extraction. By analyzing the structure of the SPP and PANet modules, it is found that the network structure contains a large number of 3 × 3 convolution layers and continuous quintic convolution structures that greatly increase the calculation amount of the model.

Figure 5a shows the process of extracting image detail features by conventional convolution, where M is the dimension of the input image, N is the number of channels filtered by the convolution kernel, and the size of the convolution kernel is 3 × 3. The dimension of the feature layer then extracted by the conventional convolution is N, and the number of parameters of the conventional convolution is $M \times N \times 3 \times 3$. The depthwise separable convolution is composed of depth convolution and 1 × 1 point convolution, as shown in Figure 5b,c. The depthwise separable convolution adopts the strategy of channel-by-channel convolution, and the extracted feature map adopts the form of point-by-point convolution to obtain a feature map with dimension N. The dimension of the input image is M, and the size of the separable convolution kernel is 3 × 3. Therefore, the number of parameters for the separable convolution operation is $M \times 3 \times 3 + 1 \times 1 \times M \times N$. Compared with the conventional convolution operation, the separable convolution is used to extract the texture features of the image, and the number of parameters is reduced, as shown in Equation (1).

$$\frac{M \times 3 \times 3 + 1 \times 1 \times M \times N}{M \times N \times 3 \times 3} = \frac{1}{N} + \frac{1}{9} \tag{1}$$

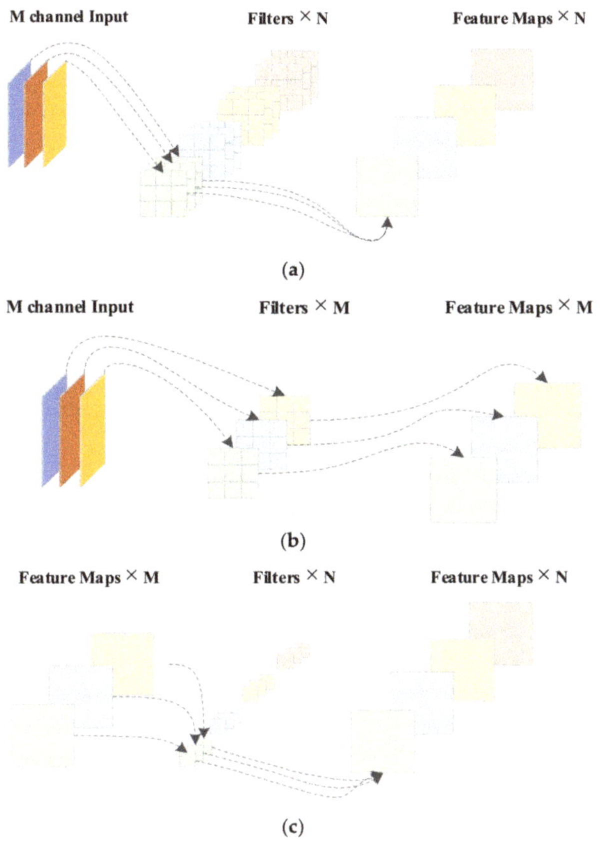

Figure 5. Depthwise separable convolution versus ordinary convolution; (**a**) general convolution filtering; (**b**) depthwise separable convolution; (**c**) point convolution operation.

The PANet module and the SPP structure contain continuous quintic convolution and continuous cubic convolution to enhance the process of image feature extraction. Equation (1) shows that the separable convolution operation greatly reduces the number of parameters compared with the conventional convolution operation. In order to further compress the network model, with the aim of maintaining channel separation, spatial convolution is realized based on separable convolution. The ordinary convolution of the SPP structure and PANet module is replaced in order to reduce the number of model parameters and memory dependence, as shown in Figure 6.

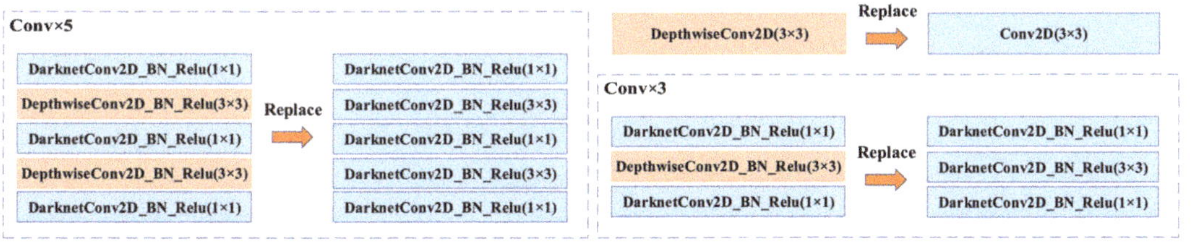

Figure 6. Depthwise separable convolution replaces ordinary convolution.

2.4. Model Inference Speed Improvements

The SwishBlock bottleneck module in the previous section uses the convolution layer and Batch Normalization layer for forward operation. The batch normalization layer can accelerate the network's convergence and control its overfitting. Although it plays an active role during training, the forward operation is added in the network forward reasoning, which takes up more memory space and affects the processing speed. Therefore, in this section, the parameters in the batch normalization layer of the SwishBlock bottleneck module and the convolutional layer are merged to improve the forward reasoning speed of the model. The original batch normalization layer is represented as follows:

$$\mu_B \leftarrow \frac{1}{m}\sum_{i=1}^{m} x_i \tag{2}$$

$$\sigma_B^2 \leftarrow \frac{1}{m}\sum_{i=1}^{m} (x_i - \mu_B)^2 \tag{3}$$

$$\hat{x}_i \leftarrow \frac{x_i - \mu_B}{\sqrt{\sigma_B^2 + \varepsilon}} \tag{4}$$

$$y_i \leftarrow \gamma \hat{x}_i + \beta \tag{5}$$

where μ_B is the mean value of the dataset, and σ_B^2 is the variance of the dataset. $\hat{x}_i$ is the normalization of the dataset, and y_i is the output result after translation and scaling through the batch normalization layer. The convolution layer is calculated as shown in Equation (6), and the output of the BN layer through the convolution layer is shown in Equation (7). Equation (6) is substituted into Equation (7) and expanded to calculate the new weight and bias term of the convolution layer, as shown in Equation (8). Thus, the new weight and bias are shown in Equations (9) and (10), respectively. The new weight and bias term are used to perform the convolution layer calculation of the SwishBlock bottleneck module, and the result is the same as that of the original convolution layer plus the batch normalization layer, while reducing the forward reasoning speed of the model.

$$out = \sum_{i=1}^{k} w_i x_i + b \tag{6}$$

$$BN = \frac{\gamma(out - \mu_B)}{\sqrt{\sigma_B^2 + \varepsilon}} + \beta \tag{7}$$

$$BN = \frac{\gamma(\sum_{i=1}^{k} w_i x_i + b - \mu_B)}{\sqrt{\sigma_B^2 + \varepsilon}} + \beta = \frac{\gamma\sum_{i=1}^{k} w_i x_i + \gamma(b - \mu_B)}{\sqrt{\sigma_B^2 + \varepsilon}} + \beta \tag{8}$$

$$w_{new} = \frac{\gamma\sum_{i=1}^{k} w_i}{\sqrt{\sigma_B^2 + \varepsilon}} \tag{9}$$

$$b_{new} = \frac{\gamma(b - \mu_B)}{\sqrt{\sigma_B^2 + \varepsilon}} + \beta \tag{10}$$

2.5. Improvements of Loss Function

The loss function of the YOLOv4 network during training consists of three parts: boundary box regression loss L_{ciou}, confidence loss L_{conf} and classification loss L_{class}, as shown in Equation (11).

$$\begin{cases} L_{conf} = & -\sum_{i=0}^{S^2}\sum_{j=0}^{B} I_{i,j}^{obj}[C_i^j \log(C_i^j) + (1 - C_i^j)\log(1 - C_i^j)] \\ & -\lambda_{noobj}\sum_{i=0}^{S^2}\sum_{j=0}^{B} I_{i,j}^{noobj}[C_i^j \log(C_i^j) + (1 - C_i^j)\log(1 - C_i^j)] \\ L_{class} = & -\sum_{i=0}^{S^2} I_{i,j}^{obj} \sum_{c \in classes} [P_i^j \log(P_i^j) + (1 - P_i^j)\log(1 - P_i^j)] \\ Loss = & L_{ciou} + L_{conf} + L_{class} \end{cases} \tag{11}$$

where S^2 and B are the feature map scale and a priori box, and λ_{noobj} is the weight coefficient. $I_{i,j}^{obj}$ and $I_{i,j}^{noobj}$ represent the target and no target at the j-th a priori box of the i-th grid, respectively. c is the diagonal distance between the predicted box and the closure area of the actual box. b, w, and h represent the center coordinates, width, and height of the prediction box, while b^{gt}, w^{gt}, and h^{gt} represent the center coordinates, width, and height of the actual box, respectively. C_i^j and $\overline{C}_i^j$ represent the confidence of the prediction box and the labeled box, and P_i^j and $\overline{P}_i^j$ represent the class probability of the prediction box and the labeled box, respectively.

When the model is used to detect cracks, the size of the cracks themselves is small, and the objects occupy a small proportion of the background. To balance the proportion of the foreground and background data samples, a modulation coefficient α is added to the cross-entropy loss function of object detection and classification using the focal loss function of the RetinaNet network for reference, as shown in Equation (12).

$$L_{class} = -\sum_{i=0}^{S^2} I_{i,j}^{obj} \sum_{c \in classes} [P_i^j \log(P_i^j) + (1 - P_i^j)^{\alpha} \log(1 - P_i^j)] \tag{12}$$

3. Experiments

A program of experimental studies is shown in Figure 7.

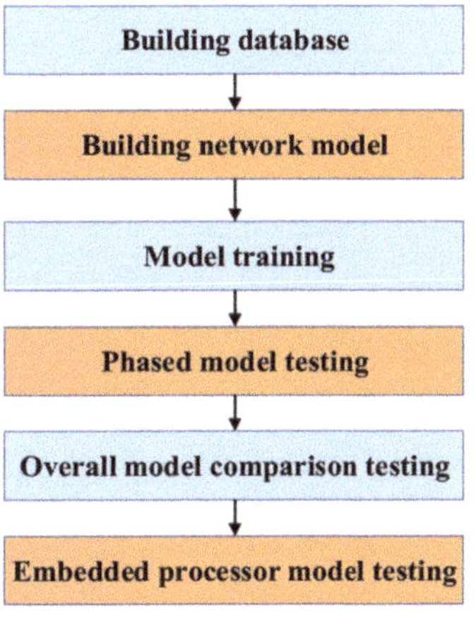

Figure 7. A program of experimental studies.

3.1. Image Database Creation

A smartphone is used for image acquisition. For the purpose of collecting images of small cracks on a concrete surface, all images are taken with a distance of 0.1 m between the smartphone and the concrete surface. Two-thousand original images with sizes of 3024 × 3024 pixels are extracted from the surfaces of concrete buildings. Each original image can be cropped to generate 139 images with sizes of 256 × 256 pixels. However, several cropped images do not include cracks. As a result, the images with cracks are meticulously selected from the cropped image set. Finally, 10,000 images conforming to the requirements are selected to create the database.

To assess the generalization ability of the improved model, the 10,000 images are divided into five parts according to the fivefold cross-validation principle, of which 80% are used to train and validate the model and the remaining 20% are used to test. More precisely, 8000 images are randomly selected from the 10,000 images, among which 7000 images are used to generate a training set and 1000 images are used to create a validation set. The remaining 2000 images not selected for training or validation are used to build a testing set.

3.2. Model Initialization

In the process of network training, in order to improve efficiency and better save computing resources and time, this paper adopts the training strategy of freezing certain layers. The entire training process is divided into two stages. In the first stage, only the backbone network structure is trained; in the second stage, the overall network structure is trained. In the training process, the Cosine Annealing learning rate strategy is adopted, and the hyperparameters are optimized according to the genetic algorithm. The initial parameter settings of the first stage and the second stage are shown in Tables 1 and 2, respectively.

Table 1. Initial parameters of the first stage of the training process.

Parameter	Value
Base_LR	10^{-3}
Batch_Size	16
Train_Epoch	100
Weight_decay	5×10^{-4}
Lr_scheduler_Max_iterations	5
Lr_scheduler_Minimum_lr	10^{-5}

Table 2. Initial parameters of the second stage of the training process.

Parameter	Value
Base_LR	10^{-4}
Batch_Size	16
Train_Epoch	500
Weight_decay	5×10^{-4}
Lr_scheduler_Max_iterations	5
Lr_scheduler_Minimum_lr	10^{-5}

3.3. Evaluation Metrics of Accuracy

Crack detection based on deep learning is quantitatively measured by objective evaluation metrics, which can measure many aspects of the quality of a restoration algorithm. There are many objective evaluation metrics commonly used in object detection, such as intersection over union (IoU), precision, recall and mean average precision (mAP). IOU is the ratio of the intersection and union between the bounding box predicted by the model and the real bounding box, which is also called the Jaccard index.

mAP is a common index used to evaluate the accuracy of algorithms in the field of object detection. In this paper, the objective evaluation index mAP is used for calculation, as shown in Equation (13), where AP is the average precision. Taking recall as the horizontal axis and precision as the vertical axis, the P-R curve can be obtained, and the AP value can then be calculated. Simply, this averages the precision values on the P-R curve. The definition of AP is shown in Equation (14).

$$mAP = \frac{1}{|Q_R|}\sum_{q=1}^{Q_R} AP(q) \tag{13}$$

$$AP = \int_0^1 p(r)dr \tag{14}$$

The construction of the P-R curve is drawn by the precision and the recall. The precision refers to the number of correct recognitions of all samples predicted to be positive. The recall reflects the missed detection rate of the model. Precision and recall are defined in Equations (15) and (16), respectively. True positive (TP) indicates that the detection category is positive and predicted to be positive, while false positive (FP) indicates that the detection category is negative and predicted to be negative. False negative (FN) indicates that the detection category is positive and predicted to be negative, and P is the number of positive samples in the testing set. The precision and recall are independent of each other. High precision means that the false detection rate is low, which can lead to a high missed detection rate.

In this paper, in addition to the mAP, the model size and computational complexity FLOPs are used to evaluate the model compression algorithm. The model's size is closely related to its parameters, which can be used to measure the simplification of the YOLOv4 model. FLOPs reflect the calculation amount of the algorithm. The unit of FLOPs is GMacs, which is short for Giga multiply–accumulation operations per second. It represents the floating-point operations per second, which can reflect the algorithm's calculation performance.

$$Precision = \frac{TP}{TP + FP} \tag{15}$$

$$Recall = \frac{TP}{TP + FN} = \frac{TP}{P} \tag{16}$$

4. Results and Discussion

The training process was implemented on a server with a high-performance GPU (NVIDIA GeForce RTX 2080 Super), 64 GB DDR4 memory, and an Intel i9-10900K CPU. The training process is based on the deep learning framework Pytorch 1.14.

4.1. Test Comparison Results of Sub-Modules of the Improved Algorithm

In order to fully verify the rationality of the various improvements to the YOLOv4 network model in this paper, a step-by-step verification experiment is performed. Objective evaluation indexes are used to qualitatively measure the quality and speed of the restoration algorithm. The commonly used objective metrics for evaluating object detection performance are mAP, parameters, and calculation amount (FLOPs).

First, the main framework of the model is improved, and the specific results are compared as shown in Table 3. It can be seen that when only the main framework of the model is improved, mAP is 96.43% of the original YOLOv4 model, and the amounts of the model parameters and algorithm calculations are reduced by 83.89% and 97.12%, respectively.

Table 3. Improved performance comparison of main model framework.

Method	mAP (%)	Parameters (M)	FLOPs (GMacs)
YOLOv4	95.50	64.00	63.92
Improved backbone	92.09	10.31	1.84

Based on the depth separable convolution structure, the ordinary convolution of the SPP structure and the PANet module is replaced by spatial convolution. Table 4 shows the performance comparison of the improved main model framework, PANet and SPP structure. It can be seen that after replacing the ordinary convolution of the SPP and PANet structure with separable convolution, the parameters are reduced from 10.31 M to 8.22 M, and the amount of calculations is reduced from 1.84 GMacs to 0.69 GMacs. The amount of model parameters and calculations are both greatly reduced.

Table 4. Performance comparison of improved main model framework, PANet and SPP structure.

Method	mAP (%)	Parameters (M)	FLOPs (GMacs)
YOLOv4	95.50	64.00	63.92
Improved backbone	92.09	10.31	1.84
Improved backbone + PANet + SPP structure	91.88	8.22	0.69

To improve the forward reasoning speed of the model, the parameters in the convolutional layer and the batch normalization (BN) layer are merged. Frames per second (FPS) is used to consider the effectiveness of the forward inference speed, and the processing speed is compared with that of the original YOLOv4 network model. The specific index FPS are shown in Table 5. It can be seen that compared with the original YOLOv4 network model; the FPS processed by the improved network model is greatly improved. After the parameters in the convolution layer and the BN layer are merged, the processing speed of a 256 × 256 pixel image is increased from 101 FPS to 112 FPS.

Table 5. Performance comparison of merge convolution layer and BN layer.

Method	FPS
	256 × 256
YOLOv4	31
Improved backbone + PANet+ SPP structure	101
Improved backbone + PANet + SPP structure + Merging convolution layer and BN layer	112

To balance the proportion of the foreground and background data samples, the specific mAP, model parameters and calculation amount of the improved loss function are shown in Table 6. It can be seen that after adding the modulation factor to the loss function of the model, the mAP of the algorithm in the test database increases from 91.89% to 94.09%.

Through a complete comparison between the original YOLOv4 model and the improved YOLOv4 model, the tested mAP is shown in Figure 8. It can be seen that when the self-made database is used for detection, the mAP of the original YOLOv4 model is 95.50%, the mAP of the improved model is 94.09% and the detection performance decreases slightly. It can be concluded from Table 6 that the image detection performance mAP of the improved Yolov4 model is 98.52% of that of the original model, and the model parameters and calculation amount are reduced by 87.43% and 99.00%, respectively.

Table 6. Comparison of performance of different algorithms using RSOD dataset for object detection.

Method	mAP (%)	Parameters (M)	FLOPs (GMacs)
YOLOv4	95.50	64.00	63.92
Improved backbone	92.09	10.31	1.84
Improved backbone + PANet + SPP structure	91.88	8.22	0.69
Improved backbone + PANet + SPP structure + Merging convolution layer and BN layer	91.89	8.04	0.64
Improved backbone + PANet + SPP structure + Merging convolution layer and BN layer + loss function	94.09	8.04	0.64

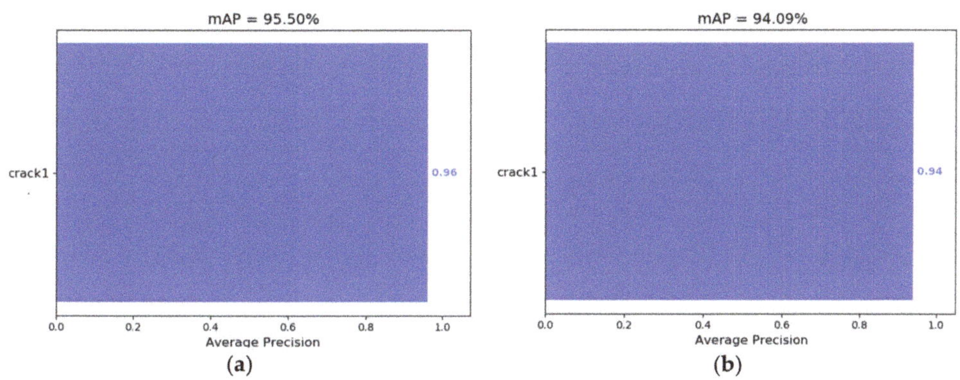

Figure 8. mAP of the original YOLOv4 model and the improved model. (**a**) mAP of the original YOLOv4 model; (**b**) mAP of the improved model.

In this study, the model modules are improved to reduce the depth and complexity of the overall network structure. Meanwhile, the separable convolution is used to realize spatial convolution, and the SPP and PANet modules are improved to reduce the model parameters. The convolutional layer and batch normalization layer are merged to improve the model inference speed. In addition, using the focal loss function for reference, the loss function of object detection network is improved to balance the proportion of the cracks and the background samples. The detection performance of the improved model is satisfactory in mAP, and the model size and calculation amount are greatly reduced.

4.2. Comparative Results of Frontier Algorithm Tests

In this paper, the self-made database is used for training and testing, and the frontier network models in the field of object detection are used for comparison. The comparison results are shown in Table 7. It can be concluded that the improved model exhibits almost no loss in mAP compared to the high-performance algorithms, but the model size and calculation amount are greatly reduced. Through comparison with the faster lightweight network models, it can be seen that the model sizes are close, but the calculation amount FLOPs are reduced, and the detection performance mAP is higher than that of the classic lightweight network models. To demonstrate the detection performance of the improved model more intuitively, images shown in Figure 9 were randomly selected from the database for testing.

Table 7. Comparison of object detection performance of different algorithms.

Method	Backbone	mAP (%)	Parameters (M)	FLOPs (GMacs)
YOLOv4	CSPDarknet	95.50	64.00	63.92
YOLOv5m	CSPDarknet	85.58	21.40	51.30
SSD	VGG-16	89.64	26.29	127.50
CenterNet	ResNet-50	92.35	32.67	35.79
YOLOv4-tiny	CSPDarknet	72.22	5.90	4.31
MobileNet-SSD	MobileNet-v1	84.28	8.85	12.40
Ours	CSPDarknet	94.09	8.04	0.64

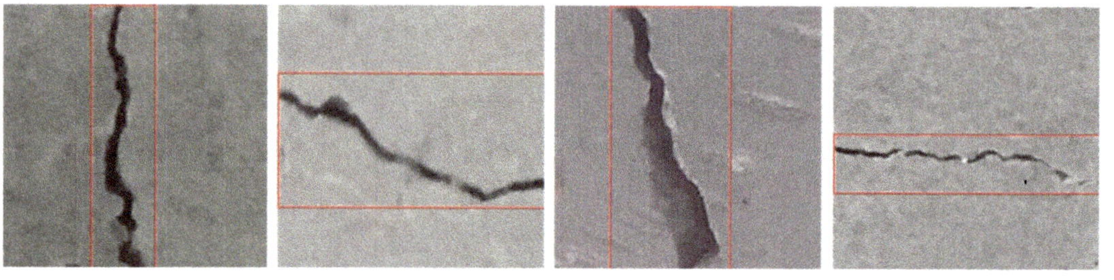

Figure 9. Detection results of the concrete surface cracks.

4.3. Experimental Results of Embedded Platform

To further verify the processing capability of the improved model in mobile devices, the trained model is deployed to the Jetson Xavier NX embedded platform for verification. The processor is small in size, low in power consumption, and strong in computing performance. The performances of the YOLOv4 network model, the YOLOv4-tiny network model and the improved model in this paper are compared in terms of the objective evaluation indicators mAP and FPS respectively, as shown in Table 8. It can be concluded that for the Jetson Xavier NX embedded platform, the input image is 256 × 256 pixels, and the YOLOv4 network model can process 16 FPS due to its complex structure, which cannot meet the needs of mobile devices for real-time crack detection. The YOLOv4-tiny network model and the improved model in this paper can process 56 and 44 FPS, which can meet the needs of real-time detection. However, the mAP of the YOLOv4-tiny network model is 72.22%, and the recognition rate is low. Compared with the YOLOv4-tiny network model, the improved YOLOv4 network model has higher accuracy and faster processing speed, which meets the requirements of accurate real-time object detection.

Table 8. Comparison of object detection performance of different algorithms.

Method	mAP (%)	FPS
YOLOv4	95.50	16
YOLOv4-tiny	72.22	56
Ours	94.09	44

5. Conclusions

A real-time concrete surface crack detection method based on the improved YOLOv4 is proposed. The improved model for concrete crack detection adopts the symmetry concept in the extraction of backbone and the design of neck and head. It is described in detail in Section 2. A smartphone is used to collect 2000 raw 3024 × 3024 pixel images from the surfaces of concrete buildings. To reduce the computation of the training process, the collected images are cropped to 256 × 256 pixels. Sets of 7000, 1000, and 2000 images are used for training, validation, and testing, respectively. The improved YOLOv4 model achieved an mAP of 94.09%, which is 98.52% of the original YOLOv4 model. The crack detection performance decreased slightly, but the parameters and calculation amount of the model are reduced by 87.43% and 99.00%, respectively. Compared with the results of the high-performance network models in object detection (such as YOLOv4, YOLOv5m, SSD, and CenterNet), it can be concluded that the improved model has almost no loss in mAP, but the model size and calculation amount are greatly reduced. In addition, compared with the detection results of the lightweight network models (such as YOLOv4-tiny and MobileNet-SSD), the model sizes are close, but the calculation amount FLOPs are reduced, and the detection performance mAP is higher. When the improved model was deployed to the Jetson Xavier NX embedded platform for testing, it achieved an mAP of 94.06% with 44 FPS. The size, accuracy, and processing speed of the model can meet the requirements of accurate real-time object detection, which can provide support for the development of mobile monitoring system. As a result, it can achieve real-time automatic vision-based crack detection on concrete surface without other equipment.

Although the improved YOLOv4 model shows good performance, there is still a long way to go before it is suitable for engineering applications. First, in the implementation of the improved method, there are many artificially adjusted hyperparameters derived from the training and verification set. Many experiments need to be conducted to explore the influence of these hyperparameters on the performance of the model. Second, a real-time mobile crack detection system (including APPs and a website) should be developed to monitor the concrete surface cracks for timely repair and protection. Lastly, we will collect more types of defect images to expand the database, such that the proposed method has greater accuracy and robustness.

Author Contributions: Conceptualization, G.Y. and Y.S.; methodology, Y.S. and X.L.; software and formal analysis, X.L.; writing—original draft preparation, Y.S.; review and editing, M.W. All authors have read and agreed to the published version of the manuscript.

Funding: This research was funded by the Fundamental Research Funds for the Central Universities (no. 2020CDJQY-A067), the National Key Research and Development Project (no. 2019YFD1101005) and the National Natural Science Foundation of China (no. 51608074).

Institutional Review Board Statement: Not applicable.

Informed Consent Statement: Not applicable.

Data Availability Statement: Please contact the corresponding author.

Conflicts of Interest: The authors declare no conflict of interest.

References

1. Gavilan, M.; Balcones, D.; Marcos, O.; Llorca, D.F.; Sotelo, M.A.; Parra, I.; Ocana, M.; Aliseda, P.; Yarza, P.; Amirola, A. Adaptive road crack detection system by pavement classification. *Sensors* **2011**, *11*, 9628–9657. [CrossRef] [PubMed]
2. Cabaleiro, M.; Lindenbergh, R.; Gard, W.F.; Arias, P.; van de Kuilen, J.W.G. Algorithm for automatic detection and analysis of cracks in timber beams from LiDAR data. *Constr. Build. Mater.* **2017**, *130*, 41–53. [CrossRef]
3. Li, S.; Yuan, C.; Liu, D.; Cai, H. Integrated Processing of Image and GPR Data for Automated Pothole Detection. *J. Comput. Civ. Eng.* **2016**, *30*, 04016015. [CrossRef]
4. Li, Q.; Zou, Q.; Zhang, D.; Mao, Q. FoSA: F* Seed-growing Approach for crack-line detection from pavement images. *Image Vis. Comput.* **2011**, *29*, 861–872. [CrossRef]
5. Zhou, Y.; Wang, F.; Meghanathan, N.; Huang, Y. Seed-Based Approach for Automated Crack Detection from Pavement Images. *Transp. Res. Rec.* **2016**, *2589*, 162–171. [CrossRef]
6. Choi, J.-I.; Lee, Y.; Kim, Y.Y.; Lee, B.Y. Image-processing technique to detect carbonation regions of concrete sprayed with a phenolphthalein solution. *Constr. Build. Mater.* **2017**, *154*, 451–461. [CrossRef]
7. Yeum, C.M.; Dyke, S.J. Vision-Based Automated Crack Detection for Bridge Inspection. *Comput. Aided Civ. Inf. Eng.* **2015**, *30*, 759–770. [CrossRef]

8. Dorafshan, S.; Thomas, R.J.; Maguire, M. Comparison of deep convolutional neural networks and edge detectors for image-based crack detection in concrete. *Constr. Build. Mater.* **2018**, *186*, 1031–1045. [CrossRef]
9. Oh, J.-K.; Jang, G.; Oh, S.; Lee, J.H.; Yi, B.-J.; Moon, Y.S.; Lee, J.S.; Choi, Y. Bridge inspection robot system with machine vision. *Automat. Constr.* **2009**, *18*, 929–941. [CrossRef]
10. Ciresan, D.; Meier, U.; Masci, J.; Schmidhuber, J. Multi-column deep neural network for traffic sign classification. *Neural Netw.* **2012**, *32*, 333–338. [CrossRef]
11. Krizhevsky, A.; Sutskever, I.; Hinton, G.E. ImageNet classification with deep convolutional neural networks. *Commun. ACM* **2017**, *60*, 84–90. [CrossRef]
12. He, K.; Zhang, X.; Ren, S.; Sun, J. Spatial Pyramid Pooling in Deep Convolutional Networks for Visual Recognition. *IEEE Trans. Pattern Anal.* **2015**, *37*, 1904–1916. [CrossRef]
13. Chen, F.-C.; Jahanshahi, M.R. NB-CNN: Deep Learning-Based Crack Detection Using Convolutional Neural Network and Naïve Bayes Data Fusion. *IEEE Trans. Ind. Electron.* **2018**, *65*, 4392–4400. [CrossRef]
14. Bang, S.; Park, S.; Kim, H.; Kim, H. Encoder–decoder network for pixel-level road crack detection in black-box images. *Comput. Aided Civ. Inf. Eng.* **2019**, *34*, 713–727. [CrossRef]
15. Ren, S.; He, K.; Girshick, R.; Sun, J. Faster R-CNN: Towards Real-Time Object Detection with Region Proposal Networks. *IEEE Trans. Pattern Anal.* **2017**, *39*, 1137–1149. [CrossRef]
16. Mei, Q.; Gül, M. A cost effective solution for pavement crack inspection using cameras and deep neural networks. *Constr. Build. Mater.* **2020**, *256*, 119397. [CrossRef]
17. Yang, Y.; Xiang, C.; Jiang, M.; Li, W.; Kuang, Y. Bridge damage identification method considering road surface roughness by using indirect measurement technique. *China J. Highw. Transp.* **2019**, *32*, 99–106.
18. Fei, Y.; Wang, K.C.P.; Zhang, A.; Chen, C.; Li, J.Q.; Liu, Y.; Yang, G.; Li, B. Pixel-Level Cracking Detection on 3D Asphalt Pavement Images Through Deep-Learning- Based CrackNet-V. *IEEE Trans. Intell. Transp.* **2020**, *21*, 273–284. [CrossRef]
19. Dung, C.V.; Anh, L.D. Autonomous concrete crack detection using deep fully convolutional neural network. *Automat. Constr.* **2019**, *99*, 52–58. [CrossRef]
20. Yang, Y.; Liang, J.; Yuan, A.; Lu, H.; Luo, K.; Shen, X.; Wan, Q. Bridge element bending stiffness damage identification based on new indirect measurement method. *China J. Highw. Transp.* **2021**, *34*, 188–198.
21. Lee, D.; Kim, J.; Lee, D. Robust Concrete Crack Detection Using Deep Learning-Based Semantic Segmentation. *Int. J. Aeronaut. Space* **2019**, *20*, 287–299. [CrossRef]
22. Sun, Y.J.; Yang, Y.; Yao, G.; Wei, F.J.; Wong, M.P. Autonomous Crack and Bughole Detection for Concrete Surface Image Based on Deep Learning. *IEEE Access.* **2021**, *9*, 85709–85720. [CrossRef]
23. Wei, F.J.; Yao, G.; Yang, Y.; Sun, Y.J. Instance-level recognition and quantification for concrete surface bughole based on deep learning. *Autom. Construction* **2019**, *107*, 102920. [CrossRef]
24. Yao, G.; Wei, F.J.; Yang, Y.; Sun, Y.J. Deep-Learning-Based Bughole Detection for Concrete Surface Image. *Adv. Civ. Eng.* **2019**, *2019*, 1–12. [CrossRef]
25. Wei, W.; Ding, L.Y.; Luo, H.B.; Li, C.; Li, G.W. Automated bughole detection and quality performance assessment of concrete using image processing and deep convolutional neural networks. *Constr. Build. Mater.* **2021**, *281*, 122576. [CrossRef]
26. Luo, R.F.; Zhang, L. Intelligent Detection Method for Internal Cracks in Aircraft Landing Gear Images under Multimedia Processing. *Symmetry* **2021**, *13*, 778. [CrossRef]
27. Yang, Y.; Cheng, Q.; Zhu, Y.; Wang, L.; Jin, R. Feasibility study of tractor-test vehicle technique for practical structural condition assessment of beam-like bridge deck. *Remote Sens.* **2020**, *12*, 114. [CrossRef]
28. Zhang, Y.X.; Lei, Y. Data Anomaly Detection of Bridge Structures Using Convolutional Neural Network Based on Structural Vibration Signals. *Symmetry* **2021**, *13*, 1186. [CrossRef]
29. Yang, Y.; Li, J.L.; Zhou, C.H.; Law, S.S.; Lv, L. Damage detection of structures with parametric uncertainties based on fusion of statistical moments. *J. Sound Vib.* **2019**, *442*, 200–219. [CrossRef]
30. Liu, B.; Zhang, Y.; He, D.J.; Li, Y.X. Identification of Apple Leaf Diseases Based on Deep Convolutional Neural Networks. *Symmetry* **2017**, *10*, 11. [CrossRef]
31. Yang, Y.; Li, C.; Ling, Y.; Tan, X.; Luo, K. Research on new damage detection method of frame structures based on generalized pattern search algorithm. *China J. Sci. Instrum.* **2021**, *42*, 123–131.
32. Liu, W.; Anguelov, D.; Erhan, D.; Szegedy, C.; Reed, S.; Fu, C.Y.; Berg, A.C. SSD: Single Shot MultiBox Detector. In Proceedings of the 2016 14th European Conference on Computer Vision (ECCV), Amsterdam, The Netherlands, 8–10 October 2016; pp. 21–37.
33. Bochkovskiy, A.; Wang, C.Y.; Liao, H. YOLOv4: Optimal Speed and Accuracy of Object Detection. In Proceedings of the 2020 IEEE/CVF Conference on Computer Vision and Pattern Recognition (CVPR), Seattle, WA, USA, 14–19 June 2020.
34. Tan, M.X.; Pang, R.M.; Le, Q.V. EfficientDet: Scalable and Efficient Object Detection. In Proceedings of the 2020 IEEE/CVF Conference on Computer Vision and Pattern Recognition (CVPR), Seattle, WA, USA, 13–19 June 2020.

symmetry

Article

Post-Processing of High Formwork Monitoring Data Based on the Back Propagation Neural Networks Model and the Autoregressive—Moving-Average Model

Yang Yang [1,2], Lin Yang [1,2] and Gang Yao [1,2,*]

[1] Key Laboratory of New Technology for Construction of Cities in Mountain Area, Ministry of Education, Chongqing University, Chongqing 400044, China; 20121601009@cqu.edu.cn (Y.Y.); 201916021069@cqu.edu.cn (L.Y.)
[2] School of Civil Engineering, Chongqing University, Chongqing 400044, China
* Correspondence: yaogang@cqu.edu.cn

Abstract: Many high formwork systems are currently equipped with health monitoring systems, and the analysis of the data obtained can determine whether high formwork is a hazard. Therefore, the post-processing of monitoring data has become an issue of widespread concern. In this paper, we discussed the fitting effect of the symmetrical high formwork monitoring data using the autoregressive–moving-average (ARMA) model and the back propagation neural networks (BPNN) combined model to process. In the actual project, the symmetry of the high formwork system allows the analysis of local monitoring results to be well extended to the whole. For the establishment of the ARMA model, the accurate judgment of the model order has a significant impact. In this paper, back propagation neural networks (BPNN) are used to simulate the ARMA process. The order of the ARMA model is estimated by determining the optimal neural network structure, which is suitable for linear or nonlinear sequences. We validated this approach from the ARMA model data simulated in Monte Carlo and compared it with the Akaike information criterion (AIC) and Bayesian information criterion (BIC). The length of the sequence, the coefficients and the order of the ARMA model are considered as factors that influence the judgment effect. Under different conditions, the BPNN always shows an accuracy rate of more than 90%, while the BIC only has a higher accuracy rate when the model order is low and the judgment efficiency of the AIC is below 50%. Finally, the proposed method successfully modeled the stress sequence and obtained the stress change trend. Compared with AIC and BIC, the efficiency of the processing time series is increased by about 50% when an order is obtained by BPNN.

Keywords: structural health monitoring; high formwork; ARMA; BPNN; stress trend prediction

Citation: Yang, Y.; Yang, L.; Yao, G. Post-Processing of High Formwork Monitoring Data Based on the Back Propagation Neural Networks Model and the Autoregressive—Moving-Average Model. *Symmetry* **2021**, *13*, 1543. https://doi.org/10.3390/sym13081543

Academic Editor: Basil Papadopoulos

Received: 31 July 2021
Accepted: 16 August 2021
Published: 23 August 2021

1. Introduction

The safety management of the high formwork is one of the important tasks of construction safety management, and the majority of failures occur due to inadequate site supervision and poor design [1]. Undoubtedly, the real-time monitoring of near-miss accidents provides an insight into possible accidents and can significantly improve safety performance by appropriate action being taken before potentially impending accidents occur [2]. In view of this, many high formwork systems have installed health monitoring systems of different sizes around the world, and these have accumulated a large amount of data over a long period of time [3,4]. Therefore, how to process this huge monitoring data accurately and timely has become the key process of the high formwork condition assessment and performance prediction [5].

Structural damage will lead to changes in the physical properties of the structure, and these changes are often reflected in the monitoring sequence [6,7]. In the previous evaluation of structural safety performance, most of its data were related to time [8–12].

The methods of dealing with stress sequences from health monitoring include the time e-frequency analysis method, modal analysis method, analytic hierarchy process, genetic algorithm, neural network analysis method, time series analysis method, and reliability theory analysis method [13–15]. Among them, the autoregressive–moving-average (ARMA) is a method widely used to model time series [16]. Two steps of the ARMA model that have an important impact on the establishment process are order estimation and coefficient estimation [17]. Coefficient estimation is based on an accurate estimate of the order. The most widely used methods of order estimation are the Akaike information criterion (AIC) [18] and the Bayesian information criterion (BIC) [19]. AIC and BIC are based on the concept of entropy and provide criteria for weighing the complexity of the estimation model against the goodness of the fitted data. In recent years, with the popularity of deep learning, ARMA models and artificial neural network associations have been used to analyze time series [20–24]. These studies prove that the combination of the two is effective for time series analysis [25]. However, in previous literature, the simulation performance of neural networks lacks comprehensive research on different models.

As widely used time series models, ARMA and BPNN are used less in structural monitoring. This paper applies these two methods to the template system for the first time. In previous studies, we have successfully applied neural networks to the safety assessment of long-span bridges and achieved good results [26]. The timeliness of the structure monitoring sequence is more obvious than the safety assessment of long-span bridges, so we model this using the ARMA model and the BPNN model [27]. Compared with the previous literature, we have improved the network structure of BPNN to a better mathematical model. The paper has mainly improved three aspects: (a) Designed and improved the BPNN structure according to the characteristics of the monitoring system; (b) Improved the judgment criteria of the ARMA model order and the model fitting effect; (c) Considering the influence of bias in BPNN modeling, the recognition of discrete data is enhanced.

In this paper, we use neural networks and ARMA to model and analyze the high formwork monitoring sequences with the goal of data-driven modeling. We use back propagation neural network (BPNN) [28] structures for the identification of the order of the ARMA model. The order estimation of the model uses the mean square error (MSE) of the neural network as the basis for the judgment of the order [29]. For neural network structures with low MSE, the structure of the input layer corresponds to the order of the ARMA model. Through the Monte Carlo simulation, a series of model simulation data is obtained. We used the defined neural network model to make a comprehensive comparative analysis of the sequences of different coefficients, sequence lengths, and orders.

This paper is organized as follows. The theoretical basis of the two methods is described in Section 2. In Section 3, the setting of neural network parameters is introduced and verified by the Monte Carlo method. In Sections 4 and 5, the methods proposed in this paper are compared with the typical existing methods, and the effect of using the actual observation data is also analyzed. Finally, the conclusions are provided in Section 6.

2. Methodology

2.1. ARMA (p, q) Model

The autoregressive–moving-average is an important simulation method of stationary time series. Before discussing the order judgment of the time series, we need to analyze the structure of the ARMA model. $\{X_t\}$ is an ARMA (p, q) process if $\{X_t\}$ is stationary and if for every t,

$$X_t - \phi_1 X_{t-1} - \cdots - \phi_p X_{t-p} = Z_t + \theta_1 Z_{t-1} + \cdots + \theta_q Z_{t-q} \tag{1}$$

where $\{Z_t\} \sim WN(0, \sigma^2)$ and the polynomials $(1 - \phi_1 z - \cdots - \phi_p z^p)$ and $(1 - \theta_1 z - \cdots - \theta_p z^p)$ have no common factors [17]. The p of the left-hand side of Equation (1) represents the order of the autoregressive (AR) process. Similarly, the q of the right-hand side is the order of the moving-average (MA) process. When they are equal, the ARMA model has mathematical symmetry.

For a mathematical model, it is an important requirement for the effectiveness to be able to fully represent information of the sequence, which is also true for the ARMA model. In Equation (1), Z_t is actually a sequence of white noise. Its characteristic is that the value of Z_t does not affect the trend of X_t. At the same time, when the parameters of Equation (1) and X_t are known, the value of Z_t can also be determined within a certain range. The characteristics of Z_t provide a basis for us to determine the order of the ARMA model through the neural network.

2.2. Back Propagation Neural Networks (BPNN)

BPNN discover intricate structures in large datasets by using the backpropagation algorithm to indicate how a machine should change its internal parameters [30]. The ARMA model in this paper is a linear time-invariant system, which can be effectively simulated by using BPNN. BPNN is a multi-layer feedforward neural network trained according to the error back propagation algorithm. The complete neural network structure is composed of a large number of neurons. A typical BPNN consists of three layers: input layer, hidden layer, and output layer. Generally, we use normalized data as the input layer. Different from the input layer, the neurons in the hidden layer and the output layer have computational functions and have similar definitions. For neurons in the hidden layer and output layer, its iteration consists of two parts, namely forward propagation and back propagation. The forward propagation of a single neuron consists of two steps. First, calculate $\{z\}$ through weight and bias, and then calculate $\{a\}$ through an activation function $g(x)$, where $\{a\}$ is the input layer of the next layer of neurons or output layer. According to the calculation result of forwarding propagation, the weight and deviation are updated through back propagation. The back propagation of the BPNN is calculated by the gradient descent method. After many iterations, the neural network can fit the data with less error. It is worth noting that the activation functions of the hidden layer and the output layer can be different. Figure 1 shows single neuron calculation and BPNN structure.

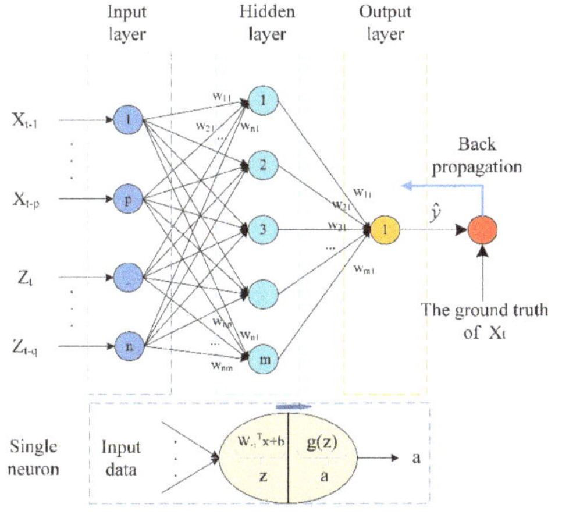

Figure 1. Single neuron calculation and BPNN structure.

When using the BPNN to simulate the ARMA model, we expect this method to be able to estimate the parameters of the ARMA model and obtain a method for determining the order of the model. Hossain et al., 2020, studied artificial neural networks (ANN) to determine the order of ARMA model, but this method did not consider the influence of bias in the process of formula derivation [20]. Therefore, we re-derive the relevant formula. Equation (1) can be rewritten in the following form

$$X_t = \sum_{i=1}^{p} \phi_i X_{t-i} + \sum_{j=0}^{q} \theta_j Z_{t-j} \qquad (\theta_0 = 1) \tag{2}$$

where X_t is the time series, Z_t is the noise sequence, ϕ_i and θ_j are coefficients of ARMA model.

Next, we compare the difference between the calculation method of BPNN and Equation (2). Figure 1 shows the processing of the input data by a single neuron. According to this, we can obtain the calculation process of the hidden layer neuron on the input data as follows

$$\begin{bmatrix} z_1 \\ z_2 \\ z_3 \\ \vdots \\ z_m \end{bmatrix} = \begin{bmatrix} w_{11} & w_{12} & w_{13} & \cdots & w_{1n} \\ w_{21} & w_{22} & w_{23} & \cdots & w_{2n} \\ w_{31} & w_{32} & w_{33} & \cdots & w_{3n} \\ \vdots & \vdots & \vdots & \vdots & \vdots \\ w_{m1} & w_{m2} & w_{m3} & \cdots & w_{mn} \end{bmatrix} \cdot \begin{bmatrix} x_1 \\ x_2 \\ x_3 \\ \vdots \\ x_n \end{bmatrix} + \begin{bmatrix} b_1 \\ b_2 \\ b_3 \\ \vdots \\ b_n \end{bmatrix} \tag{3}$$

or

$$Z^{[1]} = W^{[1]}X + b^{[1]}$$

Next,

$$\begin{bmatrix} a_1 \\ a_2 \\ a_3 \\ \vdots \\ a_m \end{bmatrix} = \begin{bmatrix} g_1(z_1) \\ g_1(z_2) \\ g_1(z_3) \\ \vdots \\ g_1(z_m) \end{bmatrix} \tag{4}$$

or

$$a^{[1]} = g_1(Z^{[1]})$$

where X is a column vector composed of input data, $W^{[1]}$ is a column vector composed of W_i^T, $b^{[1]}$ sis a column vector composed of bias, g is the activation function used, and $a^{[1]}$ is a column vector composed of activation value. The output of the proposed BPNN can be written as follows

$$[z_1] = \begin{bmatrix} w_{11} & w_{12} & w_{13} & \cdots & w_{1m} \end{bmatrix} \cdot \begin{bmatrix} a_1 \\ a_2 \\ a_3 \\ \vdots \\ a_m \end{bmatrix} + [b_1] \tag{5}$$

or

$$Z^{[2]} = W^{[2]}a^{[1]} + b^{[2]}$$

Next,

$$\hat{y} = a^{[2]} = g_2(Z^{[2]}) \tag{6}$$

where the meaning of each letter is similar to before. In this process, if we do not consider the bias and the activation function of the hidden layer, and at the same time set the activation function of the output layer to the linear activation function, we will obtain the following results

$$\hat{y} = W^{[2]}W^{[1]}X \tag{7}$$

Equation (7) is consistent with the conclusion derived by Hossain et al., 2020 [20].

Although the method described in Equation (7) can easily obtain the coefficient estimates of the ARMA model, it does not consider the influence of the bias and the nonlinear activation function on the neural network process. The existence of bias is of great significance to the operation of neural networks. It can improve the accuracy of neural network classification and reduce the noise in the evaluation process [31]. When we add bias, although the effect of neural network iteration can be improved, the coefficient estimates cannot be obtained as easily as Equation (7). This is because of the influence of the deviation column vector, the coefficient calculation of the ARMA model in the calculation of the BPNN can no longer be simply obtained through the weight matrix. Our other improvement to Equation (7) is the addition of a nonlinear activation function. This is not only because the nonlinear activation function can better exert the computational performance of the neural network but also the coefficient estimation of the ARMA model itself is a nonlinear process. In the symmetric formwork system, BPNN can overcome the shortcomings of insufficient randomness of ARMA order estimation. It is worth noting that the coefficient of the model can be better estimated by the least square method when we can accurately determine the order of the model [32,33].

3. Model Establishment

3.1. Simulation Settings

This section introduces the simulation methods separately from the simulation of data and the design of the artificial neural network. The paper mainly uses MATLAB 2017a to establish an analytical model, and the relevant calculations are also completed in the software.

In the real world, most systems can be modeled by ARMA (5, 5), so in this paper, we set the maximum value of AR and MA order to 5 (AR (1–5) and MA (0–5), the numbers represent the range of values for the corresponding order) [34]. All the simulated datasets in this paper are generated by Monte Carlo simulation. We constructed a time series (X_t) based on the random simulated noise series (Z_t) and the coefficients of the ARMA model. The expectation of the noise sequence is 0 and the variance is 1. The coefficients of the model were generated by a random method and met the conditions of causality and invertibility. The initial value of the time series was determined by the noise series.

For the neural network, we used Equation (8) to calculate the number of neurons in the hidden layer, and its value was a dynamic integer. The maximum number of epochs to train was 100, the performance goal was set to 10^{-7}, and the training was terminated when the MSE did not drop for 10 consecutive iterations. The neural network parameters were updated using Adam optimizer, and its learning rate was set to 0.01, which was chosen empirically [35]. Although the ARMA model is a linear time-invariant system, the linear unit (ReLU) activation function of hidden neurons cannot handle occasional discrete data well. Thus, we used a nonlinear activation function (Sigmoid) as the activation function [36–38].

$$M = integer(sqrt(m+1) + 15) \tag{8}$$

Based on the above conditions, we built 30 neural networks (combination of AR (1–5) and MA (0–5)) to analyze time series. For different neural networks, we converted the time series into corresponding datasets, 80% of the processed data was used as the training set and the rest as the validation set. We used the MSE of the validation set as the basis for judging the order. Figures 2 and 3 show a system identification block diagram.

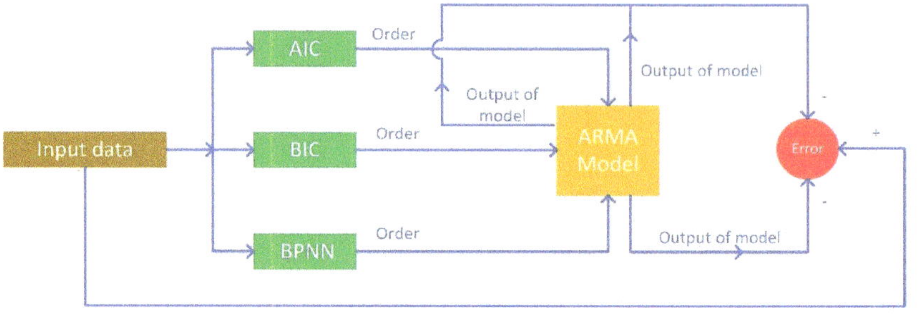

Figure 2. System identification block diagram.

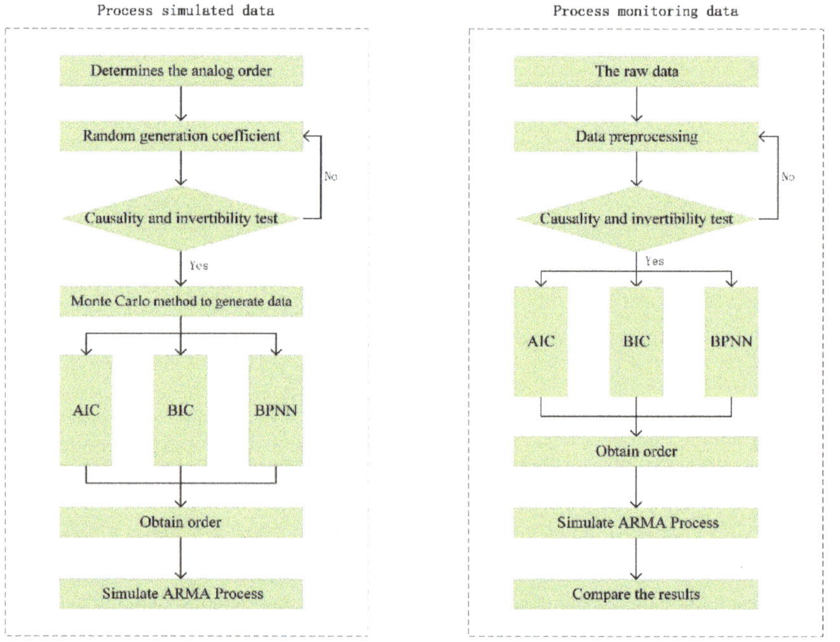

Figure 3. ARMA model-building method.

3.2. Pre-Simulation

According to the settings in Section 3.1, we verified the established model. We tested whether the neural network is suitable for the simulation of the ARMA model to verify the correctness of the derivation in Section 3.2. On the other hand, the proposed method is used for order estimation, and special cases are discussed at the same time.

Figure 4 is the MSE loss value of the time series conforming to the ARMA (3, 2) model. Figure 4a,b represent the time series by simulating the neural network of ARMA (3, 2) and ARMA (1, 2) respectively. We use this example to illustrate the judgment theory of neural networks. As shown in Figure 4a, when the time series passes the correct model, MSE has been in a downward trend and, after a sufficient period of the epoch, MSE reaches the target value (10^{-7}). In Figure 4b, when the time series passes through the mismatched model, we find that MSE reaches the optimal value (10^{-3}) at 33 Epochs and does not drop again for ten consecutive times. The reason for this phenomenon is that, for the correct neural network model, the model can approach an analytical solution after sufficient iterations. For the wrong neural network models, the value of MSE will often not drop after reaching the critical point, and the correct model can obtain a satisfactory MSE. This is also the basis for us to judge the order of the ARMA model through MSE.

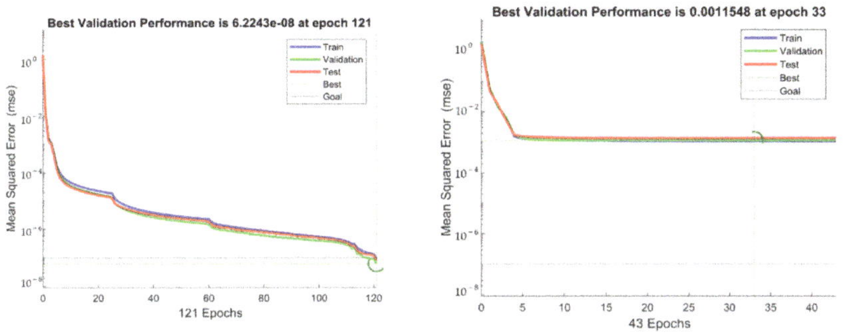

(**a**) (**b**)

Figure 4. Training neural network of ARMA (3, 2): (**a**) ARMA (3, 2), (**b**) ARMA (1, 2).

Next, the problem we needed to solve was how to determine the order of the ARMA model through the BPNN. From Figures 1 and 4, we can see that when determining the best input layer of the BPNN, the p and q of the input layer corresponded to the best order of the ARMA model. Therefore, we expected that the neural network's MSE loss function should be the smallest for the correct model order. Finally, the time series obtained MSE through 30 possible neural network structures.

Figure 5 shows the MSE calculation results of two different time series. It can be seen from Figure 5a that, as the order increases, the MSE presents an obvious downward trend, in which the red circle marks the true model orders. When the critical point is reached, the MSE will not change significantly with the increase of the order because high-level neural network models can reflect low-level changes. In the calculation, we found that the calculation result of MSE has the special case shown in Figure 5b. Figure 5b simulates a special case of the ARMA (2, 1) calculation. The three points represented by the red circle in Figure 5 may all be the value of the order, and the MSE is relatively small at these three points. Therefore, in addition to comparing MSE, we also introduce the gradient to determine the order of the model when there are multiple critical points. When the descending gradient of the critical point is the largest and the MSE is small, the point is considered to be the best value for this set of critical points. From Figure 5, we find that the asymmetric ARMA structure is more prone to result judgment difficulties because the descending gradient of its MSE is gentler near the correct values.

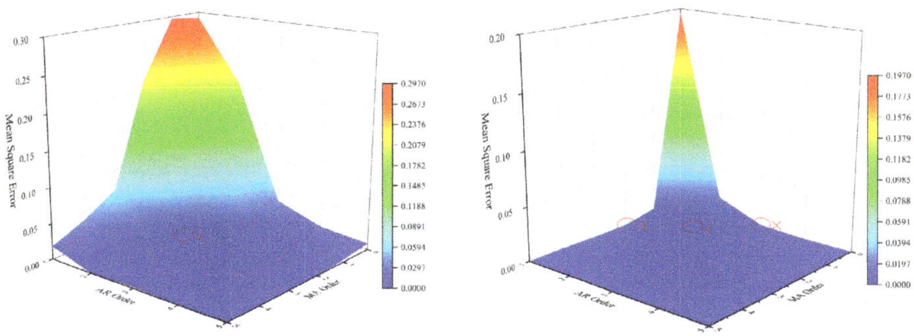

(**a**) (**b**)

Figure 5. MSE of BPNN: (**a**) ARMA (2, 2), (**b**) ARMA (2, 1).

4. Performance on Simulated Data

4.1. Different Coefficients

In the process of simulating data conforming to the ARMA model, the coefficients are restricted by many conditions. According to the definition of the ARMA model, the coefficients need to meet the requirements of causality and invertibility. In [17], the judgments of causality and invertibility are given by Equations (9) and (10).

Causality is equivalent to the condition

$$\phi(z) = 1 - \phi_1 z - \cdots - \phi_p z^p \neq 0 \text{ for all } |z| \leq 1 \tag{9}$$

Invertibility is equivalent to the condition

$$\theta(z) = 1 + \theta_1 z + \cdots + \theta_q z^q \neq 0 \text{ for all } |z| \leq 1 \tag{10}$$

where $\phi(\cdot)$ and $\theta(\cdot)$ are the pth and qth-degree polynomials. The complex z is used here since the zeros of a polynomial of degree $p > 1$ or $q > 1$ may be either real or complex. The region is defined by the set of complex z such that $|z| = 1$ is referred to as the unit circle. From Equations (9) and (10), the conditions of causality and invertibility are satisfied when the roots of $\phi(z) = 0$ and $\theta(z) = 0$ are outside the unit circle. In the calculation, we found that it is also necessary to consider whether the selected coefficients can effectively reflect the characteristics of the model in addition to causality and invertibility. In order to improve the sensitivity of the model, we set the minimum absolute value of the coefficient to 0.1. This avoids the fact that the coefficients are too small to make the polynomial difficult to identify.

In this section, we have simulated three ARMA models, each of which used 25 sets of different coefficients to simulate time series. Since the symmetrical ARMA model was less prone to result judgment difficulties, we used the asymmetrical ARMA model here. The three models were ARMA (1, 2), ARMA (2, 3), and ARMA (4, 2). Table 1 is the model coefficients we obtained through the random method, and Figure 6 is the verification of causality and invertibility.

In Figure 6, the roots of $\phi(z) = 0$ and $\theta(z) = 0$ for each model are shown in a different color. All roots are outside the unit circle, which shows that the coefficients meet the requirements of causality and invertibility. For the coefficients in Table 1, we simulated 30 different realizations of the system's response with time series lengths of 400. From these overdetermined ARMA model orders, the goal was to determine the correct ARMA model order using BPNN and compare its results with AIC and BIC. Another purpose was to study whether the effects of different order determination methods under random parameters were consistent. Model identification using the AIC and BIC was performed using functions in MATLAB R2017a.

Figure 7 is a stacked area diagram of the order estimation results. Figure 7 shows that the order estimation accuracy of BPNN is above 90%, AIC is below 10%, and the results of BIC are unstable. Comparing Figure 7a,b, the judgment results of BPNN and AIC are relatively stable, while the judgment efficiency of BIC criteria is significantly reduced and affected by the change of coefficients. For Figure 7c, the judgment effect of the BIC is basically the same as that of the AIC, and there is no obvious change in the BPNN. For the same model, the correct rate of BPNN is the highest, and the AIC is the lowest. The BIC is somewhere in between, but it is more sensitive to changes in model coefficients. For different ARMA models, the order estimation results of BPNN under different coefficients are relatively stable and accurate. In addition, the accuracy of the BIC is significantly reduced when the ARMA order is higher. For example, in ARMA (4, 2), its judgment effect is almost the same as that of the AIC. The influence of the change of order on different judgment criteria is analyzed in detail in Section 4.3. On the whole, the judgment result of BPNN has obvious advantages.

Table 1. ARMA coefficient.

(a) ARMA (1, 2)			
Number	ϕ_1	θ_1	θ_2
1	0.68598	0.52501	0.30802
2	0.64442	0.31632	0.25568
3	0.6858	0.52559	0.43188
4	0.69234	0.16724	0.36293
5	0.64172	0.40218	0.26578
6	0.69125	0.10683	0.69129
7	0.67355	0.13828	0.39175
8	0.58189	0.44822	0.29035
9	0.65346	0.5068	0.37852
10	0.65751	0.46853	0.31898
11	0.57612	0.55828	0.1989
12	0.5549	0.5677	0.36828
13	0.57328	0.16801	0.40967
14	0.62506	0.11917	0.45852
15	0.64303	0.12413	0.6465
16	0.66412	0.18832	0.58978
17	0.52794	0.59022	0.20596
18	0.65484	0.54687	0.44928
19	0.50026	0.14956	0.65269
20	0.57071	0.65167	0.23093
21	0.51364	0.23927	0.27787
22	0.58889	0.18203	0.58252
23	0.5552	0.19295	0.54223
24	0.66197	0.22495	0.50503
25	0.69481	0.31173	0.66661

(b) ARMA (2, 3)					
Number	ϕ_1	ϕ_2	θ_1	θ_2	θ_3
1	0.56187	0.11088	0.31731	0.10306	0.44403
2	0.29368	0.22157	0.17051	0.27467	0.32838
3	0.19265	0.66323	0.48151	0.26584	0.21114
4	0.28469	0.26203	0.227	0.20943	0.46187
5	0.12039	0.54207	0.12816	0.50777	0.13457
6	0.50685	0.18235	0.11274	0.22457	0.43534
7	0.25565	0.66183	0.12575	0.16069	0.66979
8	0.24022	0.26447	0.26283	0.27753	0.25457
9	0.30066	0.25647	0.25259	0.10025	0.44024
10	0.23385	0.27877	0.29301	0.36222	0.33182
11	0.52257	0.24083	0.26788	0.29762	0.27343
12	0.37619	0.15621	0.30297	0.13064	0.25225
13	0.29519	0.36147	0.11577	0.21734	0.46065
14	0.41579	0.42125	0.18491	0.10473	0.568
15	0.25602	0.57018	0.13555	0.14648	0.5147
16	0.37881	0.45223	0.18419	0.21338	0.48166
17	0.33787	0.31189	0.20047	0.27474	0.24118
18	0.40362	0.45338	0.2618	0.2151	0.3823
19	0.15286	0.38683	0.51148	0.15672	0.16223
20	0.39763	0.16128	0.14097	0.29433	0.26854
21	0.32716	0.51292	0.1052	0.14028	0.63252
22	0.22749	0.28874	0.16985	0.16377	0.46749
23	0.41344	0.43832	0.42688	0.23569	0.27483
24	0.26962	0.64428	0.50579	0.25514	0.10056
25	0.62415	0.11642	0.21241	0.23787	0.27413

Table 1. *Cont.*

(c) ARMA (4, 2)						
Number	ϕ_1	ϕ_2	ϕ_3	ϕ_4	θ_1	θ_2
1	0.1173	0.14669	0.11194	0.28898	0.26858	0.27696
2	0.10356	0.38866	0.27642	0.22115	0.34628	0.58485
3	0.19003	0.10435	0.16407	0.19478	0.35867	0.35484
4	0.19558	0.17231	0.18333	0.40822	0.52755	0.18906
5	0.19987	0.1165	0.12641	0.47265	0.52957	0.18639
6	0.16881	0.14111	0.4436	0.1778	0.26526	0.59525
7	0.1385	0.1794	0.1131	0.42791	0.35514	0.29305
8	0.12084	0.26884	0.13637	0.38152	0.37562	0.20007
9	0.14009	0.1314	0.14363	0.50568	0.35714	0.21889
10	0.11697	0.17643	0.19267	0.19714	0.57655	0.28871
11	0.32149	0.14979	0.13027	0.35725	0.19231	0.45819
12	0.11596	0.1146	0.29422	0.22075	0.6898	0.30744
13	0.19825	0.2323	0.111	0.12595	0.107	0.58112
14	0.15196	0.10187	0.22066	0.45185	0.23608	0.36163
15	0.27141	0.1968	0.15257	0.28422	0.57048	0.39562
16	0.11062	0.22939	0.37438	0.25847	0.68981	0.19552
17	0.14397	0.10805	0.39279	0.15033	0.1266	0.458
18	0.19712	0.20237	0.28527	0.12375	0.23074	0.68393
19	0.18443	0.1625	0.10434	0.32789	0.33492	0.2104
20	0.4223	0.19247	0.11811	0.21622	0.22415	0.35038
21	0.11076	0.26585	0.42486	0.11908	0.16995	0.58259
22	0.15044	0.19221	0.14957	0.19805	0.38759	0.4243
23	0.11706	0.13743	0.14761	0.14352	0.10387	0.51243
24	0.17383	0.1815	0.17688	0.30087	0.55079	0.35127
25	0.23135	0.23691	0.10951	0.37275	0.13111	0.42879

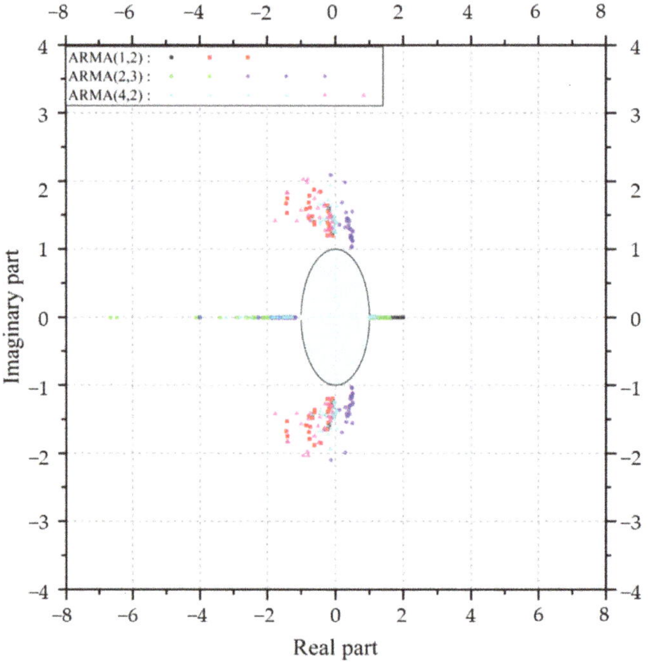

Figure 6. The verification of causality and invertibility.

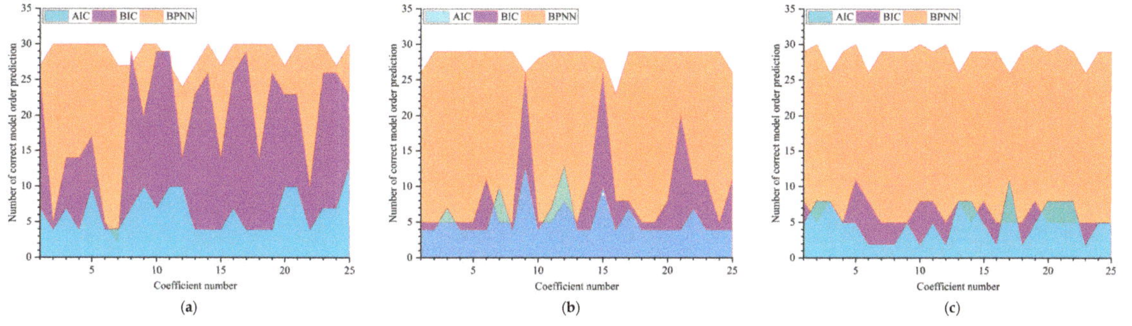

Figure 7. Comparison of the correct number of estimation methods for each order under different coefficients: (**a**) ARMA (1, 2), (**b**) ARMA (2, 3), (**c**) ARMA (4, 2).

4.2. Different Length

In Section 4.1, we discussed the effect of each judgment criterion under different coefficients. Next, we study the effect of the length of the time series on the accuracy. We selected a set of representative coefficients from the three models in Section 4.1, where ARMA (1, 2) selected the 13th group, ARMA (2, 3) selected the 21st group, and ARMA (4, 2) selected the 9th group. Then, we simulated 100 different responses with time series lengths of 200, 400, 600, 800, and 1000. The expressions of the three models are shown in Equation (11).

$$
\begin{aligned}
&ARMA(1,2) : X_t = Z_t + 0.5733X_{t-1} + 0.1680Z_{t-1} + 0.4097Z_{t-2} \\
&ARMA(2,3) : X_t = Z_t + 0.3272X_{t-1} + 0.5129X_{t-2} + 0.1052Z_{t-1} + 0.1403Z_{t-2} + 0.6325Z_{t-3} \\
&ARMA(4,2) : X_t = Z_t + 0.1401X_{t-1} + 0.1314X_{t-2} + 0.1436X_{t-3} + 0.5057X_{t-4} + 0.3571Z_{t-1} + 0.2189Z_{t-2}
\end{aligned} \tag{11}
$$

The model order estimation results are presented in Figure 8. From Figure 8, we can find that the BPNN can provide accurate order estimates for each length of the signal in models of different orders and the accuracy rate is above 90%. The accuracy of AIC and BIC has a certain upward trend with the increase of sequence length, while the estimated result of BIC can reach 90% under certain models. In Figure 8a,b, the judgment effect of the BIC and the BPNN is basically the same in the case of a long sequence. From Figure 8c, we can clearly find that the judgment result of the BPNN is much more accurate than the AIC and the BIC under the high-order model. In Figure 8c, the accuracy of AIC and BIC is below 40%. Similar to Section 4.2, the BIC performs better under low-order models than high-order models. In addition, we find that the accuracy of the AIC is low, but the effect is relatively stable under different models.

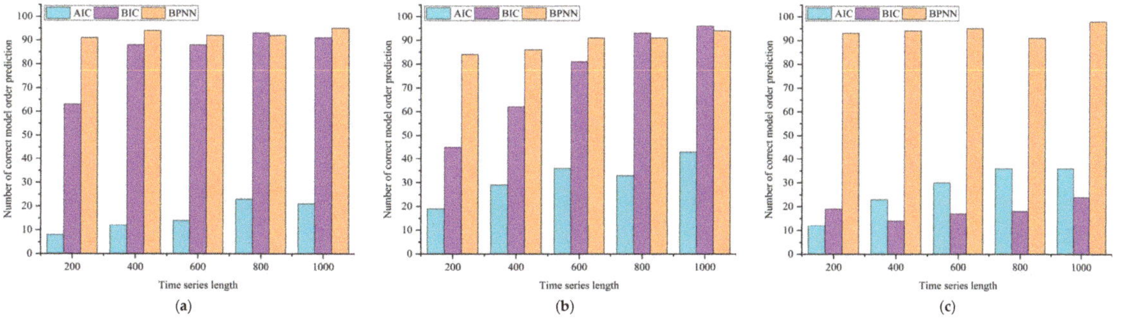

Figure 8. Comparison of the correct number of estimation methods for each order under different sequence length: (**a**) ARMA (1, 2), (**b**) ARMA (2, 3), (**c**) ARMA (4, 2).

4.3. *Different Order*

In this section, we fixed the maximum AR and MA orders at 5 (AR (1–5) and MA (0–5)). We performed 100 Monte Carlo simulations on these 30 models and used the AIC, BIC, and BPNN to estimate the order. In Section 4.1, we found that the coefficient affects the estimate of the order. In order to make the results more representative, we randomized the coefficients of each model and met the conditions of causality and invertibility. In Section 4.2, we already knew that the length of the time series would affect the estimation of the order, so we set the length of the series to 1000 for better performance of all the methods. The setting of coefficient randomization can better study the effect of BPNN order estimation under different models. The length of the time series can make the accuracy of the AIC and the BIC higher, so as to better compare with the estimation results of the BPNN.

Figure 9 shows the order estimation results of different models. In Figure 9, the accuracy of BPNN is generally above 90%. The accuracy of the BIC is above 70% when the order is small, but it does not work well under higher-order models. Although the accuracy of the AIC is below 30%, the result is relatively stable. Combined with the present conditions, the order estimation of BPNN can have a prominent performance under random coefficient and different model orders. Another point worth noting is that when the order of the model is higher, only the BPNN can obtain satisfactory estimation results. From this example, it can be found that BPNN still has an excellent estimation effect, even though the ARMA model has mathematical symmetry.

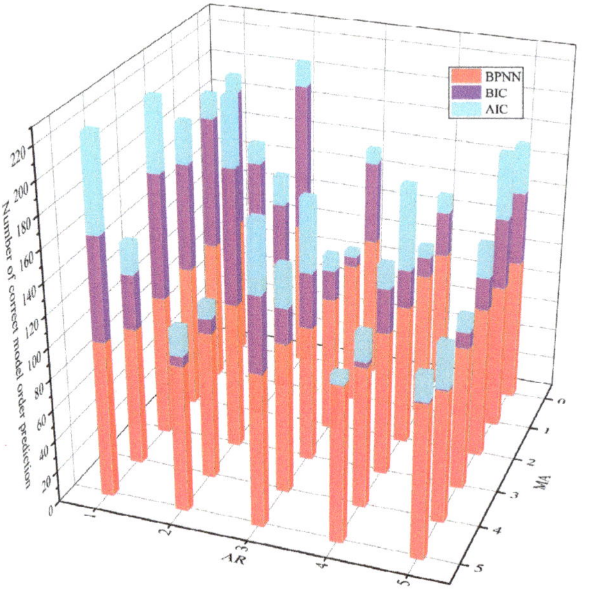

Figure 9. Comparison of the correct number of estimation methods for each order under different orders.

Hossain et al., 2020, simulated physiological systems through ARMA and BPNN. As with their findings, the BPNN always shows an accuracy rate of more than 90% under different conditions. However, we found that AIC and BIC accuracy were low in our study, which may be due to a different coefficient selection. This may be because we did not have too much human intervention in the choice of coefficients.

5. Application

5.1. High Formwork Safety Monitoring System

The high formwork safety monitoring system is a solution for real-time automatic safety monitoring of many major safety risk points during the pouring construction process of the tall formwork support system. The system uses wireless automatic networking, high-frequency continuous sampling, real-time data analysis and on-site sound and light alarms. There are four main components of the high formwork safety monitoring system: collector, analyzer, cloud platform, and client. The collector is responsible for sampling and uploading sensor data; the analyzer networks the collector, transfers the data, alerts, and uploads the data to the cloud platform; the cloud platform is responsible for data storage, display, early warning, data analysis, and other functions on the server; the client mainly implements the remote configuration of the data display and monitoring system on the cloud platform. The structure monitored in this paper has obvious symmetry, and the arrangement of the measuring points is also symmetrical and orderly. The composition of the high formwork safety monitoring system is shown in the Figure 10. Figure 11 shows the 3D model of the high formwork and the installation scheme of the instrument. Table 2 lists the instrument-related parameters.

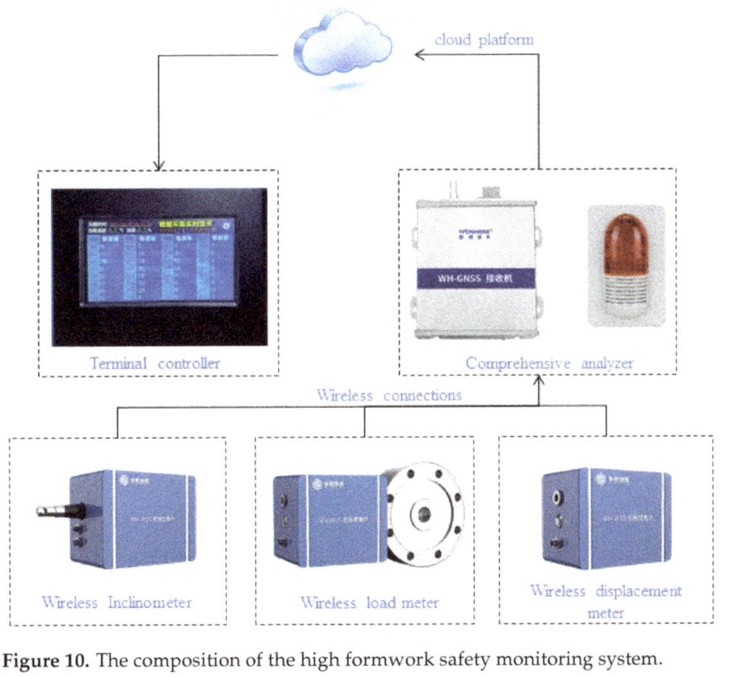

Figure 10. The composition of the high formwork safety monitoring system.

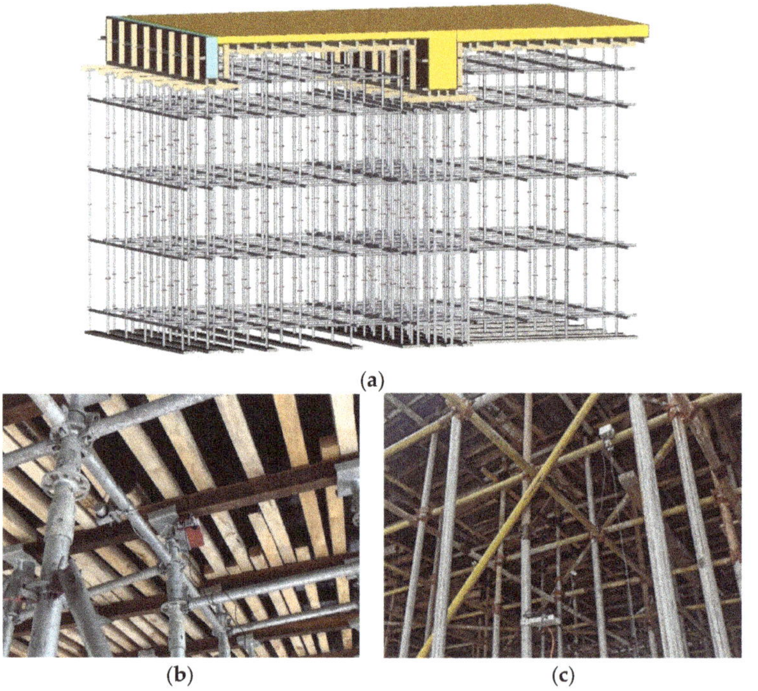

Figure 11. Monitoring arrangement of the high formwork system: (**a**) the 3D model of the high formwork; (**b**) the installation of the collector; (**c**) the installation of the analyzer.

Table 2. The instrument-related parameters.

	Wireless Receiver	Wireless Inclinometer	Wireless Load Meter	Wireless Displacement Meter
Version	WH-GNSS	WH-WTS	WH-WLS	WH-WDS
Communication interface	10 M, RS232	NBTOT	NB-IOT/LORA	NBTOT
Power consumption	<20 uA	<60 uA	<60 uA	<60 uA
Positioning accuracy	Plane: ±3 mm Elevation: ±5 mm	±0.01°	0.05	0.2%F·S
Operating temperature	−30~70 °C	−40~85 °C	−20~65 °C	−20~65 °C
Producer	HUAHEIOT	HUAHEIOT	HUAHEIOT	HUAHEIOT
Country	China	China	China	China

5.2. Application in Stress Sequence

In the past, the ARMA model was often used to remove the noise of the time series to obtain the trend of the series [39–41]. These documents select a number of different ARMA models to process the time series and obtain the optimal solution by comparing the results. Selecting a model through the results often takes more time when the amount of data is large, and an accurate estimation of the order can reduce the workload and facilitate the batch processing of data. In this subsection, we demonstrate that the proposed model order selection method based on BPNN can be used to analyze stress sequences. The time series we analyzed comes from a part of the stress change of the high formwork system. According to the loading status of the system, the stress change of the high formwork can be divided into the loading phase and load stabilization phase. For the stress time series in the use phase, the ARMA model generally satisfies the requirements of the series causality. If the time series contains data in the loading phase, then we often need to perform nonlinear processing (differential) on the data to meet the requirements of causality. The accurate processing of monitoring data results is an indispensable part of the high formwork safety monitoring system. Based on the initial data collected by the high formwork safety monitoring system in the actual project, this paper discusses the specific application of the proposed method in the post-processing of monitoring data. In the following, we use two examples to verify the effect of the order selection method proposed earlier.

Example 1. *This example considers a time series that meets the requirements of causality, and the data is obtained from engineering field measurements. The data includes stress changes at 37 positions, and the sequence used in this example is one of the data. Since the increase in the length of the sequence has no adverse effects, we intercepted the sequence with a length of 5000, which is the stress change at the steady stage of the load. We use AIC, BIC, and BPNN to estimate the best order of the model. As before, regarding the setting of the BPNN, we use 80% of the data as the training set and the rest as the test set. The parameter settings are also the same as before. Finally, the ARMA parameters were estimated using the least-squares method for the model order estimated by the BPNN, the AIC and the BIC.*

Figure 12a plots the original data and the sequence obtained by different processing methods, and the residuals of the different methods are shown in Figure 12b. In order to better describe the distribution of the data, we draw the envelope of the obtained sequence and calculate the average width between the upper and lower envelopes. Compared with the AIC and the BIC, the BPNN method reduces the average width of the envelope by 83.08% and 9.16%. It can be found that the BPNN and BIC have similar and accurate judgments on sequence trends for the data in this example, while the AIC has a higher degree of dispersion. In addition, there is no obvious difference in residuals of the three judgment methods.

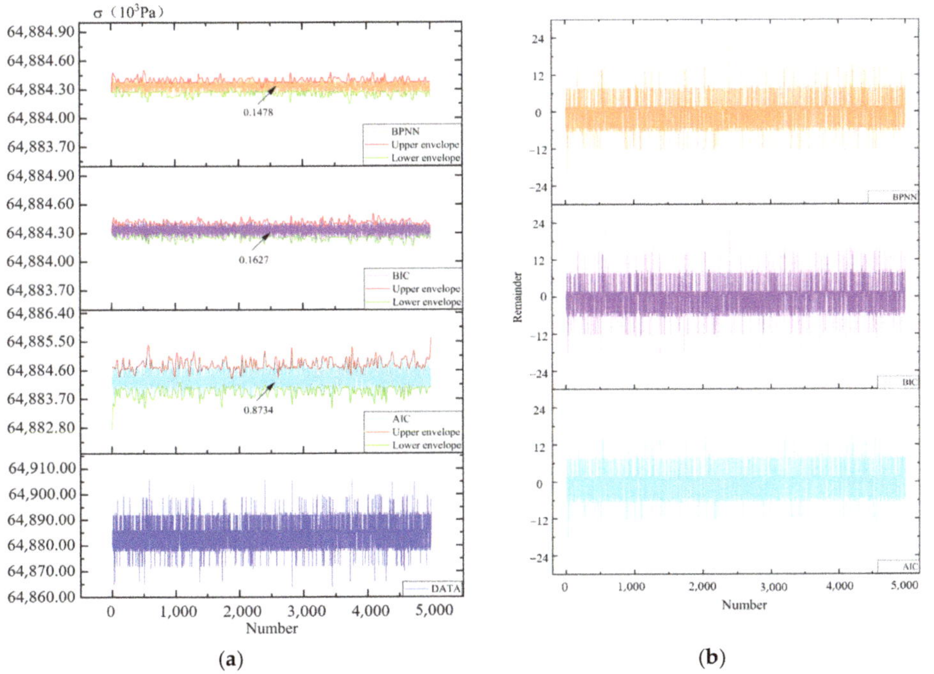

Figure 12. Simulation results of different methods: (**a**) Sequence trend (**b**) Residual sequence.

Example 2. *The data used in this example includes the loading phase and the load stabilization phase, and the length of the sequence is 29,610. The setup of the neural network is the same as in Example 1. The difference from Example 1 is that this time series does not meet the requirements of causality, so we have performed different processing on it. The differentiated sequence meets the requirements of causality. Similar to before, we estimated the order of the differenced sequence and used the ARMA process for fitting. Finally, we restored the sequence [42].*

Figure 13 shows the processing results of different judgment methods. The obvious difference between Figure 12a and 13a is that the sequence of the former is stable, while the latter has an obvious rising stage, which is why the latter needs to be processed by difference. Compared with the AIC and the BIC, the BPNN method reduces the average width of the envelope by 51.91% and 52.14%. Therefore, we find that the BPNN's analysis of data trends is more compact than the AIC and the BIC. This shows that the model obtained by the BPNN for order estimation has a better effect on noise extraction.

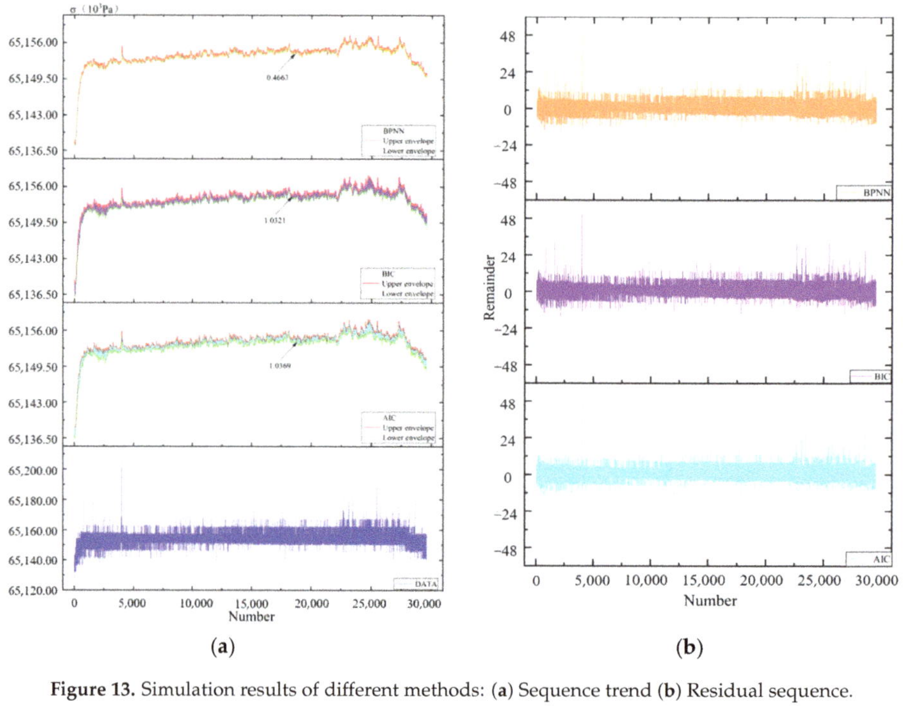

Figure 13. Simulation results of different methods: (**a**) Sequence trend (**b**) Residual sequence.

6. Conclusions

This paper proposes a modeling method for the high formwork monitoring data. The details of establishing the ARMA model and BPNN have been presented, and the algorithm of model order estimation by BPNN is introduced. Through the method of Monte Carlo simulation, we studied the accuracy of the three methods under different coefficients, different sequence lengths, and different model orders. For the actual measured stress series, we used three methods to estimate the model order and then used the least square method to estimate the model coefficients. Finally, we applied the established model to the symmetrical high formwork monitoring data. According to the simulation and application results, the following conclusions can be made.

- For each system, the accuracy rate of the proposed model order selection method is above 90%, and both show better performance than the AIC or the BIC. At the same time, the BIC criterion is better than the AIC when the model order is lower;
- In the Monte Carlo simulation, changing the model's coefficients will affect the accuracy of the BIC judgment, the instability increases significantly in the higher-order model. However, the order judgment method of BPNN still has an accuracy rate of more than 90%;
- The mathematically symmetric ARMA model is more likely to make errors in the BPNN method, so this type of model needs to be judged in conjunction with the MSE descent gradient.
- The judgment efficiency of AIC and BIC will increase as the length of the time series increases. The proposed BPNN order judgment method is not sensitive to the change of sequence length and has a relatively high accuracy rate;
- For changes in the order of the model, both AIC and BIC are more sensitive. In particular, the BIC cannot be judged correctly when the model order is high. The BPNN still maintains a good judgment effect;
- For the measured data used in the paper that meets the requirements of causality, the judgment effect of BPNN is not significantly different from that of the BIC, but the AIC is obviously inferior. When the time series does not meet the causality requirements, we will transform it into a stationary series. The analysis result shows that the processing result of BPNN increased by about 50%;
- The stress sequence of the high formwork can be processed by the ARMA process to obtain its change trend and noise sequence. This is feasible for obtaining effective information on the stress sequence.

Due to the influence of computing resources, the model of all orders is not calculated when considering sequence coefficients and sequence length. In addition, the model established in this paper does not consider the impact of accidental factors. We will discuss the effects of accidental factors on model-building in future studies, and further study the methods of predicting periodic data.

Author Contributions: Conceptualization, Y.Y. and G.Y.; methodology, Y.Y. and G.Y.; software, Y.Y. and L.Y.; formal analysis, Y.Y. and L.Y.; writing—original draft preparation, Y.Y. and G.Y.; writing—review and editing, Y.Y. and G.Y. All authors have read and agreed to the published version of the manuscript.

Funding: This work was funded by National Key R&D Program of the Ministry of Science and Technology (No. 2019YFD1101005-4), Fundamental Research Funds for the Central Universities (2020CDJQY-A067) and the 111 project of the Ministry of Education and the Bureau of Foreign Experts of China (No. B18062).

Institutional Review Board Statement: Not applicable.

Informed Consent Statement: Not applicable.

Data Availability Statement: Please contact the corresponding author.

Conflicts of Interest: The authors declare no conflict of interest.

References

1. Beale, R.G. Scaffold research—A review. *J. Constr. Steel Res.* **2014**, *98*, 188–200. [CrossRef]
2. Yang, Y.; Cheng, Q.; Zhu, Y.; Wang, L.; Jin, R. Feasibility Study of Tractor-Test Vehicle Technique for Practical Structural Condition Assessment of Beam-Like Bridge Deck. *Remote Sens.* **2020**, *12*, 114. [CrossRef]
3. Wu, W.; Yang, H.; Chew, D.A.S.; Yang, S.-H.; Gibb, A.G.F.; Li, Q. Towards an autonomous real-time tracking system of near-miss accidents on construction sites. *Autom. Constr.* **2010**, *19*, 134–141. [CrossRef]
4. Huang, Y.L.; Chen, W.F.; Chen, H.J.; Yen, T.; Kao, Y.G.; Lin, C.Q. A monitoring method for scaffold-frame shoring systems for elevated concrete formwork. *Comput. Struct.* **2000**, *78*, 681–690. [CrossRef]
5. Entezami, A.; Sarmadi, H.; Behkamal, B.; Mariani, S. Big Data Analytics and Structural Health Monitoring: A Statistical Pattern Recognition-Based Approach. *Sensors* **2020**, *20*, 2328. [CrossRef] [PubMed]

6. Yang, Y.; Yang, Y.B.; Chen, Z.X. Seismic damage assessment of RC structures under shaking table tests using the modified direct stiffness calculation method. *Eng. Struct.* **2017**, *131*, 574–586. [CrossRef]
7. Yang, Y.; Xiang, C.; Jiang, M.; Li, W.; Kuang, Y. Bridge Damage Identification Method Considering Road Surface Roughness by Using Indirect Measurement Technique. *China J. Highw. Transp.* **2019**, *32*, 99–106.
8. Bai, J.; Chen, H.; Zhao, J.; Liu, M.; Jin, S. Seismic design and subassemblage tests of buckling-restrained braced RC frames with shear connector gusset connections. *Eng. Struct.* **2021**, *234*, 112018. [CrossRef]
9. Chen, H.; Bai, J. Seismic performance evaluation of buckling-restrained braced RC frames considering stiffness and strength requirements and low-cycle fatigue behaviors. *Eng. Struct.* **2021**, *239*, 112359. [CrossRef]
10. Qin, Y.; Shu, G.P.; Zhou, G.G.; Han, J.H. Compressive behavior of double skin composite wall with different plate thicknesses. *J. Constr. Steel Res.* **2019**, *157*, 297–313. [CrossRef]
11. Yao, G.; Wu, C.; Yang, Y. Scientometric analysis for mechanical performance on broken-line long-span steel structure in construction considering geometric nonlinearity. *Symmetry* **2021**, *137*, 1229. [CrossRef]
12. Wang, Y.H.; Liu, Y.J.; Zhou, X.H. Seismic behavior of steel coupling beam with different buckling constraint materials. *Constr. Build. Mater.* **2017**, *149*, 111–129. [CrossRef]
13. Huang, N.E.; Shen, Z.; Long, S.R.; Wu, M.L.C.; Shih, H.H.; Zheng, Q.N.; Yen, N.C.; Tung, C.C.; Liu, H.H. The empirical mode decomposition and the Hilbert spectrum for nonlinear and non-stationary time series analysis. *Proc. R. Soc. Lond. Ser. A Math. Phys. Eng. Sci.* **1998**, *454*, 903–995. [CrossRef]
14. Bellahsene, H.; Taleb-Ahmed, A. ARMA order model detection using minimum of Kurtosis: Application on seismic data. *Arab. J. Geosci.* **2018**, *11*, 1–7. [CrossRef]
15. Yang, Y.; Li, C.; Ling, Y.; Tan, X.; Luo, K. Research on new damage detection method of frame structures based on generalized pattern search algorithm. *Chin. J. Sci. Instrum.* **2021**, *42*, 123–131.
16. Momin, B.; Chavan, G. Univariate Time Series Models for Forecasting Stationary and Non-stationary Data: A Brief Review. *Inf. Commun. Technol. Intell. Syst.* **2018**, *84*, 219–226.
17. Brockwell, P.J.; Davis, R.A.; Brockwell, P.J.; Davis, R.A. ARMA Models. In *Introduction to Time Series and Forecasting*, 3rd ed.; Springer Texts in Statistics; Springer: Cham, Switzerland, 2016; Volume 3, pp. 73–96.
18. Akaike, H. A Bayesian analysis of the minimum AIC procedure. *Ann. I Stat. Math.* **1978**, *30*, 9–14. [CrossRef]
19. Akaike, H. Statistical predictor identification. *Ann. I Stat. Math.* **1970**, *22*, 203–217. [CrossRef]
20. Hossain, M.-B.; Moon, J.; Chon, K.H. Estimation of ARMA Model Order via Artificial Neural Network for Modeling Physiological Systems. *IEEE Access* **2020**, *8*, 186813–186820. [CrossRef]
21. Hang, J.; Zhang, J.; Cheng, M. Fault diagnosis of wind turbine based on multi-sensors information fusion technology. *IET Renew. Power Gener.* **2014**, *8*, 289–298. [CrossRef]
22. Qian, X.; Iop. Parameter prediction based on Improved Process neural network and ARMA error compensation in Evaporation Process. In *IOP Conference Series: Earth and Environmental Science*; IOP Publishing: Bristol, UK, 2018; Volume 108, p. 022078. [CrossRef]
23. Kocak, C.; Dalar, A.Z.; Yolcu, O.C.; Bas, E.; Egrioglu, E. A new fuzzy time series method based on an ARMA-type recurrent Pi-Sigma artificial neural network. *Soft Comput.* **2020**, *24*, 8243–8252. [CrossRef]
24. Khashei, M.; Bijari, M. An artificial neural network (p, d, q) model for timeseries forecasting. *Expert Syst. Appl.* **2010**, *37*, 479–489. [CrossRef]
25. Chon, K.H.; Cohen, R.J. Linear and nonlinear ARMA model parameter estimation using an artificial neural network. *IEEE Trans. Biomed. Eng.* **1997**, *44*, 168–174. [CrossRef] [PubMed]
26. Yang, Y.; Lin, Y.; Bo, W.; Gang, Y.; Hang, L.; Soltys, R. Safety Prediction Using Vehicle Safety Evaluation Model Passing on Long-Span Bridge with Fully Connected Neural Network. *Adv. Civil Eng.* **2019**, *2019*, 1–12. [CrossRef]
27. Yao, G.; Guo, H.; Yang, Y.; Xiang, C.; Robert, S. Dynamic Characteristics and Time-History Analysis of Hydraulic Climbing Formwork for Seismic Motions. *Adv. Civil Eng.* **2021**, *2021*, 1–17.
28. Zipser, D.; Andersen, R.A. A back-propagation programmed network that simulates response properties of a subset of posterior parietal neurons. *Nature* **1988**, *331*, 679–684. [CrossRef]
29. Xu, L. Least mean-square error reconstruction principle for self-organizing neural-nets. *Neural Netw.* **1993**, *6*, 627–648. [CrossRef]
30. LeCun, Y.; Bengio, Y.; Hinton, G. Deep learning. *Nature* **2015**, *521*, 436–444. [CrossRef]
31. Yao, X.; Liu, Y. A new evolutionary system for evolving artificial neural networks. *IEEE Trans. Neural Netw.* **1997**, *8*, 694–713. [CrossRef]
32. Graupe, D.; Krause, D.J.; Moore, J.B. Identification of autoregressive moving-average parameters of time series. *IEEE Trans. Autom. Control* **1975**, *AC20*, 104–107. [CrossRef]
33. Martinelli, G.; Orlandi, G.; Burrascano, P. ARMA estimation by the classical predictor. *IEEE Trans. Circuits Syst.* **1985**, *32*, 506–507. [CrossRef]
34. Lee, K.C.; Oh, S.B. An intelligent approach to time series identification by a neural network-driven decision tree classifier. *Decis. Support Syst.* **1996**, *17*, 183–197. [CrossRef]
35. Kingma, D.P.; Ba, J. Adam: A Method for Stochastic Optimization. In Proceedings of the 3rd International Conference on Learning Representations, San Diego, CA, USA, 7 May 2015.

36. Stanley, K.O.; Clune, J.; Lehman, J.; Miikkulainen, R. Designing neural networks through neuroevolution. *Nat. Mach. Intell.* **2019**, *1*, 24–35. [CrossRef]
37. Guenther, F.; Fritsch, S. Neuralnet: Training of Neural Networks. *R J.* **2010**, *2*, 30–38. [CrossRef]
38. Gu, J.; Wang, Z.; Kuen, J.; Ma, L.; Shahroudy, A.; Shuai, B.; Liu, T.; Wang, X.; Wang, G.; Cai, J.; et al. Recent advances in convolutional neural networks. *Pattern Recognit.* **2018**, *77*, 354–377. [CrossRef]
39. Wang, K.; Wu, Y.; Gao, Y.; Li, Y. New methods to estimate the observed noise variance for an ARMA model. *Measurement* **2017**, *99*, 164–170. [CrossRef]
40. Zhang, Y.; Kim, C.-W.; Zhang, L.; Bai, Y.; Yang, H.; Xu, X.; Zhang, Z. Long term structural health monitoring for old deteriorated bridges: A copula-ARMA approach. *Smart Struct. Syst.* **2020**, *25*, 285–299.
41. De Livera, A.M.; Hyndman, R.J.; Snyder, R.D. Forecasting Time Series With Complex Seasonal Patterns Using Exponential Smoothing. *J. Am. Stat. Assoc.* **2011**, *106*, 1513–1527. [CrossRef]
42. Valipour, M.; Banihabib, M.E.; Behbahani, S.M.R. Comparison of the ARMA, ARIMA, and the autoregressive artificial neural network models in forecasting the monthly inflow of Dez dam reservoir. *J. Hydrol.* **2013**, *476*, 433–441. [CrossRef]

symmetry

Article

Force Analysis of Self-Anchored Suspension Bridges after Cable Clamp Slippage

Hongfeng Li *, Yancheng Liu, Chunwei Li, Hao Hu and Quansheng Sun

School of Civil Engineering, Northeast Forestry University, Harbin 150040, China; yanchengliu@nefu.edu.cn (Y.L.); lichunwei@nefu.edu.cn (C.L.); huhao@nefu.edu.cn (H.H.); sunquansheng@nefu.edu.cn (Q.S.)
* Correspondence: lihongfeng@nefu.edu.cn

Abstract: The slippage of cable clamps during the long-term operation of suspension bridges is a common and detrimental phenomenon. From an experimental point of view, the cable clamp slippage of a suspension bridge was investigated to reveal the effect of this sliding on the force acting on the full bridge. The forces acting on the bridge before and after the slippage were analyzed using a finite element model. The calculation results showed that the cable clamp slippage directly affects the cable forces of the hangers. The hanger cable force decreased by 19.2% when the slippage reached 10.2 cm, while the maximum increase in the cable force of adjacent hangers was 147.7 kN, an increase of 7.25%. The variation of forces in the hanger cable disrupted the force balance of the main girder, thereby producing a torque effect at the corresponding position in the girder, i.e., increased torque. Meanwhile, the slippage affected the axial tension in the main cable and the main girder. The impact of the tower internal force was less than 1%. Hence, the study concluded that the effect of cable clamp slippage is better understood, ensuring the safety of the suspension bridge.

Keywords: self-anchored suspension bridge; cable clamp; slippage; force analysis

Citation: Li, H.; Liu, Y.; Li, C.; Hu, H.; Sun, Q. Force Analysis of Self-Anchored Suspension Bridges after Cable Clamp Slippage. *Symmetry* **2021**, *13*, 1514. https://doi.org/10.3390/sym13081514

Academic Editors: Yang Yang, Ying Lei, Xiaolin Meng and Jun Li

Received: 22 July 2021
Accepted: 16 August 2021
Published: 18 August 2021

1. Introduction

In suspension bridges, cable clamps are key nodes connecting the main cables and hangers. With the progress and development of suspension bridges, cable clamps have gradually evolved from eyebars and pin-type hinge structures to riding straddle and pin-type structures. The choice is closely related to the type of hanger used: riding straddle cable clamps are commonly used for wire rope hangers, whereas pin-type cable clamps are used for parallel wire hangers. In the U.S. and Japan, preference is given to riding straddle cable clamps, whereas pin-type cable clamps are more common in China and Europe [1]. Figure 1 shows the different types of cable clips.

During the construction and operation, the cable clamp will no longer be firmly anchored because of the loosening of the fastening bolts, the main cable has a thinner section under long-term loading, and the linear expansion coefficient is inconsistent under the influence of temperature. These factors are important inducements that cause slippage of the suspension bridge cable clamps. Therefore, the cable clamp slippage is a common phenomenon that is detrimental to suspension bridges. In the inspection manual for suspension bridges developed by the Honshu–Shikoku Bridge Construction Corporation, Japan, cable clamp slippage is listed as a key inspection item [2]. In the United States, the cable inspection and strength assessment guidelines provide detailed inspection requirements for hangers and cable clamps [3]. There are strict regulations on cable clamp slippage in the Chinese Highway Bridge Technical Condition Assessment Standard [4].

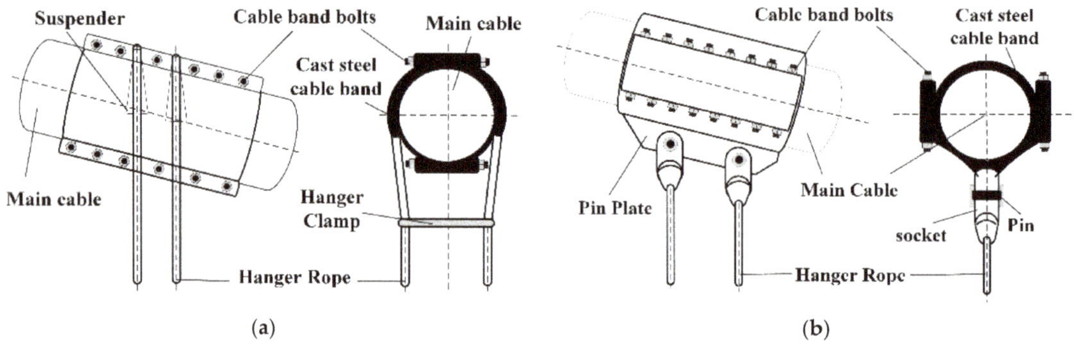

Figure 1. Different types of cable clips: (**a**) abridged general view of a riding straddle cable clip; (**b**) abridged general view of a pin-type cable clamp.

Cable clamps with poor anti-sliding performance tend to slide along the main cable, as shown in Figure 2, which directly causes a loss of stress in the entire structure and major safety hazards in the structural system. Insufficient anti-skid performance of the cable clamp will cause the cable clamp to slide down the main cable, which will lead to the redistribution of the internal force of the principal force-bearing components such as the main cable, hanger, and main beam. The internal force deviates from the design value and the safety reserve of the structure is cut. In practice, engineering problems due to the poor anti-sliding performance of cable clamps may force a project to be terminated or even rebuilt. Hence, scholars have extensively studied the problem of cable clamp slippage.

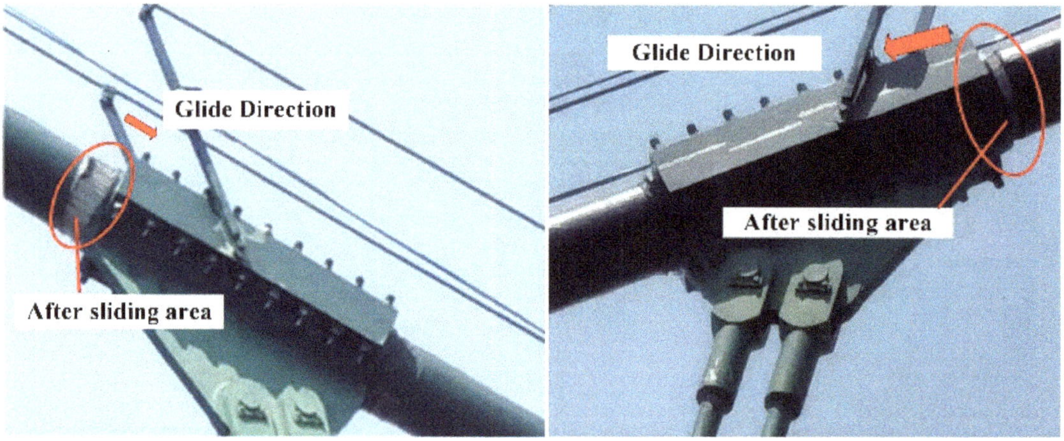

Figure 2. Actual images of a cable clip after slipping.

In 2000, Ji et al. analyzed the additional force due to the hanger pull acting on the screws in the hanger, with an objective to improve the cable clamp design of the Kurima Bridge, Japan. The Liaison Bridge Corporation proposed the application of pin-type cable clamps for the first time [5]. Since 2009, scholars have started paying attention to cable clamp slippage in suspension bridges. Ren et al. analyzed the anti-skid friction coefficient of a cable clamp and the main cable void ratio by conducting an anti-sliding test on the main cable and cable clamp [6]. Huang et al. studied the variations in the lateral displacement of the main cable and the lateral deflection angle of the cable clamp under the effect of hanger tension in the main cable [7]. In 2013, Li et al. determined the upper limit of the transverse elastic modulus based on the macroscopic force of an ideally arranged main cable [8]. In 2014, Ruan et al. studied and analyzed the spatial effect and stress distribution characteristics of a cable clamp [9]. Ma et al. conducted jack push tests on the main cable of a suspension bridge to analyze the friction coefficient of the cable clamp against sliding and the internal and external void ratios of the cable clamp [10]. Zhuge et al. studied the friction factor of a carbon fiber reinforced plastic (CFRP) cable–cable clamp interface [11]. Li et al. proposed the Tsai–Hill failure criterion for composite materials [12]. In 2015, Zhou et al. calculated the anti-sliding friction coefficient and the internal and external void ratios of a steel wire main cable clamp coated with a zinc–aluminum alloy [13]. In 2016, Sun et al. derived a simplified calculation formula to determine the increase in the elevation of the main cable mid-span control point and the average increment in the hanger cable force [14]. In 2018, Shen et al. analyzed the change laws of the ultimate anti-sliding friction resistance of a cable clamp under the action of the cable force, and the contact force between the main cable and the cable clamp surface [15,16]. In 2019, Ruan et al. proposed a theoretical model considering transversely isotropic materials based on the generalized Hooke's law. Through anti-slipping performance tests, they determined the actual ultimate sliding resistance and comprehensive friction coefficient of a cable clamp [17]. In 2020, Miao et al. proposed a new slippage-based failure criterion based on an analytical model [18] and briefly revised the original slippage resistance formula using the Coulomb friction law.

In summary, most studies have been based on the interface performance of the main cable and the cable clamp, focusing on the analysis of the effects of the clamping force of the cable clamp and the friction coefficient on the anti-slipping performance. Studies on the total force acting on suspension bridges are lacking. The integrity of the overall structure of a full bridge following cable clamp slippage has rarely been studied. In practice, although cable clamp slippage is minor, it is a key factor causing a change in the forces acting on the entire bridge. Therefore, this study mainly analyzed the overall force acting on suspension bridges following cable clamp slippage.

In this paper, the influence of pin-type cable clamp slippage on a self-anchored suspension bridge is reported. Considering the geometric nonlinear influence of the bridge, a finite element simulation was conducted. First, we selected a large-span self-anchored bridge as an example to describe the geometry of self-anchored suspension bridges and the magnitude of the cable clamp slippage in detail. Subsequently, a method for simulating the cable clamp slippage was proposed and verified through examples. Additionally, the total force acting on the suspension bridge before and after cable clamp slippage was analyzed, and recommendations for operation and maintenance was provided.

2. Background

A self-anchored suspension bridge with double towers and double cable planes (see Figure 3) was selected as the research object. The main span, side span, auxiliary span, cable sag ratio, and sag-span ratio of this bridge are 248 m, 108 m, 46 m, 49.6 m, and 1/5, respectively. The main tower is a double-tower portal structure system with a height of 80.5 m; the main cable is a parallel steel wire; the hanger is a parallel steel wire strand; and the stiffening girder is a steel–concrete composite girder system.

The construction of this self-anchored suspension bridge was initiated in 2011. A recent structural inspection revealed that the cable clamp close to the main tower has caused severe cable clamp slippage, as shown in Figure 4. Hence, this bridge was thoroughly checked for cable clamp slippage.

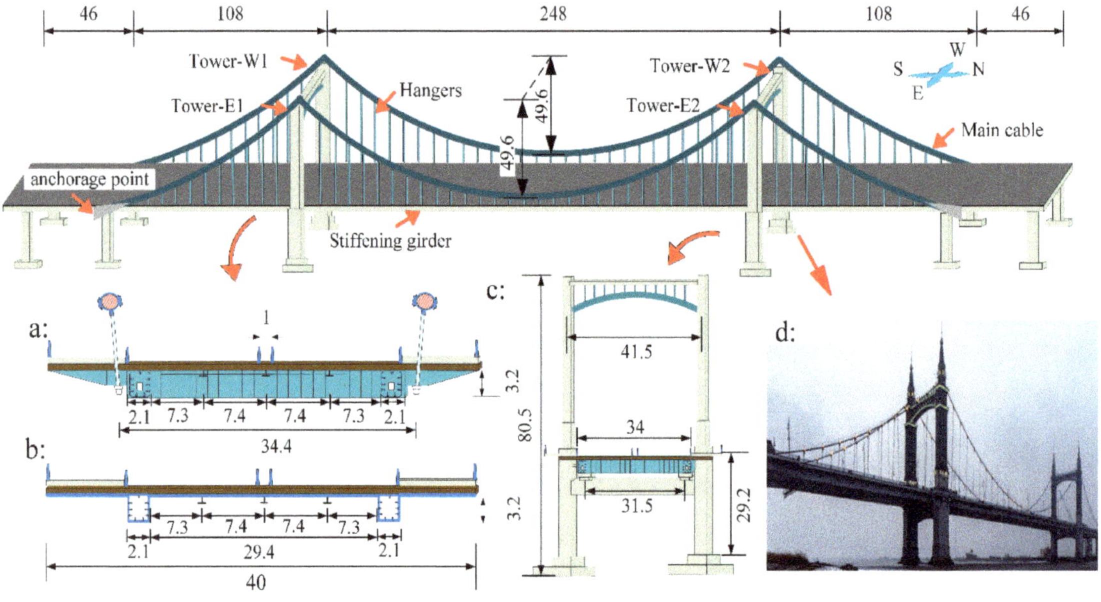

Figure 3. Overview of a self-anchored suspension bridge (m): (**a**) standard cross section at the hanger; (**b**) standard cross section without sling; (**c**) standard cross section of the main tower; (**d**) photo of a self-anchored suspension bridge.

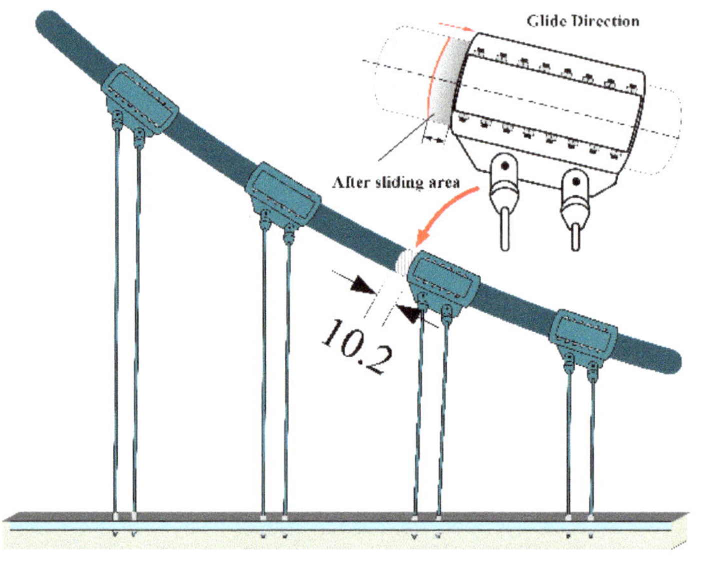

Figure 4. Image and schematic of a cable clamp after slipping.

As showed in Figure 4, the sliding direction of the cable clamp is along the radial direction of the main cable. After inspection, the radial sliding distance of all cable clamps was measured in detail, and the number of each sliding cable clamp and the amount of slippage was recorded. In the establishment of the finite element model, adjusting the position of the cable clamp simulates the actual bridge stress condition. In order to emphasize the influence of the change in the position of the cable clamp on the whole bridge, the cable clamp with the largest slip amount is mainly simulated during the simulation.

During the inspection of the suspension bridge, it was found that the cable clamp that belonged to the third hanger in the mid-span direction of the W1 main tower was the most serious. A detailed inspection of the remaining cable clamps revealed that a total of 20 cable clamps in the whole bridge had slipped. Among them, there were 11 sliding positions of the west cable clamp, and 9 sliding positions of the east cable clamp. Figure 5 shows the downward positions of the cable clamps.

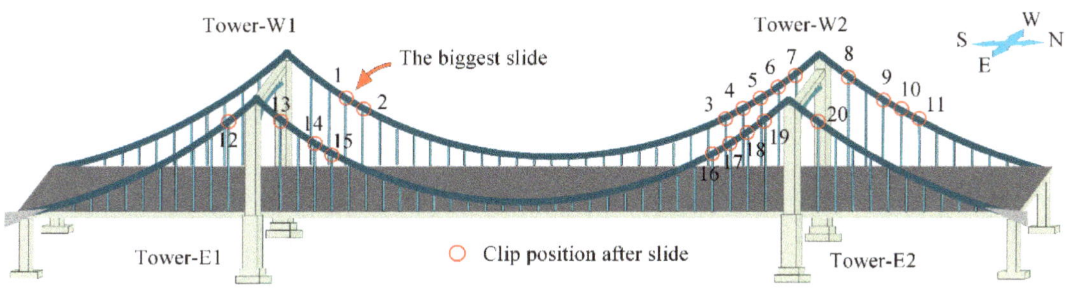

Figure 5. Positions of the cable clamps after slipping.

The measurements showed that five cable clamps slipped beyond 5 cm, and the maximum slippage was found to be 10.2 cm. Table 1 summarizes the slippage of the cable clip.

Table 1. Slippage of the cable clip.

Sample	**1**	**2**	**3**	**4**	**5**	**6**	**7**	**8**	**9**	**10**
Slippage (cm)	10.2	5.1	2.3	5.4	3.3	5.7	3.2	2.2	1.7	1.3
Sample	11	12	13	14	15	16	17	18	19	20
Slippage (cm)	2.4	2.6	2.1	1.8	1.2	3.3	2.4	6.7	3.1	2.5

Because of the slipping of several cable clamps, the variation in the structural force acting on the full bridge is complicated. To eliminate the numerous interferences and obtain the most representative situation, the slipping of a single cable clamp on one side is employed for the simulation. We established a model for the slipping of this clamp. In addition to considering the weight of the main cable, stiffening beams, main tower, and hangers, it is necessary to consider auxiliary components, such as the bridge deck paving, sidewalks, and guardrails, and convert them into constant loads for the simulation analysis. The dead load distribution of the bridge deck paving asphalt, sidewalks, and guardrails and other ancillary components is 132 kN/m.

3. Simulation of Cable Clamp Slippage

3.1. Finite Element Modelingsubsection

MIDAS Civil was used to establish a finite element model for the theoretical calculations. The bridge is first assembled from a free cables state to a finished bridge state and then changed from a finished bridge state to a slip state. In a structural analysis, the equilibrium equation of the force should be established based on the geometric position of the structure after deformation, and the relationship between the force and the deformation is nonlinear. Therefore, the model cannot be established directly from the slip state. When building the model, the finished bridge state model must be established first, and then the slip state should be established based on the finished bridge state. The models currently employed can mainly be divided into nondamaged bridge completion stage models and single-hanger slip whole bridge models for calculation and comparison. The main girder and the main tower of the whole bridge model are simulated by beam elements, and the main cables and hangers are mainly stimulated by cable elements. Table 2 lists the main modeling material parameters.

Table 2. Specifications of the main components.

Structural Part	Material	Modulus of Elasticity (MPa)	Design Compressive Strength (MPa)	Design Tensile Strength (MPa)
Main cable	Galvanized high-strength steel wire	2 × 105	—	1670
Hanger	Galvanized high-strength steel wire	2 × 1015	—	1770
Concrete main girder Tower column and lower girder of bridge tower	C50 concrete	3.45 × 104	22.4	1.83
Bridge deck	C55 concrete	3.55 × 104	24.4	1.89
Composite girder steel girder	Q370qE	2 × 105	22.4	335

To ensure that the force condition of the bridge is simulated correctly and is consistent with reality, when the model is established, the boundary conditions are as follows: consolidate the bottom nodes of the main tower; rigidly connect the top of the main cable to the main tower and then release the full fixation in the rest of the direction of the bridge displacement; rigidly connect the bottom of the main cable to the main girder and constrain all direction; elastic support simulation is adopted at each support of the main beam. The entire bridge is divided into 4706 elements and 5409 nodes. Among them, the main cable has 106 elements, and only 98 elements of the hanger are used as tensile members. Figure 6 shows the model diagram.

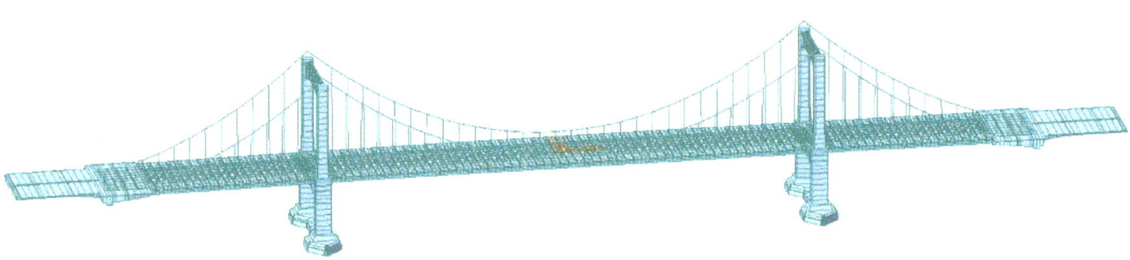

Figure 6. Integral finite element model of a bridge.

3.2. Simulation Steps for Cable Clamp Slippage

For cable clamp slippage, in this study, the method of double-cable element replacement with common nodes is adopted. A new construction stage after the last construction stage of the original bridge type is established and defined as the cable clamp slip stage. We compared the two stages and observed the variations in the structural parameters after cable clamp slippage. In the slipping stage, the original main cable element is passivated, and the adjusted main bridge state cable element is activated to ensure that the model follows the original design. In the sliding stage of the cable clamp, the original main cable element, without slipping of the cable clamp, is passivated, and the main cable element after the cable clamp slips is activated correspondingly. The model takes the original mechanical state when the bridge is erected under the original design without the stress cable length as the initial state, the adjusted cable element is used to re-calculate iteratively, and finally the model converges to the slip state to complete the cable clamp slip. In the simulation of the moved finite element model, the cable element of the aforementioned main cable is adjusted by modifying the unstressed length parameters of the main cable element. The original main cable element is controlled by the unstressed cable length given by the design, and the bridge is in the final state of the bridge with the original design cable length under the working force. The replaced cable element changes its unstressed length relative to the original cable element.

The modifications are as follows: Using the measured or known slippage of the cable clamp in the final slipping state, the stress-free length change in the main cable section before and after the slipping of the cable clamp node is calculated, and the stress-free length of the main cable element, before and after the slipping of the cable clamp node, is manually adjusted. The total unstressed length of the main cable element before and after the adjustment should be kept constant.

3.3. Verification of Measured Cable Force Against Model Cable Force

The hanger is the force transmission member that transfers the stiffening girder's own weight and external load to the main cable. It is the link between the stiffening girder and the main cable that bears the axial tension. The magnitude of the constant load axial force in the hanger determines the behavior of the main cable in the suspension bridge. The overall linearity in the completed state of the bridge also determines the magnitude and distribution of the dead load bending moment of the stiffening girder; therefore, the hanger is key to studying the finished state of the bridge after completion of the suspension bridge. The cable force is the most intuitive measurement standard that reflects the overall force of the hanger [19]. The force distribution, between the main cable and the main girder of the actual spatial self-anchored suspension bridge, is mainly determined by the sling force, and the load on the main girder is transmitted by the hanger. Therefore, the load on the main beam is directly related to the weight of the main beam and the cable force of the sling.

To ensure that the finite element simulation is consistent with the actual force trend in the actual completed state, the hanger cable force was used to verify the model. The verification method was used to measure the hanger frequency using the frequency method when the bridge was completed in 2011, and the result was then compared with the cable force obtained from the finite element method. The suspension bridge structure is a symmetrical structure, so only the cable forces of the model and actual bridge on one side of the structure are compared. As shown in Figure 7, the cable force distribution in the model is consistent with the actual cable force distribution trend of the bridge in 2011. The maximum cable force in the finite element model is 2590 kN, and the maximum cable force measured in the bridge in 2011 was 2629.8 kN. The difference is 39.8 kN, and the positions are all close to the hanger of the main tower.

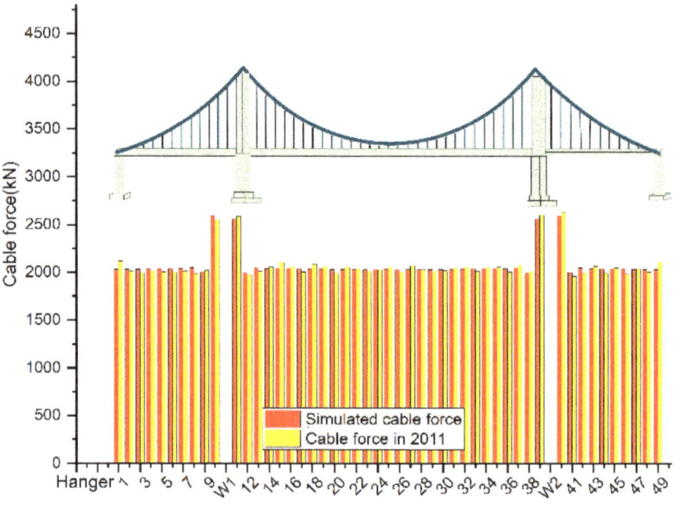

Figure 7. Verification of measured and simulated cable forces.

As shown in Figure 8, the difference between the cable force provided by the finite element model and the measured cable force of the bridge in 2011 is within the range of 60.3 to 86.5 kN, with the difference being <5%. The maximum rate of change on both sides of the hanger is 4.26%. This is because the length of the hanger on both sides is relatively short. In the actual test, the error of the cable force value calculated by the frequency method will be relatively high; nevertheless, the test result and overall trend in the model are in line with those observed in 2011 when the bridge was completed. Therefore, from the above comparison, it can be seen that the proposed model yields the same force state as that of the bridge in 2011. The model can be used to simulate the single point slip of the suspension bridge for a detailed analysis and calculation owing to the reliable accuracy.

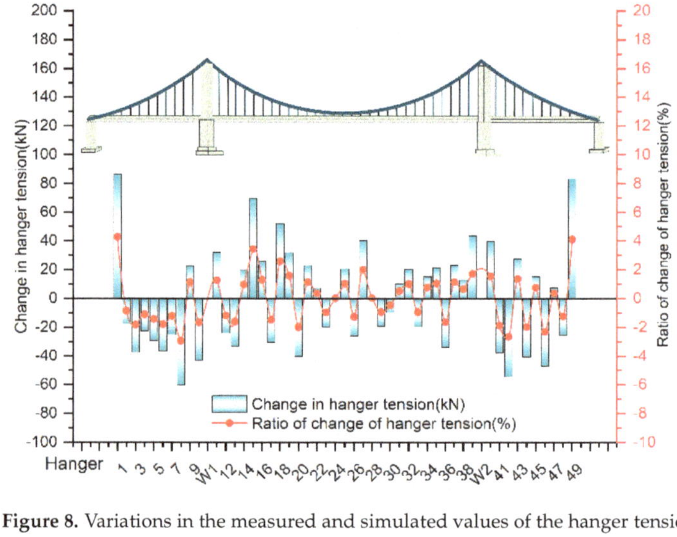

Figure 8. Variations in the measured and simulated values of the hanger tensions.

To further verify the accuracy of the clamp slipping model simulation, the hanger cable force after clamp slippage in the model is compared with the measured cable force of the suspension bridge. As shown in Figure 9, the trend in the simulated hanger tension after single-clamp slippage and the measured cable force is the same. The greater difference in the individual cable force can be attributed to the reallocation of the hanger tension after repeated cable clamp slippage during actual measurement. Nevertheless, the simulated hanger tension after the most severe clamp slippage coincides with the trend in the measured data. Hence, the simulation of the single-clamp slippage by the finite element model is accurate.

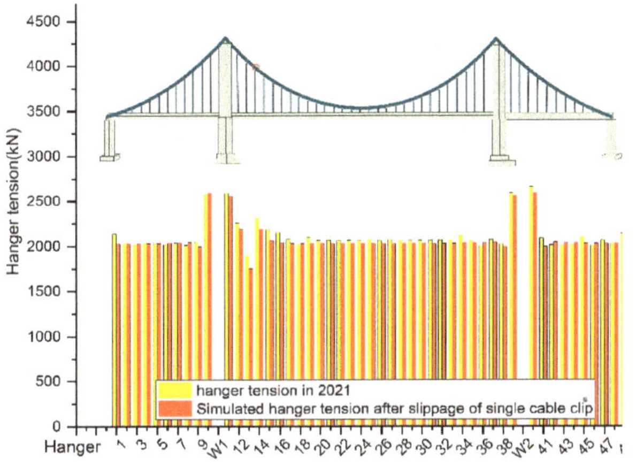

Figure 9. Cable forces after simulated slip of a single cable clip and measured hanger tension in 2021.

As shown in Figure 10, the overall cable force after the measured slip in 2021 is greater than the finite element model simulation result. The overall difference between the measured cable force and the simulated cable force is <10%, where the maximum difference in the measured and simulated cable forces after clamp slippage is 137.1 kN, and the simulated cable force is 7.26% lower than the measured cable force. The above comparison shows that the proposed finite element model yields the same result as the force state measured in 2021, indicating that the accuracy of the proposed suspension bridge model is reliable and can be used for further analysis and calculation.

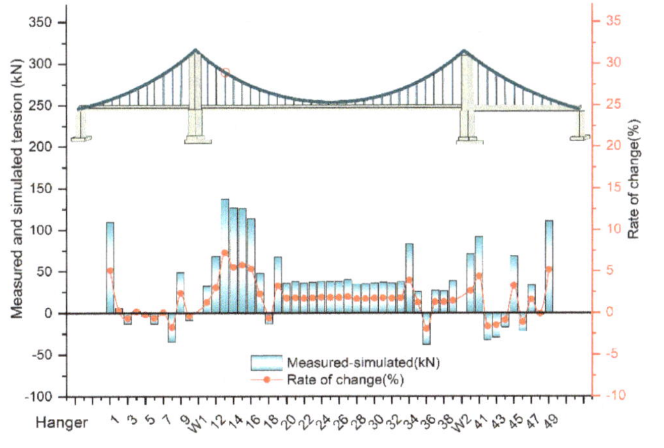

Figure 10. Variations in the simulated and measured hanger tensions after the simulated slip of a single cable clamp in 2021.

4. Results of Finite Element Calculation

In the analysis to clearly illustrate the impact of cable clamp slippage on the overall structure of the self-anchored suspension bridge, this study conducted a single-point slip analysis to simulate and analyze the impact on the structure. The most serious clamp slipping location on this bridge is selected for the slippage simulation of the single cable clamp.

4.1. Hanger Cable Force before and after Cable Clamp Slippage

In the study of cable clamp slippage, it is found that the most direct result of slippage is the change in the magnitude of the cable force. As shown in Figure 11, the cable force before and after the slip of a single cable clamp is simulated using the finite element model, and it is found that the slippage decreases the cable force of this hanger, while the cable force of the two adjacent hangers relatively increases.

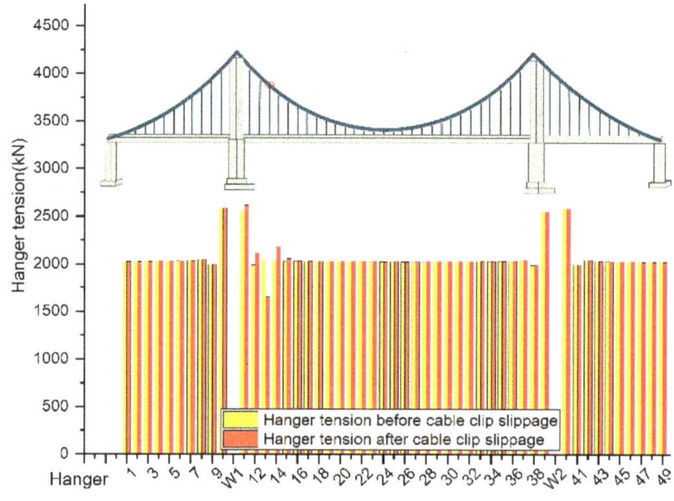

Figure 11. Diagram of hanger tension before and after the slip of a simulated single cable clip.

In combination with the cable force of the entire bridge hanger, the single-cable clamp slippage affects about two hanger positions at the front and rear, and the cable force of the hanger at the farthest distance is <1%. From the overall position, because of the stiffness of the main tower, the slippage in the middle span affects the cable force of the side span hanger by less than 1%.

Figure 12 focuses on the impact of single-cable clamp slippage on the cable force of adjacent hangers on the same side. As observed, the cable force at the slipping position is reduced by 392.6 kN, the hanger cable force drops by 19.2% compared with that before the slip, and the cable force of the four hangers before and after the slip is increased by 62.8, 123.3, 147.7, and 26.2 kN. Compared with pre-slip, the cable force growth rates are 2.45%, 6.19%, 7.25%, and 1.29%. The impact of the remaining hanger cable force is <1%. The single-clamp slippage affects other hanger cable forces. The impact is only in the range of the two hangers before and after.

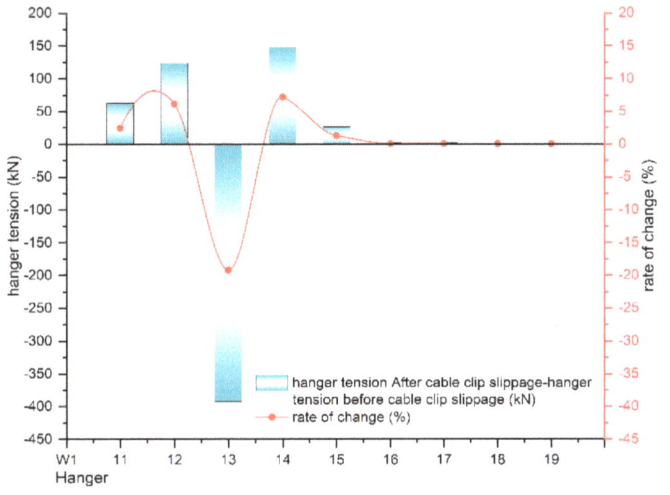

Figure 12. Variation in the hanger tension before and after cable clip slippage.

4.2. Force Acting on the Main Girder before and after Cable Clamp Slippage

The suspension bridge is a flexible structure, wherein the stiffening girder mainly provides the torsional stiffness and load acting surface and transmits the load to the hanger. The hanger connects the main cable and the stiffening girder and distributes the load from the stiffening girder to the main cable. Therefore, the change in the hanger tension directly affects the force state of the main girder, and the effect of single-clamp slippage on the main girder is discussed here.

As shown in Figure 13, the maximum variation in the bending moment at the single-clamp slipping position is 118 kN, and the variation rate is 1.28%. This figure also shows that the magnitude of the effect decreases with the increase in the slipping cable clamp position distance, and the effect of mid-span slippage on the side-span main girder bending moment is <1%.

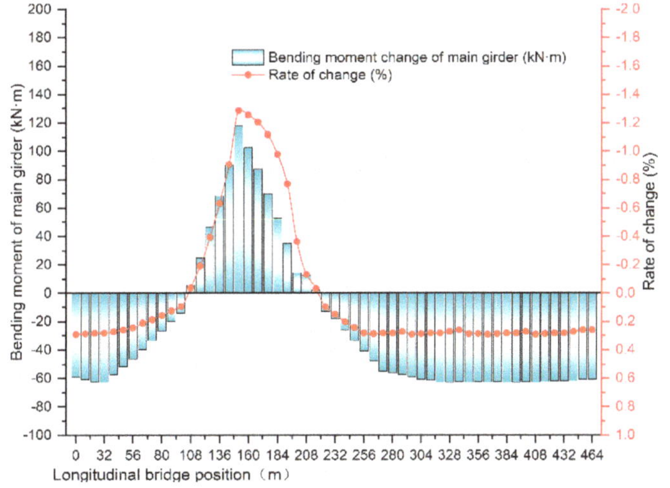

Figure 13. Variation in the bending moment of the main girder.

After cable clamp slippage, because of the change in the cable force of the horizontal symmetric hanger, the force of the main girder is no longer on the same horizontal plane. As a result, the change in the cable force causes a local torsion effect on the stiffened girder. To clearly analyze the influence range of the torque change in the main girder, the main girder at the cable clamp slipping position is taken as the 0 point, and the distance is used to represent the influence range. The change in the internal force of the main girder after slipping was carefully studied, and it was found that the influence of the slippage in the middle span was only within the middle span, and the influence on the side span was low.

From Figure 14, the biggest change in the torque of the main girder after the single cable clamp slips is the main girder between the W1 tower and W2 tower, while the main girder torque change of the side span is smaller. From a local analysis, the slippage of the cable clamp only has an impact on the torque near the hanger, and the range of the four hangers is approximately its range of influence. Through distance analysis, the influence range of the single cable clamp on the main girder torque after slipping is 24 m in total. Among them, the slip of the single cable clamp has a greater impact on the torque of the main beam close to the W1 main tower, with an impact range of 16 m, and an impact range of 8 m in the mid-span direction.

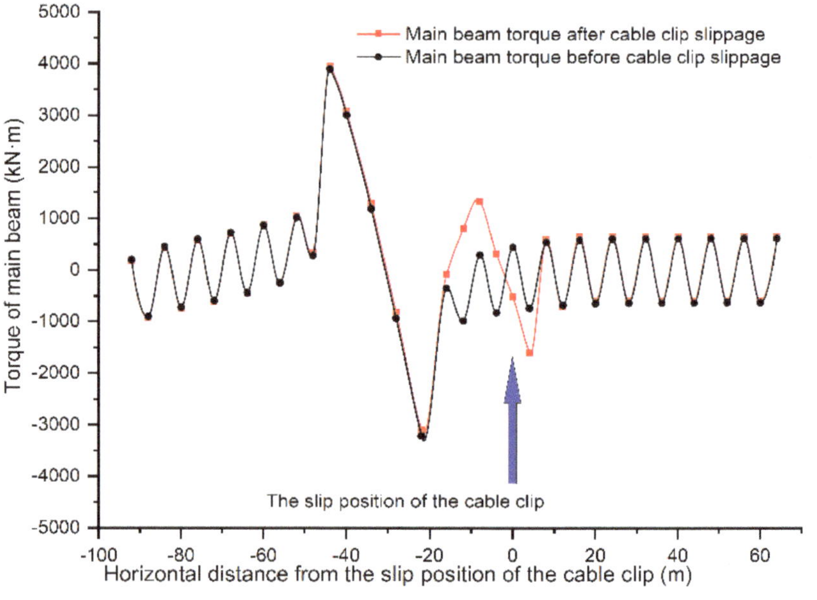

Figure 14. Torque of the main girder.

As shown in Figure 15, cable clamp slippage has a significant impact on the local torque of the main girder. The torque after slippage is 3.5 times that before slippage, and the maximum torque change after slippage is 1785 kN. In terms of the long-term torque effect, this is unfavorable to the main girder in terms of the force.

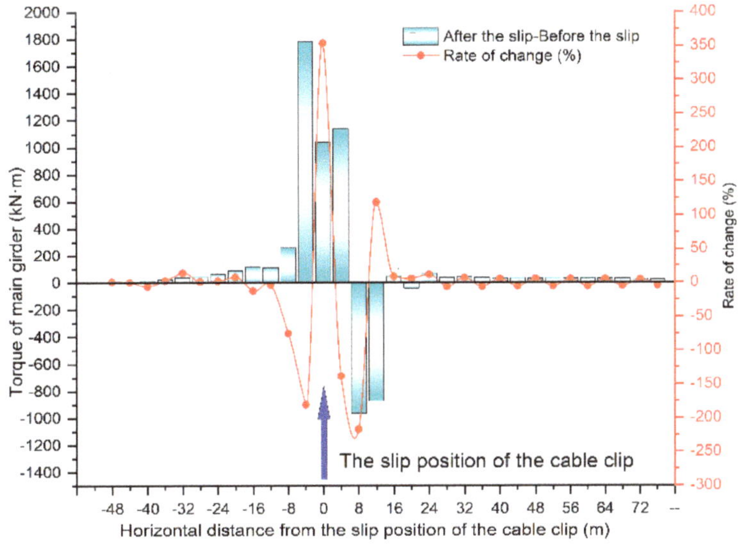

Figure 15. Variations in the main girder torque before and after cable clamp slippage.

4.3. Forces Acting on the Main Cable before and after Cable Clamp Slippage

The main cable is the main load-bearing member of the structural system and is a geometrically varying body, mainly subjected to tension. The main cable has a high initial tensile force under constant loading, providing "geometric stiffness" for the subsequent structure, not only through its own elastic deformation, but also through the geometry to affect the system equilibrium. Hence, cable clamp slippage will have a corresponding effect on the tensile force.

As shown in Figure 16, the maximum effect of single-clamp slippage on the main cable force is at the slippage location, while the maximum main girder force is 527.6 kN, a change of 1.21% before and after slippage. The maximum change is 1.21%. The remaining positions of the main cable force change is <1%, and the middle-span clamp slippage has a certain effect on it. The main tower stiffness is higher on the mid-span side of the main cable, and the force change on the side span side of the main cable tension is <1%.

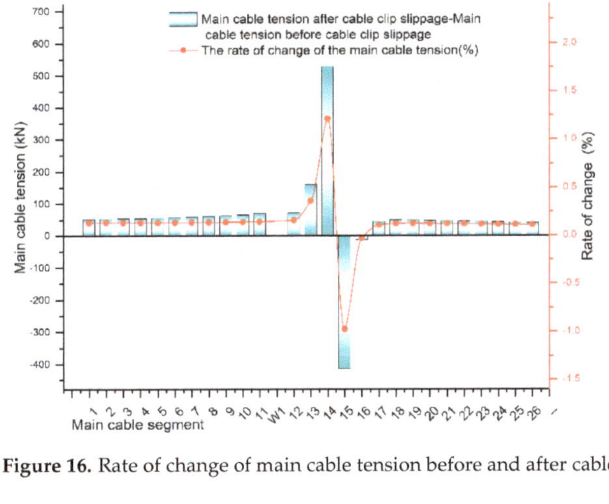

Figure 16. Rate of change of main cable tension before and after cable clip slippage.

4.4. Forces in the Main Tower before and after Cable Clamp Slippage

In a suspension bridge, the main tower is a compression-bending member. In the vertical direction, the main tower bears the vertical force of the main cable mainly through the constant load. In the longitudinal direction, the horizontal force of the main cable is balanced on both sides of the tower by the effect of the constant load, and no bending moment is generated on the tower. After cable clamp slippage, the unbalanced state of the load causes an unbalanced tension on both sides of the main tower, resulting in a horizontal displacement of the tower top. The single-clamp slippage at the mid span causes the main tower to produce a longitudinal top displacement of 0.242 mm.

After the horizontal displacement of the top of the tower, the horizontal force of the main cable on both sides of the main tower is rebalanced, the main tower is subjected to an unbalanced main cable tension, and the bending moment is changed. As shown in Figure 17, the trend in the main tower bending moment after slippage is consistent with that before slippage. Due to the single-clamp slippage, the maximum change value of the main tower is 0.0558 kN·m.

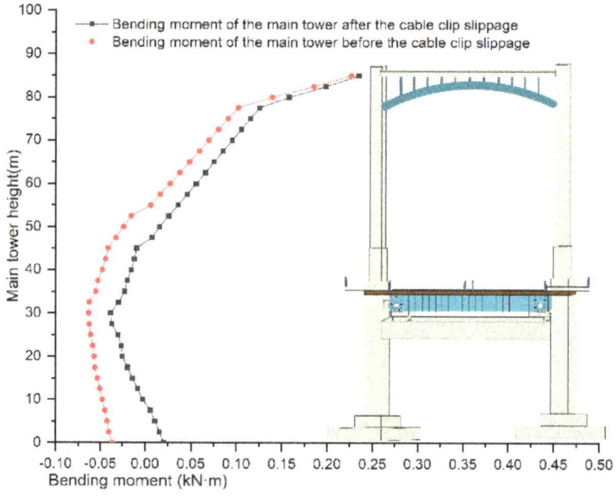

Figure 17. Bending moment diagram of main tower before and after cable clip slippage.

4.5. Suggestions for Handling Cable Clamp Slippage

A force analysis of self-anchored suspension bridges after single-clamp slippage revealed that cable clamp slippage, which typically brings about a minor change, directly affects the force of each main component. Practically, cable clamp slippage is not a fixed state, but a state that continues to deteriorate with the operational phase of the bridge. As more clamps slip, the slipping distance gradually increases, and the structural forces of the bridge tend to develop unfavorably. If this problem is ignored, the safety of the bridge will be compromised. Therefore, if slipping of cable clamps in suspension bridges is detected during structural inspection, a detailed investigation should be conducted immediately, and for cable clamps with slippage <5 mm, measures, such as re-torqueing of the bolts, should be taken to restore the slip resistance of the clamps to prevent further slippage. For cable clamps with a slippage >5 mm, immediate restoration and adjustment should be performed to ensure structural safety [20].

5. Conclusions

This study conducted an analysis of cable clamp slippage, a common phenomenon in actual bridges. To clarify the force acting on a self-anchored suspension bridge before and after cable clamp slippage, the slipping of a single clamp was used for the analysis.

We focused on the forces acting on the hangers, main girder, main cable, and main tower before and after cable clamp slippage in the studied bridge. The main research results are as follows:

(1) Simulations of the hanger cable force after single-clamp slippage revealed that the slippage directly affects the hanger cable force, and its influence range is at the two hanger positions before and after the change. After slippage, the corresponding hanger cable force was reduced by 19.2%, while the cable force of the adjacent hanger increased sharply (up to 7.25%).

(2) The bending moment and torque of the main girder before and after single-clamp slippage revealed that the change in the bending moment of the main girder was up to 1.28%. Meanwhile, the local forces on the main girder induced by single-clamp slippage led to a torsion effect. As a result, the torque increased by 250%.

(3) An analysis of the tension in the main cable before and after single-clamp slippage revealed a re-distribution of the main cable tension after the slippage. Specifically, the change in the main cable tension at the single-clamp slippage position was 1.21%, while the changes at the remaining positions were <1%.

(4) After single cable clamp slippage, the top of the main tower exhibited a longitudinal displacement of 0.242 mm. Because of the greater rigidity of the main tower, the single-clamp slippage had a small effect on the bending moment of the main tower (up to 0.0558 kN).

In the study of cable clamp slips, it was found that the most direct effect after the cable clamp slips is the change in the force of the hanger. From the perspective of the whole bridge, the slip of the cable clamp will directly lead to the imbalance of the force on the main girder, and this problem will cause the main girder to be subjected to torsion for a long time. Changes caused by the slippage of the cable clamp are often not conducive to the safe use of the bridge. Future research will continue to focus on this area, focusing on the slippage of the cable clamp and the limit of the slippage of the cable clamp. These studies are of great importance to the safety performance of the bridge.

Author Contributions: Conceptualization, H.L. and Y.L.; methodology, Y.L.; software, C.L.; experiment, Y.L. and H.H.; data curation, Y.L.; writing—original draft preparation, Y.L.; writing—review and editing, H.L. and Q.S. All authors have read and agreed to the published version of the manuscript.

Funding: The research in this paper was supported by the Science and Technology Project of the Department of Transportation of Heilongjiang Province (2021HLJ01K42) and the Science and Technology Project of the Department of Transportation of Heilongjiang Province (2020HLJ018).

Institutional Review Board Statement: Not applicable.

Informed Consent Statement: Not applicable.

Data Availability Statement: The data that support the findings of this study are available upon reasonable request.

Conflicts of Interest: The authors declare that they have no conflict of interest.

References

1. Xu, G. *Design of Large-Span Railway Suspension Bridges*; Shanghai Science and Technology Press: Shanghai, China, 2020.
2. Bridge Research Office, Japan Institute of Territorial Technology Policy. *Damage Cases of Road and Bridge with Tension Materials and Matters Needing Attention*; Japan National Land Technology Policy Research Institute: Tokyo, Japan, 2018.
3. National Academies of Sciences, Engineering, and Medicine. *Guidelines for Inspection and Strength Evaluation of Suspension Bridge Parallel Wire Cables*; The National Academies Press: Washington, DC, USA, 2004. [CrossRef]
4. Institute of Highway Research, Ministry of Transport. *Standards for Technical Condition Evaluation of Highway Bridges*; JTG/T H21-2011; Ministry of Transport of the People's Republic of China, Transportation Department: Beijing, China, 2011.
5. Ji, L.; Li, H. Laidao Bridge Hanger and Clamp Design. *Road Traffic Technol.* **2000**, *2*, 22–26. [CrossRef]
6. Ren, H.; Zhao, L.; Bai, F. Sliding resistance test and stress test analysis of main cable clamp of suspension bridge. *North Traffic* **2009**, *3*, 74–76. [CrossRef]

7. Huang, H.; Zhang, J.; Liu, A.; Mei, L.; Li, Y. Experimental study on lateral displacement of main cable and lateral deflection angle of cable clamp of self-anchored suspension bridge with space cable plane. *Highw. Eng.* **2009**, *34*, 41–44. [CrossRef]
8. Li, Z.; Jiang, J. Analysis and Research on Upper and Lower Clamp Structures of Suspension Bridges. *Bridge Constr.* **2013**, *43*, 60–65.
9. Ruan, K. Research on spatial stress analysis of cable clamp of suspension bridge. *Highw. Automob. Transp.* **2014**, *2*, 170–173. [CrossRef]
10. Ma, W.; Liu, S.; Wang, C.; Bao, Z.; Geng, J. Experimental study on void ratio of main cable and anti-slip of cable clamp of Liujiaxia Bridge. *World Bridge* **2014**, *5*, 59–62. [CrossRef]
11. Zhang, Z.; Wang, S.; Ding, Y.; Qiang, S. Anti-slip Performance of Interface between Carbon Fiber-Reinforced Plastic Main Cable and Cable Clamp for Large-Span Suspension Bridges. *J. Southwest Jiaotong Univ.* **2014**, *2*, 208–212. [CrossRef]
12. Li, Y. CFRP cable suspension bridge clamp design calculation method. *China J. Highw.* **2015**, *28*, 67–75.
13. Zhou, Z.; Yuan, Q.; Zhou, C.; Shu, S. Experimental study on skid resistance of zinc-aluminum alloy coated steel wire main cable clamp of suspension bridge. *World Bridge* **2015**, *5*, 40–43. [CrossRef]
14. Sun, Y.; Zhang, L.; Li, Z. Analysis of the influence of cable clamp on the finished state of self-anchored suspension bridge. *J. Tongji Univ. (Nat. Sci. Ed.)* **2016**, *44*, 24–28. [CrossRef]
15. Shen, R.; He, K.; Miao, R. Limit anti-skid friction analysis of pinned cable clips based on multi-scale models. *Bridge Constr.* **2018**, *48*, 20–24. [CrossRef]
16. Shen, R.; He, K.; Huang, Z. Analysis model of nonlinear relationship between main cable steel wires at cable clamp of suspension bridge. *J. Archit. Sci. Eng.* **2018**, *35*, 111–119. [CrossRef]
17. Ruan, Y.; Luo, B.; Ding, M.; Huang, L.; Guo, Z. Theoretical and Experimental Study on the Anti-sliding Performance of Casting Steel Cable Clamps. *Adv. Civ. Eng.* **2019**, *2019*, 1–18. [CrossRef]
18. Miao, R.; Shen, R.; Wang, L.; Bai, L. Theoretical and numerical studies of the slip resistance of main cable clamp composed of an upper and a lower part. *Adv. Struct. Eng.* **2020**, 136943322096527. [CrossRef]
19. Xiang, H. *Theory of Higher Bridge Structure*; China Communications Press: Beijing, China, 2007.
20. Yuan, A.; Yang, T.; Xia, Y.; Qian, L.; Dong, L.; Jin, X. Replacement Technology of Long Suspenders of Runyang Suspension Bridge. *China J. Highw. Transp.* **2021**, *34*, 289–297. [CrossRef]

symmetry

MDPI

Article

Time–Frequency Extraction Model Based on Variational Mode Decomposition and Hilbert–Huang Transform for Offshore Oil Platforms Using MIMU Data

Jian Wang [1,2,*], Xu Liu [1,2], Wen Li [1,2], Fei Liu [2] and Craig Hancock [3]

[1] School of Geomatics and Urban Spatial Informatics, Beijing University of Civil Engineering and Architecture, Beijing 102616, China; liuxucrystal@163.com (X.L.); liwen2375@163.com (W.L.)

[2] Research Center for Urban Big Data Applications, Beijing University of Civil Engineering and Architecture, Beijing 100044, China; pntrc@cumt.edu.cn

[3] School of Architecture, Building and Civil Engineering, Loughborough University, Loughborough, Leicestershire LE11 3TU, UK; c.m.hancock@lboro.ac.uk

* Correspondence: wangjian@bucea.edu.cn; Tel.: +86-188-1095-5832 or +86-188-5505-8097

Abstract: Time–frequency extraction is a key issue to understand structural symmetry of dynamic responses of offshore oil platforms for early warning during drilling operations. Current popular methods for signal characteristics extraction can only obtain the attributes with a single dimension or poor precision. To solve this, a combined Hilbert–Huang transform (HHT) and variational mode decomposition (VMD) method is proposed to extract multidimensional dynamic response characteristics of time, frequency, and energy of offshore oil platforms. Based on the extracted time–frequency–energy information, the frequency-domain integration approach (FDIA) can be applied to calculate the displacement using accelerometer in the micro inertial measurement unit (MIMU). A complementary filtering algorithm was designed to measure the torsion angle of platforms using six degrees of freedom data from the MIMU to obtain the torsion angle information. The performance of the proposed method was validated using a series of simulation shaking-table tests and a field test conducted on an offshore oil platform at Dongying City, Shandong Province, China. During the field test, seven out of eight collisions were detected in the frequency range 5 Hz to 12 Hz. The intensity of the fifth collision was the highest, and the maximum displacement obtained by the accelerometer was 6 mm. In addition, the results show a correlation between the axes of the accelerometer and gyroscope, and their combination can measure a torsion angle up to 1.1°.

Keywords: time–frequency extraction; micro inertial measurement unit (MIMU); variational mode decomposition (VMD); Hilbert–Huang transform (HHT); frequency-domain integration approach (FDIA); torsion angle calculation; offshore oil platform

Citation: Wang, J.; Liu, X.; Li, W.; Liu, F.; Hancock, C. Time–Frequency Extraction Model Based on Variational Mode Decomposition and Hilbert–Huang Transform for Offshore Oil Platforms Using MIMU Data. *Symmetry* **2021**, *13*, 1443. https://doi.org/10.3390/sym13081443

Academic Editors: Yang Yang, Ying Lei, Xiaolin Meng and Jun Li

Received: 11 July 2021
Accepted: 1 August 2021
Published: 6 August 2021

1. Introduction

With the rapid growth in world energy demand, the number of offshore oil platforms keeps increasing gradually with progressively upgrading potential security hazards. Statistics show that several structural damage accidents in offshore oil platforms happen every year, and nearly half of them are caused by severe weather, such as typhoons, hurricanes, tsunami, earthquakes, etc. Furthermore, with the increasingly complex offshore environment, ships or unidentified objects may hit the platform occasionally [1]. Once an accident occurs, it will lead to heavy casualties, property loss, and environmental pollution. Hence, the monitoring of dynamic responses of offshore oil platforms is of vital importance to safety in the offshore oil industry.

Time and frequency data are essential in investigating the dynamic responses of offshore oil platforms. To obtain high-precision dynamic displacement, extracting frequency dominant information is necessary, which can also act as a principal method to analyze dynamic response characteristics and structural symmetry. The fast Fourier transform (FFT) was traditionally applied to extract frequencies and amplitudes from monitoring datasets [2]. Nevertheless, FFT can neither extract local frequencies nor process nonstationary signals that constantly change [3]. The short-time Fourier transform (STFT) has been proposed to overcome such defects, allowing local characteristics of frequencies and amplitudes to be obtained using a moving window. STFT is also suitable for nonstationary signals as signals intercepted by the moving window can be regarded as linear [4]. However, neither FFT nor STFT can detect the relationships between time and frequency domains for time-varying signals.

To analyze frequency characteristics of nonstationary signals precisely in the time domain, the wavelet transform method was suggested because of its excellent local time-frequency properties [5]. As the wavelet transform method was of poor adaptive ability due to its fixed wavelet basis [6–8], the empirical mode decomposition (EMD) was subsequently proposed to overcome these limitations by decomposing the wavelet function adaptively [9]. Later, EMD was widely used as a time–frequency analysis method for its good performances in adaptive decomposition and nonlinear signal analysis [10,11]. However, several limitations of the EMD method have been identified, such as mode mixing (difficulty in separating modes effectively according to the time scale) and the endpoint effect (a problem of signal divergence caused by the repeated use of cubic spline interpolation) [12]. Improvements to the EMD have since been made and been applied as the ensemble empirical mode decomposition (EEMD) [13,14] and complementary ensemble empirical mode decomposition (CEEMD) [15]. However, the above methods cannot fundamentally remedy the inherent defects of EMD, primarily the phenomenon of mode mixing.

To avoid the limitations of the signal processing methods mentioned above, a variety of integration approaches can be employed, such as the EMD–wavelet and the Hilbert–Huang transform (HHT) [16,17]. The latter has been widely used to analyze dynamic responses of signals since its first application in information extraction from seismic waves [18,19]. The major advantages of HHT are as follows: (i) it is suitable for processing non-linear and nonstationary signals; (ii) it can modify the time scale adaptively; (iii) it can obtain 3D information consisting of time, frequency, and energy [20]. HHT is made up of Hilbert transform and EMD, in which EMD is the core. Accordingly, it is essential to improve the performance of EMD. Various approaches combining improved EMD and HHT are proposed, such as EEMD–HHT and CEEMD–HHT [21,22]. These approaches reduce the defects of mode mixing and endpoint effect of EMD to a certain extent.

In this study, variational mode decomposition (VMD) was first verified to avoid mode mixing and end effects [23]. Then, a combined model of VMD and Hilbert transform (VMD–HHT) was established to extract time–frequency–energy characteristics. Meanwhile, we efficiently eliminated the noise component from accelerometer data according to extracted frequency range using VMD–HHT. Thus, displacement responses were calculated using the frequency-domain integration approach (FDIA) using the accelerometer data. Additionally, torsion angles were obtained by the complementary filtering algorithm based on six degrees of freedom data from the MIMU. A series of simulation shaking-table tests were performed using accelerom in the MIMU to verify the reliability of the VMD–HHT model and displacement reconstruction method. Finally, a field test in the offshore oil platform was conducted. The results prove that VMD–HHT can notably help extract time–frequency–energy characteristics and symmetric information of offshore oil platforms wholly and accurately and improve displacement calculation by FDIA. Moreover, dynamic displacement responses and torsion angle information of the offshore oil platform can be calculated by MIMU alone.

2. A VMD–HHT Approach for Extracting Time–Frequency–Energy Characteristics of Dynamic Responses

2.1. *Hilbert–Huang Transform*

In 1998, N.E. Huang proposed the Hilbert–Huang transform (HHT) [9]. HHT can process nonstationary and nonlinear signals adaptively and is made up of EMD and

Hilbert transform. First, EMD decomposes the original signal into several intrinsic mode functions (IMFs) adaptively. Secondly, the instantaneous frequency is obtained by the Hilbert transform to each intrinsic mode function (IMF). The formula of Hilbert transform for each IMF is as follows:

$$H[c_i(t)] = \frac{1}{\pi}\int_{-\infty}^{\infty}\frac{c_i(\tau)}{t-\tau}d\tau \quad (1)$$

where $c_i(t)$ is the intrinsic mode component, $i = 1, 2, \ldots n$, and $H[\,]$ is the symbol of HHT, τ.

The analytic signal of each IMF is computed by Equation (2).

$$z_i(t) = c_i(t) + jH[c_i(t)] = a_i(t)e^{j\varphi_i(t)} \quad (2)$$

The amplitude function and phase function are defined as

$$a_i(t) = \sqrt{c_i^2(t) + H[c_i(t)]^2} \quad (3)$$

$$\varphi_i(t) = arctan\frac{H[c_i(t)]}{c_i(t)} \quad (4)$$

where $a_i(t)$ is the amplitude function, and $\varphi_i(t)$ means the phase function.

The instantaneous frequency can be obtained by the differential processing of phase function as follows:

$$\omega_i(t) = \frac{d\varphi_i(t)}{dt} \quad (5)$$

Then, the Hilbert spectrum of signal $x(t)$ can be indicated as

$$H(\omega, t) = \begin{cases} Re\sum_{i=1}^{n} a_i(t)e^{j\int \omega_i(t)dt}\omega_i(t) = \omega \\ 0 \qquad\qquad\qquad other \end{cases} \quad (6)$$

Even as the core of Hilbert–Huang transform, EMD is limited in the application because of its drawbacks, such as mode mixing, the endpoint effect, uncertain center frequency, and bandwidth of intrinsic mode. Furthermore, since the iterative calculation of EMD is very complicated, the computational efficiency of HHT is extremely low, failing to satisfy real-time project applications [12].

2.2. Variational Mode Decomposition

Variational mode decomposition (VMD) was put forward by Dragomiretskiy, K. and Zosso, D.in 2014 as a new method of time–frequency analysis based on traditional EMD [23]. The basic concept of VMD is to determine the center frequency and bandwidth of each IMF by searching for the optimal solution of the optimal function iteratively and decompose each IMF of frequency self-adaptively [24]. Research has shown that VMD could overcome the uncertainty of EMD in the center frequency and bandwidth and fundamentally reduce the endpoint effect and mode mixing. The basic steps of constructing the VMD objective function are as follows:

Analytic signals of each IMF are obtained by Hilbert transform to obtain a one-sided spectrum.

$$\left(\delta(t) + \frac{j}{\pi t}\right)\cdot\mu_k(t) \quad (7)$$

where $\mu_k(t)$ is the kth IMF, and $\delta(t)$ stands for the pulse signal.

To modulate the spectrum of each mode to baseband, the exponential term $e^{-j\omega_k t}$ is added to the analytical signal of each IMF.

$$\left[(\delta(t) + \frac{j}{\pi t}\right)\cdot\mu_k(t)\Big]e^{-j\omega_k t} \quad (8)$$

where ω_k denotes the estimated center frequency of the kth IMF.

Calculate the gradient of the 2-norm of the above functions and construct the following constraint model:

$$\begin{cases} \min_{\{u_k\},\{\omega_k\}} \left\{ \sum_k \left\| \partial_t \left[\left(\delta(t) + \frac{j}{\pi t} \right) u_k(t) \right] e^{-j\omega_k t} \right\|_2^2 \right\} \\ s.t. \quad \sum_k u_k = f \end{cases} \tag{9}$$

where $\{\mu_k\}$ is the set of each IMF, and $\{\omega_k\}$ indicates the center frequency of each IMF. Continuous iteration is carried out to optimize the above constraint model, and each center frequency and bandwidth are searched to self-adaptively decompose signal frequency.

To solve the optimal solution of the constraint model, the constrained variational problem can be transformed into an unconstrained variational problem by introducing quadratic penalty factor α and Lagrange multiplier $\lambda(t)$, the extended Lagrange expression is as follows:

$$L(\{u_k\},\{w_k\},\lambda) = \\ \alpha \sum_k \left\| \partial_t \left[(\delta(t) + \frac{j}{\pi t}) \cdot \mu_k(t) \right] e^{-j\omega_k t} \right\|_2^2 + \left\| f(t) - \sum_k u_k(t) \right\|_2^2 + \langle \lambda(t), f(t) - \sum_k u_k(t) \rangle \tag{10}$$

where α is the quadratic penalty factor, which can ensure signal accuracy in the presence of Gaussian noise. $\lambda(t)$ is the Lagrange multiplier, which can maintain strict constraint conditions. An alternate direction multiplier algorithm is used to solve the extended Lagrange function by alternating updates u_k, w_k, λ to find "saddle point" of extended Lagrange expression [23].

2.3. VMD–HHT Model

Since VMD could weaken the inherent end effect and mode aliasing of EMD, a new approach of VMD–HHT time–frequency analysis is suggested in this paper by replacing EMD with VMD. First, the acceleration data were decomposed using VMD. Secondly, the power spectrum density (PSD) was used to extract each mode frequency. As a result, the frequency range of dynamic responses could be defined, and the noise modes were eliminated. Then, the time–frequency–energy spectrum could be gained by employing HHT to extracted modes. Lastly, assisted by the spectrum to adjust and determine feature modes, characteristic information of dynamic responses with high accuracy could be obtained in all directions. Figure 1 shows the process of the VMD–HHT-based characteristic extraction model.

To verify the VMD–HHT-based characteristic extraction model for dynamic responses, a set of data was simulated to compare the performance between traditional HHT and VMD–HHT in extraction. The analog data consisted of five sinusoidal signals with different amplitudes and frequencies (the frequencies were 1 Hz, 5 Hz, 10 Hz, 15 Hz, and 35 Hz, respectively) at a sampling frequency of 100 Hz, and the effect of random noise was also added, as shown in Figure 2.

As shown in Figure 3, HHT cannot extract characteristic information of accelerometer signals. In contrast, the VMD–HHT model can accurately separate frequencies of analog signals into each band (separation frequency is consistent with setting frequency) and qualitatively analyze frequency distribution of energy based on color. Moreover, the analog data were made up of time-invariant stationary signals with constant statistics along the time axis. According to previous analyses and research, even for accessible composite signals, HHT cannot separate characteristic information effectively. Its algorithm EMD would cause mode aliasing, while the VMD–HHT would not. Therefore, this paper proposes an HHT method using VMD instead of EMD, from now on referred to as VMD–HHT time–frequency analysis model.

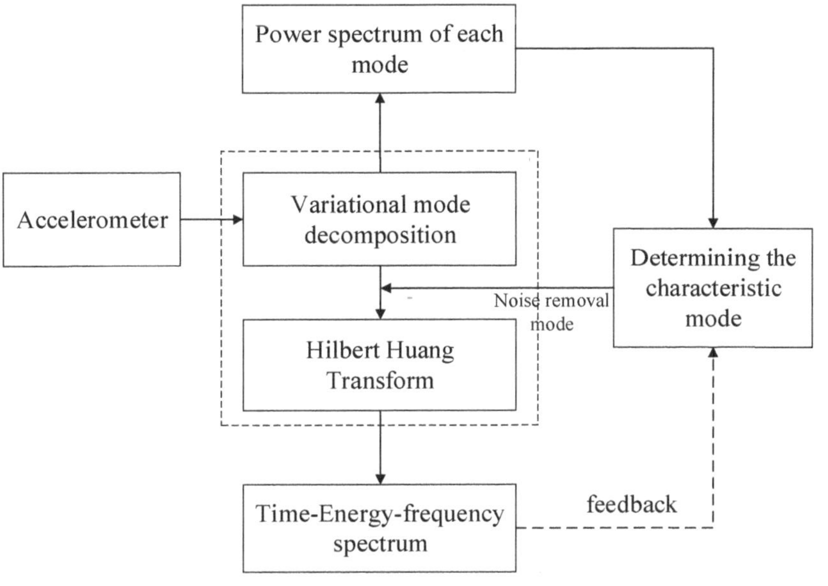

Figure 1. VMD–HHT-based characteristic extraction model.

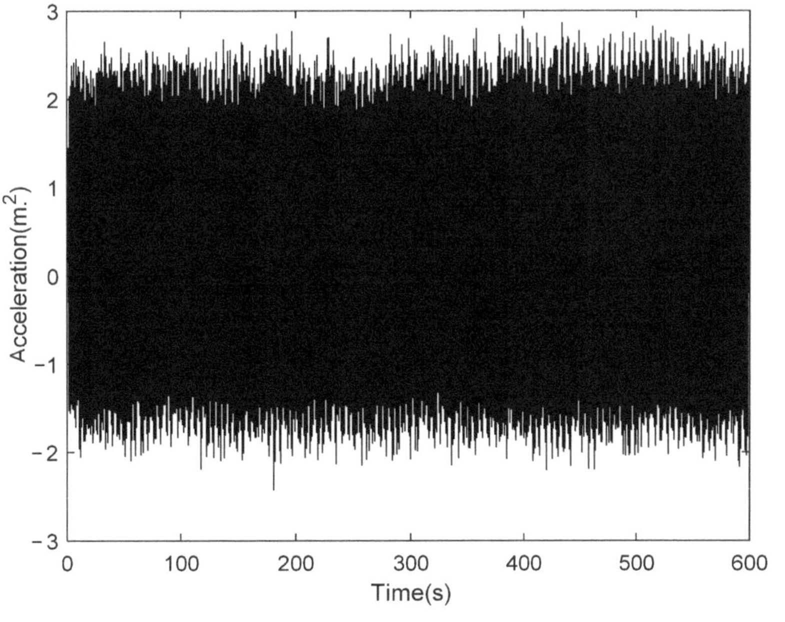

Figure 2. Analog time series of acceleration.

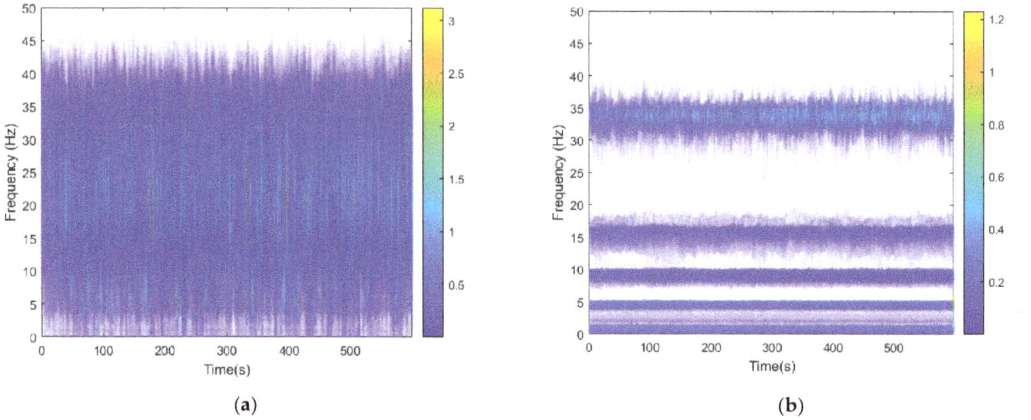

Figure 3. Comparison between HHT and VMD–HHT: (**a**) characteristic information extraction of analog signals using HHT; (**b**) characteristic information extraction of analog signals using VMD–HHT.

3. Dynamic Responses Monitored by MIMU

It is appropriate to monitor dynamic response using MIMU due to the high sampling rate, which can reach up to 100 Hz–200 Hz. The displacement, velocity, and torsion angle responses can be calculated by mechanical arrangement using MIMU. However, low-cost MIMU has a relatively low measurement accuracy with a large gyroscope drift. By processing and analyzing the observed time-series data of accelerometer and gyroscope, real-time dynamic responses of offshore oil platforms can be provided. The monitoring accuracy can also be improved.

3.1. Accelerometer-Derived Displacement Reconstruction

The frequency-domain integration approach (FDIA) is based on Fourier transform, which derives from its integral property as follows:

$$\mathcal{F}\left[\int_{-\infty}^{t} a(t)dt\right] = \frac{1}{j\omega}\cdot\mathcal{F}[a(t)] \quad (11)$$

where $\mathcal{F}$ represents the Fourier transform symbol, $a(t)$ denotes the acceleration signal, j is the imaginary number, and ω means the frequency. The above formula illustrates that the Fourier transform of acceleration signal integration is equivalent to the Fourier transform of the signal divided by the factor $i\omega$, simplifying the complex integration into division.

In FDIA, the signal is transformed into a frequency-domain signal by Fourier transform, and the operation is completed in the frequency domain. The velocity and displacement information in the time domain is obtained by inverse Fourier transform. The calculation procedure is represented in Figure 4.

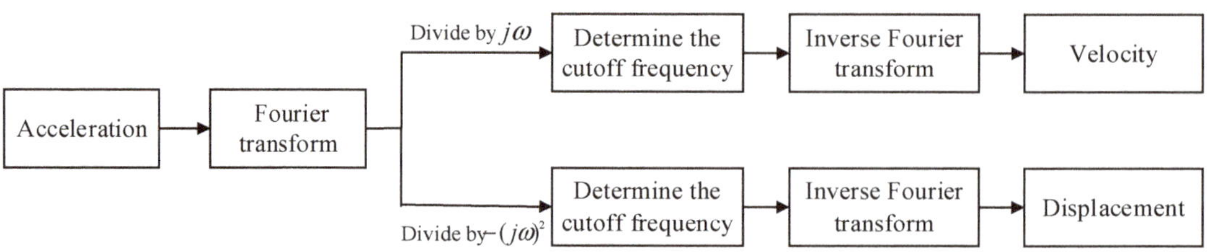

Figure 4. Algorithm of frequency-domain integration.

The frequency spectrum of the acceleration signal after Fourier transform can be expressed by

$$A(k) = \sum_{n=0}^{N-1} a(n)e^{-j(2\pi nk/N)} \tag{12}$$

where N is the number of acquisition points, $a(n)$ means the discrete expression of $a(t)$, and n and k stand for positive integers.

According to the formula, the single integration is obtained as follows:

$$V(n) = \frac{A(k)}{j\omega} = \sum_{k=0}^{N-1} \frac{1}{j2\pi k\Delta f} H(k)a(n)e^{-j2\pi nk/N} \tag{13}$$

The results of double integration are given by

$$X(n) = -\frac{A(k)}{\omega^2} = \sum_{k=0}^{N-1} \frac{1}{-(j2\pi k\Delta f)^2} H(k)a(n)e^{-j2\pi nk/N} \tag{14}$$

As for $H(k) = \begin{cases} 1\ (f_d \le k\Delta f \le f_u) \\ 0\ (other) \end{cases}$, Δf is the frequency resolution while f_d and f_u are the upper and lower cutoff frequencies, respectively. After all Fourier components of different frequencies are calculated according to the frequency-domain relationship, time-domain signals can be obtained by inverse Fourier transform.

3.2. Gyro-Derived Torsion Reconstruction

The Mahony complementary filter has been routinely used to calculate torsion angle after its first successful trial in estimating the torsion angle of UAV accurately [25]. Since it is easy to understand and less time consuming, this algorithm is widely used in engineering applications. The basic idea behind the Mahony complementary filter is that when the carrier is at a static state, the theoretical value of the gravity acceleration component is $[0\ 0\ g]^T$ in a geographical coordinate system. Then, the matrix of $[0\ 0\ g]^T$ is transformed under the carrier coordinate system to compare with the measurements of the accelerometer. Thus, this deviation is the error between the torsion angle integrated by gyroscope and measured accelerometer, respectively. The steps of the algorithm are as follows:

(1) The accelerometer is normalized to a single vector $\frac{g_b}{\|g_b\|}$.

(2) The acceleration data in the geographical coordinate system are converted to the carrier coordinate system, and the estimation in the carrier coordinate system is given by

$$\hat{v} = \begin{bmatrix} 2(\hat{q}_1\hat{q}_3 + \hat{q}_0\hat{q}_2) \\ 2(\hat{q}_2\hat{q}_3 + \hat{q}_0\hat{q}_1) \\ \hat{q}_0^2 - \hat{q}_1^2 - \hat{q}_2^2 + \hat{q}_3^2 \end{bmatrix} \tag{15}$$

where q is quaternions.

(3) The deviation between the acceleration estimation $\hat{v}$ and the measurements by accelerometer $\overline{v}$ is the error item between the integrated torsion angle of the gyroscope and the torsion angle measured by the accelerometer. The value *errror* can be expressed by the cross product.

$$errror = \hat{v} \times \overline{v} \tag{16}$$

(4) The corrected torsion angle can be obtained based on a PI controller using the results from the previous step,

$$\hat{\Omega} = \overline{\Omega}_b + K_p error + K_I \int error \tag{17}$$

where $\overline{\Omega}_b$ is the torsion angle obtained by gyro integration, while K_p and K_I are the PI control param.

(5) The quaternion differential equation can be solved by using the corrected torsion angle $\hat{\Omega}$ (See Formula (8)), and the quaternion can be updated to calculate the theoretical estimation of the accelerometer (transfer to Equation (2)).

$$\dot{q} = \frac{1}{2} q \otimes p(\hat{\Omega}) \tag{18}$$

The algorithm flowchart of Mahony complementary filtering is presented in Figure 5.

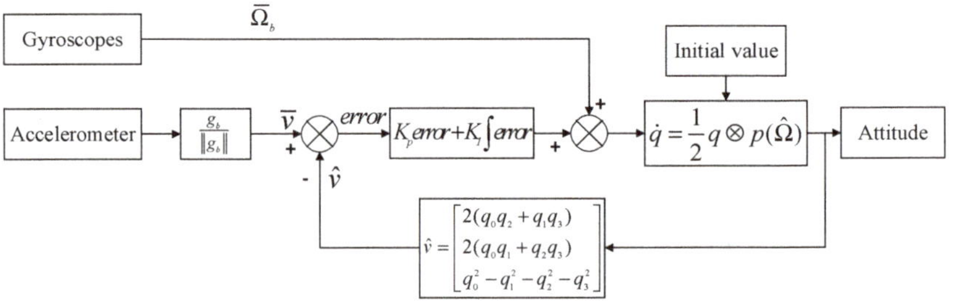

Figure 5. Algorithm of Mahony complementary filtering.

3.3. The Schematic to Monitor Dynamic Responses Based on VMD–HHT Characteristic Extraction Model

A multidimensional characteristic extraction method based on VMD–HHT is introduced to monitor dynamic responses of offshore oil platforms. First, VMD was applied to extract the frequency component of the dynamic response, and the noise component could thus be eliminated. Then, HHT was used to extract multidimensional dynamic response characteristics of time, frequency, and energy using the cleaned accelerometer data. Finally, dynamic displacement responses were calculated by the FDIA based on the cleaned data. To assess the reliability of dynamic displacement responses obtained by accelerom, GNSS data are given for comparison. In addition, to obtain torsion angle changes of offshore oil platforms, the complementary filtering algorithm was applied using six-degree of freedom of MIMU. The flowchart for dynamic responses based on VMD–HHT using MIMU is summarized in Figure 6.

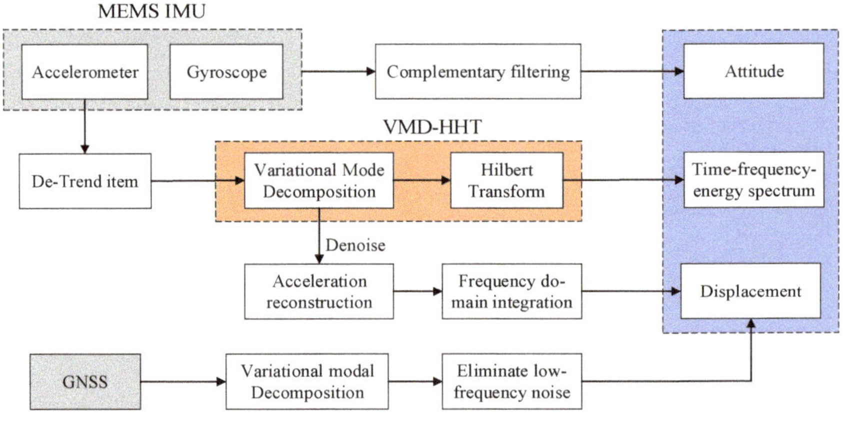

Figure 6. Flowchart for dynamic responses based on VMD–HHT using MIMU.

4. Simulation Shaking-Table Tests

4.1. Simulation Shaking Table

To verify the reliability of the VMD–HHT model and displacement reconstruction method, a series of simulation shaking-table tests were performed using accelerometer in the MIMU. The simulation shaking table (see in Figure 7) comprised a platform framework, steel plates for equipment, two mutually perpendicular screw rods, a servo motor, a control panel, and a power box. The platform frame is 1.2 m long, 1.2 m wide, and 0.5 m high, and the steel plates for equipment are 0.3 m long, 0.1 m wide, and 0.02 m thick. To install the GNSS antenna, a screw hole was drilled through the steel plates; moreover, the screw rods were driven by the servo motor when the steel plates move horizontally with screw rods. In addition, the control system of the shaking table adjusted the vibration amplitude and frequency through the control panel.

Figure 7. Shaking table.

4.2. Data Collection

On 31 December 2020, the experimental measurements of dynamic displacement monitoring were carried out on the platform located in the south square of Beijing University of Civil Engineering and Architecture (BUCEA). The equipment used in the tests includes the shaking table, a GNSS deformation monitoring system independently developed by BUCEA, the ADIS16505 IMU (inertial measurement unit) produced by Analog Devices, Inc. (containing an accelerometer and gyroscope, but only the accelerometer data were used in this test) and laptops. The param of GNSS and accelerometer are shown in the following Table 1.

Two sets of equipment, the GNSS and accelerometer, were attached to the shaking table using a metal plate. The connecting line between the centers of MIMU and GNSS antenna was directed toward north–south. Since the accelerometer was located toward the northern GNSS antennas at the distance of 10 cm using a northeast down system, the installation direction of the X-axis would point to the north. The layout of monitoring systems and monitoring points are shown in Figure 8.

Table 1. Principal specifications of GNSS and accelerometer.

Equipment	Performance	
GNSS	Signal Tracking	BDS: B1/B2; GPS:L1/L2 GLONASS: L1/L2; GALILEO:E1/E5b
	RTK (RMS)	Horizontal: ±8 mm + 1 ppm Vertical: ±15 mm + 1 ppm
	Updating Frequency	5 Hz
Accelerometer	Dynamic Range	$\pm 78.4\ m/s^2$
	Bias Stability	$26.5 \times 10^{-6} m/s^2$
	Noise Density	$167 \times 10^{-6} m/s^2/\sqrt{Hz}rms$
	Updating Frequency	100 Hz

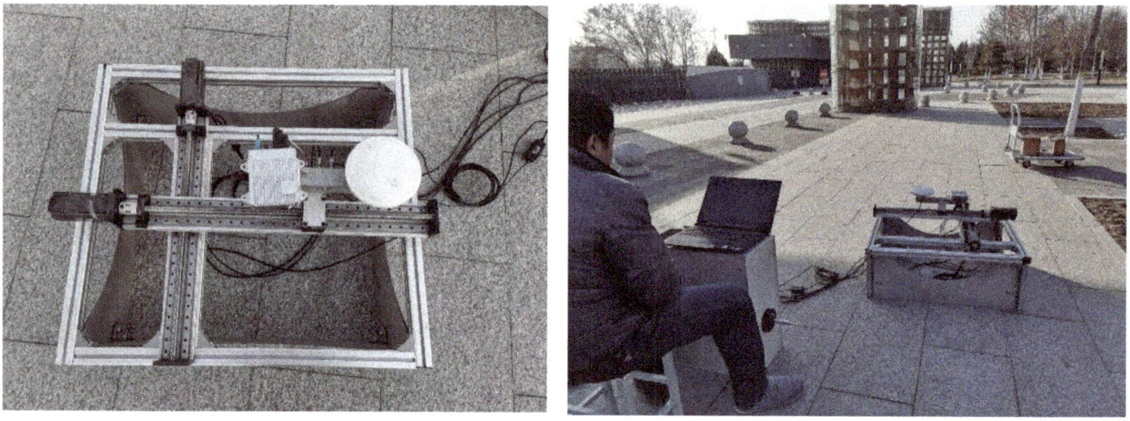

(**a**) (**b**)

Figure 8. Instrument setup for deformation monitoring systems: (**a**) the layout of monitoring systems; (**b**) the monitoring point.

4.3. Comparison of PSD, HHT, and VMD–HHT

To illustrate the advantages of the VMD–HHT model in the time domain and energy domain compared with PSD and HHT, three groups of tests were designed with time sequence amplitudes of 10 mm, 30 mm, and 50 mm. A GNSS receiver using a 5 Hz sampling rate and an accelerometer using a 100 Hz sampling rate were used to obtain the three original data sets for comparison.

The shaking table tests were conducted for 10 min (each test lasted approximately 3 min). The test site had an open view of the sky, thus avoiding multipath effects. Additionally, during the experiments, there was no wind to prevent wind-induced frequency noise. Figure 9 gives the frequency extract from three vibration events using the method of PSD; as seen from Figure 9, the frequency response peaked at 1 Hz is consistent with the vibration frequency of the shaking table. However, the time of the three groups of tests with amplitudes of 10 mm, 30 mm, and 50 mm is unknown from Figure 9 due to the frequency of all tests are 1 Hz.

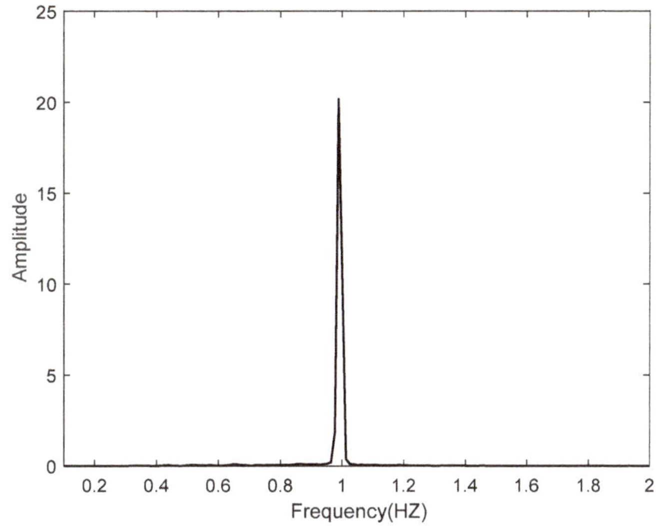

Figure 9. Vibration frequencies extracted by PSD.

HHT and VMD–HHT methods can separate each event in the time dimension and reflect the intensity of each event (Figure 10). It can be seen that the frequency range is about 1 Hz, and the intensity of three events is gradually strengthening, corresponding to amplitudes at 10 mm, 30 mm, and 50 mm. Comparing Figures 10 and 11 show that the HHT spectrum is noisier than the VMD–HHT spectrum. We can also observe that the intensity displayed in the HHT spectrum is smaller than the VMD–HHT spectrum, and the latter is more clearly detected. However, the frequency values at the beginning and end of each signal also reach higher values, especially the signal with amplitude at 50 mm. It is probably caused by endpoint effect. Overall, VMD–HHT can extract the characteristics of time, frequency, and energy more reliably than HHT.

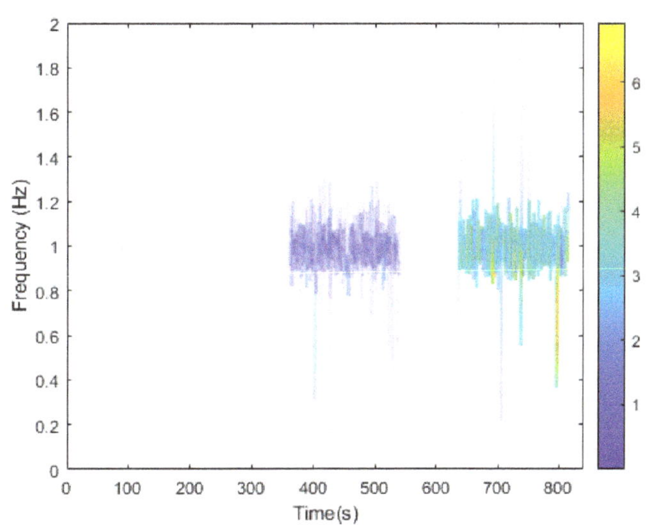

Figure 10. Time–frequency–energy extracted by HHT.

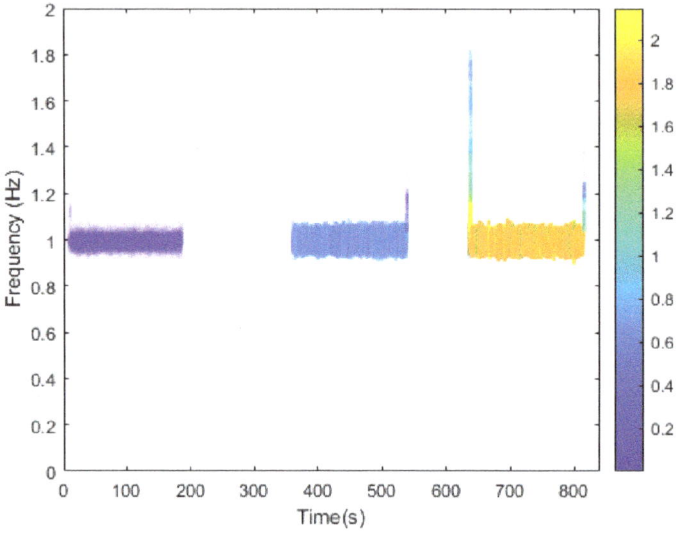

Figure 11. Time–frequency–energy extracted by VMD–HHT.

4.4. Reconstruction of Dynamic Displacement Using FDIA Based on the VMD–HHT Model and GPS

To assess the effectiveness and accuracy of FDIA based on the VMD–HHT model, displacements measured by GNSS are given for comparison. Firstly, the VMD approach was applied to remove residual and high-frequency noise from accelerometer data. Then, FDIA was used to obtain vibration displacements based on the cleaned data proposed by VMD. Figure 12 shows the displacement comparison of GNSS and FDIA; as seen from Figure 12, there is a high correlation between GNSS and FDIA. It is evident that the vibration amplitude is about 10 mm, and frequency is about 1 Hz, which is consistent with the set param. It can be concluded that VMD–HHT-based FDIA has successfully calculated the vibration displacements.

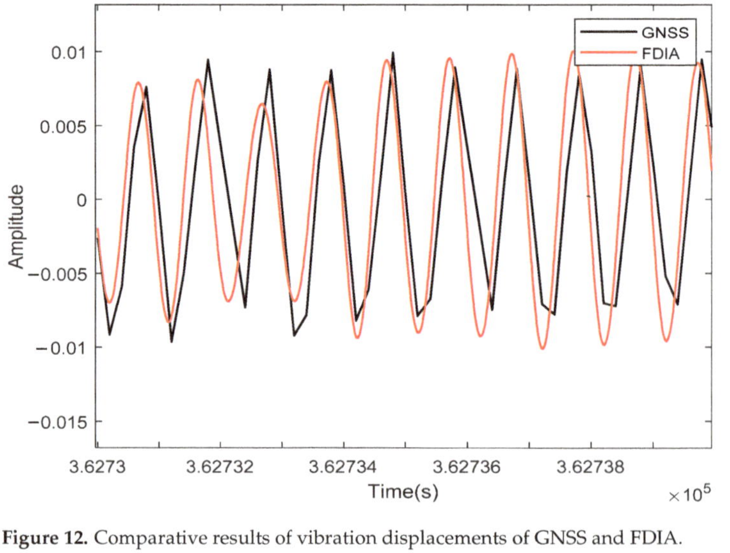

Figure 12. Comparative results of vibration displacements of GNSS and FDIA.

5. Trial Analysis of CB4A Offshore Oil Platform

Located at 38°8′39″ north latitude and 118°50′38″ east longitude, the CB4A offshore oil platform lies in the Yellow River Delta along the coast of the Bohai Sea. It is a fixed offshore oil platform in Shengli Oilfield at Dongying City with a total area of 700 m^2 (see in Figure 13). The sea area near the platform is windy all year round, affected by the southeast Pacific monsoon and the northwest monsoon of the Asian continent. More specifically, as frequently affected by the monsoon of the Asian continent, the offshore oil platform also suffers from heavy wind all year round. In addition, being close to the maritime transportation hub, the platform is under heavy traffic. As a result, ship collisions occur from time to time, and thus, the monitoring of offshore oil platforms becomes exceptionally urgent.

Figure 13. The view of the offshore oil platform.

5.1. *Equipment for Monitoring Dynamic Responses*

MIMU used in the dynamic responses monitoring trial were from Mti300 inertial sensors by Xsens Technologies B.V. of the Netherlands (as shown in the upper right corner of Figure 14b). The specifications of Mti300 are given in Table 2. Furthermore, to demonstrate the accuracy and reliability of dynamic displacement responses by MIMU, a GPS monitoring system independently developed by BUCEA was installed on CB4A offshore oil platform. GPS RTK technology was used in this trial. The specifications of GPS are given in Table 3. The layout of the monitoring systems and the monitoring point for dynamic responses are indicated in Figure 14. Two sets of equipment were positioned at the northwest corner of the platform roof and attached to the roof railing by a sheet iron. The connecting line between the centers of MIMU and GPS antenna was directed toward the north–south. The MIMU was placed 20 cm to the north of the GPS antenna navigated by the NED coordinate system, and its X-axis was installed to the north.

(a) (b)

Figure 14. Instrument setup for monitoring systems of dynamic responses of offshore oil platforms: (**a**) the layout of monitoring systems; (**b**) the monitoring point of dynamic responses.

Table 2. Performance index for Mti300.

Index Item	Gyroscope	Accelerom
Standard full range	±450 °/s	±20 g
Initial bias error (one year)	0.2 °/s	5 mg
In-run bias stability	10 °/h	15 μg
Bandwidth (−3 dB)	415 Hz	375 Hz
Noise density	0.01 °/s/$\sqrt{}$Hz	60 μg/$\sqrt{}$Hz
g-sensitivity (calibrated)	0.003 °/s/g	N/A
Nonorthogonality	0.05 deg	0.05 deg
Nonlinearity	0.01%	0.1%

Table 3. Performance index for GPS.

Tracking Signal	BDS B1/B2/B3
	GPS L1/L2/L5
	GLONASS /L1/L2
	GALILEO E1/E5a/E5b
	QZSS L1/L5
	SBAS L1
Single(RMS)	Plane: ±1.5 m; Altitude: ±3 m
DGPS(RMS)	Plane: ±0.4 m; Altitude: ±0.8 m
RTK(RMS)	Plane: ±8 mm + 1 ppm Altitude: ±15 mm + 1 ppm
Sampling rate	5 Hz

5.2. Data Acquisition

The monitoring trial of the CB4A offshore oil platform was carried out in Dongying city, Shandong Province, on 3 December 2019. The MIMU and GPS data sampling frequencies were set at 100 Hz and 5 Hz, respectively, with a 4.5 min sampling time (GPS time: 200088.0–200358.0). We planned to design an experiment for researching the dynamic response of offshore oil platforms induced by wind and waves from the beginning. However, as CB4A offshore oil platform was a stable rigid body at the height of only 7 m, it would be difficult to produce dynamic responses caused by slight winds or waves. Unfortunately, the weather was suitable, with no strong winds and/or waves during the tests. Thus, wind and waves were replaced with ship collision events. The ship collision experiment was conducted on a calm afternoon, which would avoid the influence of wind and waves. A small ship, about 5 m long, successive hit the leg of the platform traveling from south to north at a speed of 20 km per hour. When the small ship started to hit the leg of the platform, the experimenter recorded the current Beijing standard time. The continuous collision lasted for about 4 min. Eight collisions were made, and the time interval of each impact was less than 1 min. The time of each impact was recorded manually and is shown in Table 4.

Table 4. Record schedule of ship collision.

Collision	First	Second	Third	Fourth	Fifth	Sixth	Seventh	Eighth
Time	15:34:37	15:35:28	15:36:00	15:36:58	15:37:30	15:37:51	15:38:17	15:38:46
GPS Period	200094.6	200145.6	200177.6	200235.6	200267.6	200290.8	200314.6	200343.8

5.3. Dynamic Responses Time–Frequency Extraction

As the X-axis installation of the inertial sensor and ship collision were both directed toward the north, which means, the X-axis was the main direction of the dynamic response. Therefore, the paper only analyzed the X-axis data for dynamic response monitoring. As shown in Figure 15, the original X-axis data of the accelerometer were collected, and changes of acceleration were probably caused by multiple ship collisions; the corresponding collision time was estimated to be at the 7th s, 59th s, 89th s, 147th s, 198th s, 225th s, 237th s, and 252nd s, respectively (the second and third collisions were not obvious to be identified directly). They were basically consistent with manual records without abnormality; hence, subsequent analyses could be carried out.

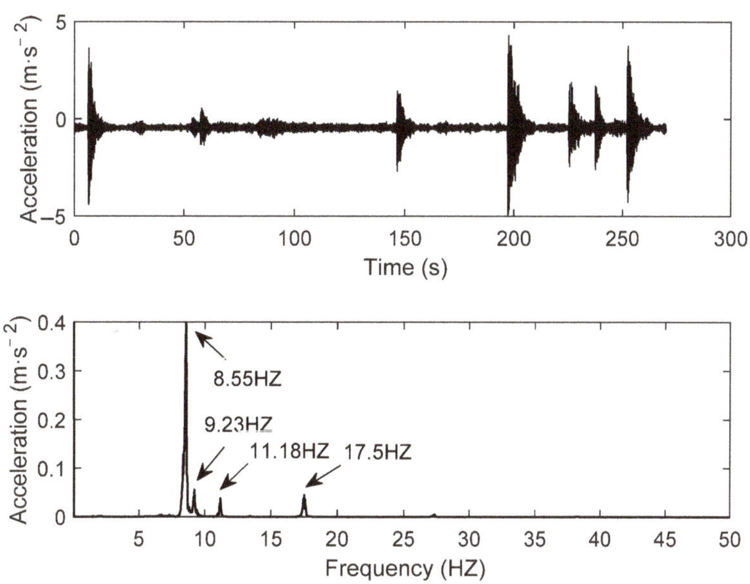

Figure 15. Acceleration and corresponding spectrum densities: raw signal of accelerometer (**top**) and power spectral density (**bottom**).

PSD was utilized to extract frequencies of dynamic responses of ship collisions, as represented in the bottom of Figure 15, where four peaks can be seen at 8.55 Hz, 9.23 Hz, 11.18 Hz, and 17.50 Hz separately, with the largest peak at 8.55 Hz. It could be concluded that much information was missing in the spectrum as (i) the sequence and time of collisions were unable to be identified corresponding to the four frequencies without the time-domain information in the spectrum and (ii) only four peaks were produced by eight ship collisions, which indicated the possibility of a common frequency phenomenon or the covering of high-power spectral density on the low-power ones. Therefore, PSD could not extract complete and reliable frequencies of dynamic responses for the ship collisions.

To further study the frequency characteristics of collisions in the time domain, the VMD–HHT method proposed in this paper was adopted to carry out a three-dimensional synchronous analysis of time–frequency energy. First, five IMFs were produced using self-adaptive VMD, as illustrated in Figure 16. All those functions contained dynamic responses of ship collisions, indicating the accelerometer's high accuracy, which was unaffected by high- and low-frequency noise and could be directly reconstructed without denoising. Later, following the Hilbert transform of the reconstructed acceleration data, the Hilbert time–frequency spectrum was obtained, as shown in Figure 11. It was evident that seven collisions, along with each corresponding time, frequency range, and energy (collision intensity), were displayed synchronously, except that the frequency range of the fifth collision was found to be between 3 Hz and 15 Hz, while others were between 5 Hz and 12 Hz. Referring to the qualitative analysis of energy, the third collision (89 s) was invisible, indicating that its intensity was the smallest. Among the seven visible collisions (Figure 17), the intensity of the second collision was the lowest, consistent with the analysis results of the original data; the intensity of the fifth collision was the highest, and its enlarged view is reflected in Figure 18, while the collision lasted for about 7 s, and the intensity changed from high to low. In summary, this VMD–HHT-based method could accurately extract frequency ranges of collisions in the time domain and clearly characterize the intensities of each collision.

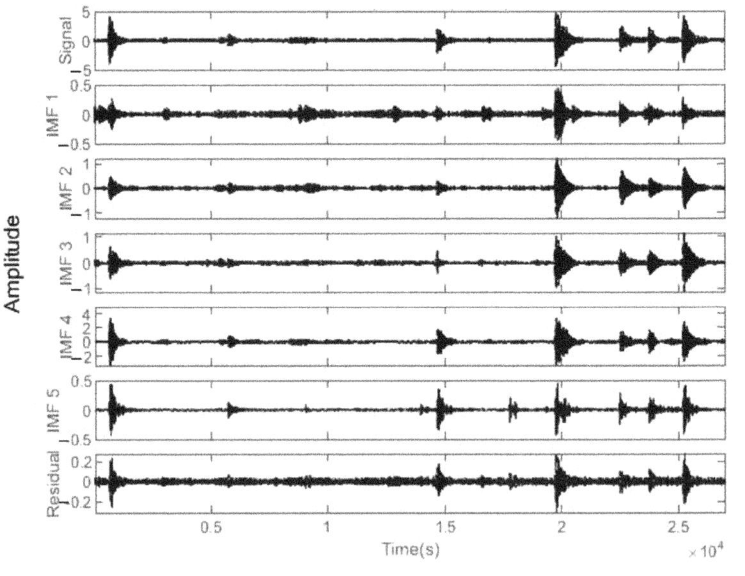

Figure 16. IMF components of accelerometer-recorded acceleration based on VMD.

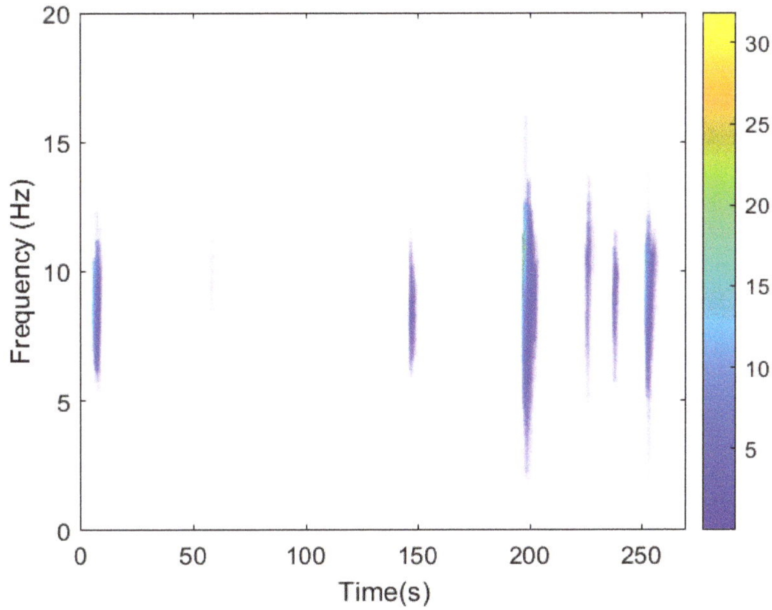

Figure 17. HHT spectrum of acceleration.

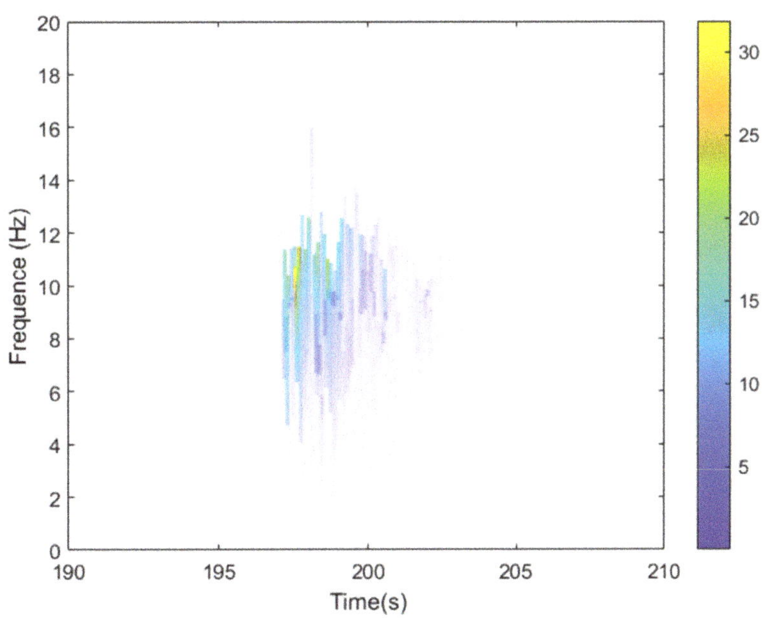

Figure 18. HHT spectrum of the fifth collision.

5.4. Reconstruction of Dynamic Displacement Based on VMD–HHT Using Accelerometer

The frequency-domain integral equation is a powerful approach to process acceleration data using a double integral to reconstruct dynamic displacement information. According to the frequency range obtained through the VMD–HHT method, the minimum cutoff frequency was set as 1 Hz, and the maximum was set as 20 Hz. Figure 19 shows the accelerometer-derived displacement after double integration. We can observe that the displacement of the third collision (89th s) was minimal. Thus, the third collision on the platform was weak. Referring to the manual record, original data, and the HHT spectrum, the unknown displacement occurred at 180th s without any collision; the displacements of eight collisions were within 6 mm. Compared with related deformation data of bridges, high-rise buildings, and dams, deformation of the offshore oil platform collided by ships was smaller [1,9]. The reason for this is that the offshore oil platform was a stable rigid body, and the colliding ship was small when anticollision rubbers were tied to the platform pile.

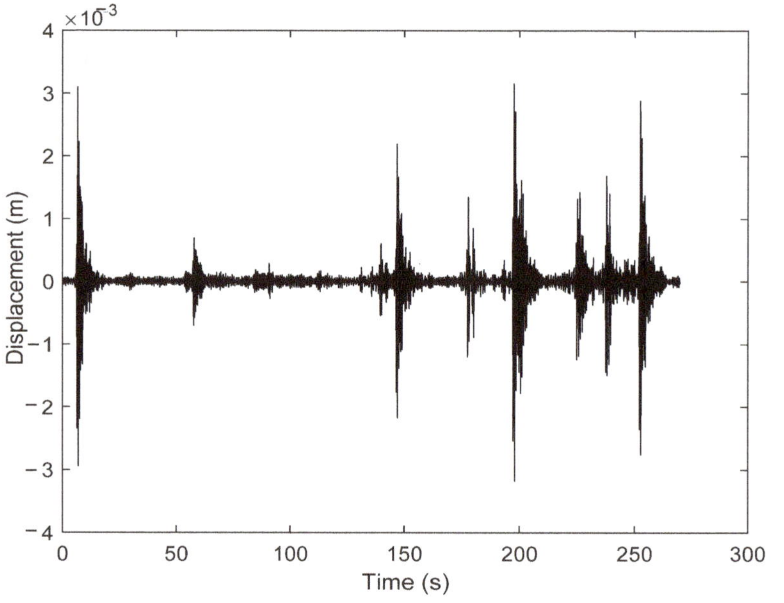

Figure 19. Accelerometer-derived displacement by frequency-domain integration.

5.5. Evaluation of the Torsion Angle Based on Mahony Complementary Filter Using MIMU

To obtain torsion angle responses caused by ship collisions, the gyroscope data of MIMU were analyzed by Mahony complementary filter. Figure 20 shows the three-axis angular rate data correctly without bias, suggesting that the Y-axis of the gyroscope was influenced by the collision most. From the above analyses, it can be concluded that X-axis was the main direction of displacement response. When the ship stroked the north of the platform (X-axis), displacement responses would be caused in the same direction, while the platform was found out to incline towards Y-axis. A visual representation of the process is given in Figure 21.

Figure 22 (bottom) shows the PSD in correspondence to angular rate output in the Y-axis of the gyroscope, which included five peaks whose corresponding frequencies were 8.57 Hz, 9.23 Hz, 11.18 Hz, 13.45 Hz, and 15.00 Hz, respectively. The main peak frequency was 8.57 Hz. Referring to Figure 2, the frequencies extracted from the X-axis of the accelerometer were 8.55 Hz, 9.23 Hz, 11.18 Hz, and 17.50 Hz independently, while the main peak frequency was 8.55 Hz. By comparing two sets of data, it is evident that collision frequencies extracted from the X-axis of accelerometer and Y-axis of gyroscope were very similar. To be more specific, the distribution was consistent, but the amplitude corresponding to each frequency had a high coherence. To summarize, the frequencies of dynamic responses could be extracted from gyroscope data of MIMU.

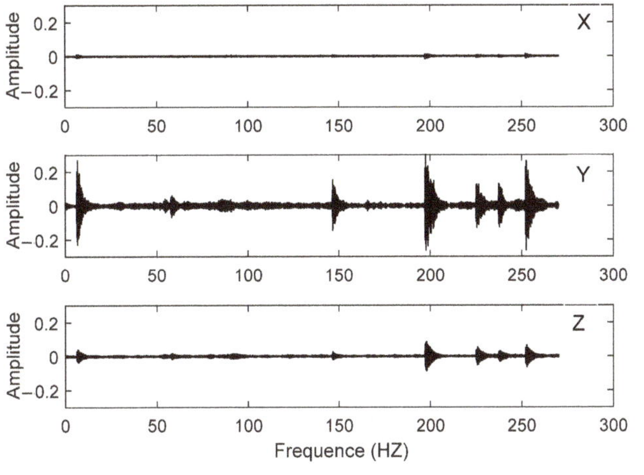

Figure 20. Three-axis raw signals recorded by gyroscope.

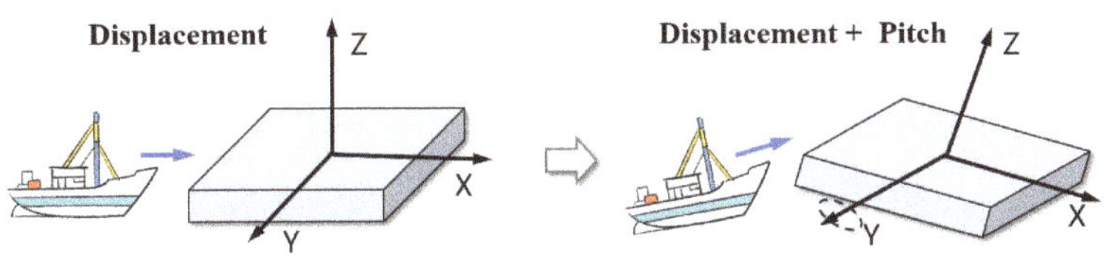

Figure 21. Schematic of dynamic displacement and pitch caused by the ship collision.

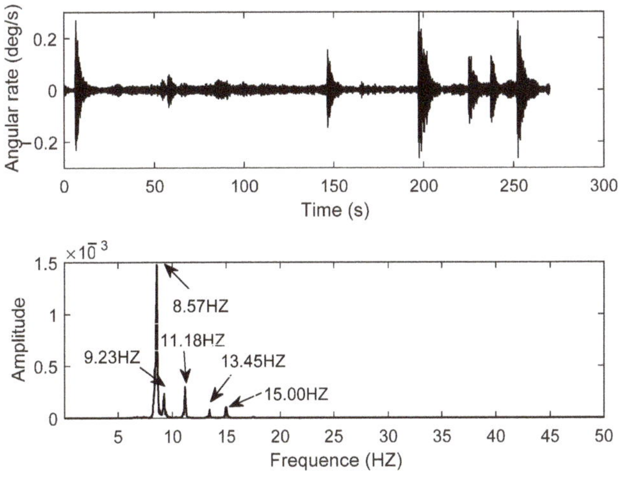

Figure 22. Gyroscope-recorded signal and corresponding spectrums: raw signal at the direction of Y (**top**) and power spectral density (**bottom**).

To further explore the relationship between the accelerometer and gyroscope axis system during collisions, three-axis PSD of accelerometer and gyroscope were given, as indicated in Figures 23 and 24. Amplitudes processed by PSD through accelerometer from high to low were X-axis $\geq$ Z-axis $\geq$ Y-axis. Those of gyroscope were Y-axis $\geq$ Z-axis $\geq$ X-axis, which means the event had the most significant impact on the X-axis of the accelerometer and the Y-axis of gyroscope, followed by their Z-axes. By examining the frequency distribution, the similarity of PSD was high among X-axes and Z-axes of the accelerometer and Y-axis of the gyroscope. According to the results, when the ship struck the platform from X-direction (north direction), a displacement response in the Z-axis would be caused by the displacement response of the X-axis and the rotation of the Y-axis; the PSD of the Y-axis of the accelerometer was similar to that of Z-axis of the gyroscope (as X-axis of the gyroscope was seriously affected by the noise; thus, no comparative analysis would be performed). Hereby, two conjectures are proposed as follows: (i) Since the collision direction was not completely in the X-direction and there was a component on the Y-axis, it means that a small-angle oblique impact has occurred to cause the rotation of the Z-axis, and the horizontal distortion of the platform is illustrated in Figure 25; (ii) the X-axis of the installed inertial sensor did not coincide with the north direction, that is, the installed axis deviated from the local geographical coordinate system, resulting in the component on the Y-axis to cause a small-angle oblique collision and Z-axis' rotation.

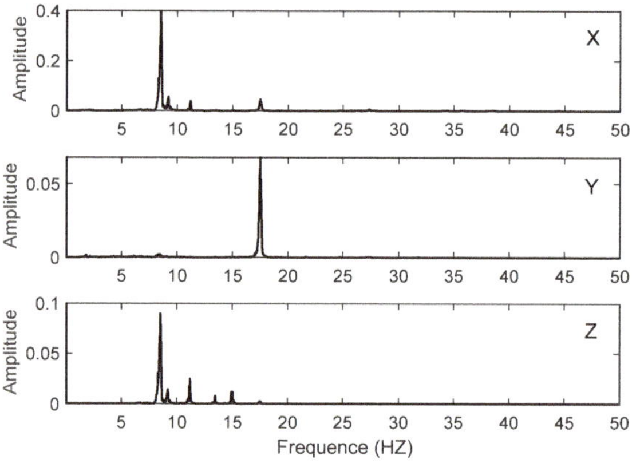

Figure 23. PSD of the acceleration recorded by the accelerometer in three directions.

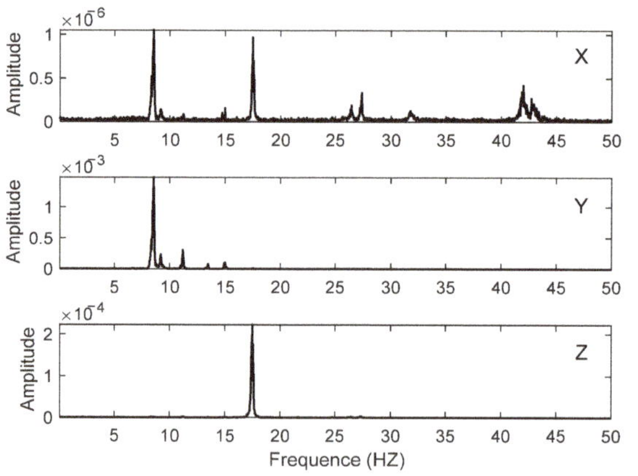

Figure 24. PSD of the angular velocity recorded by the gyroscope in three directions.

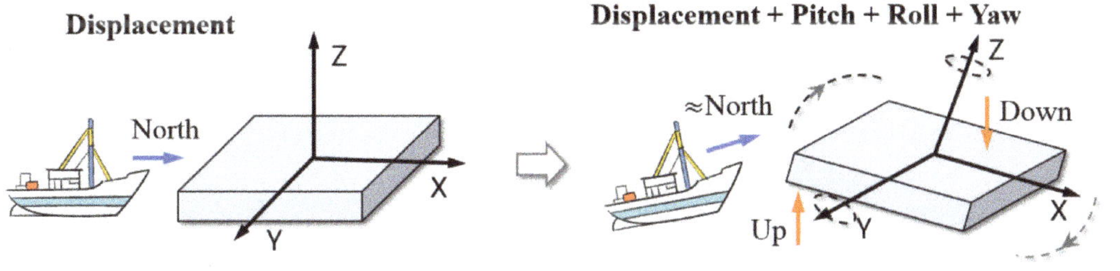

Figure 25. Schematic of dynamic displacement and torsion angle caused by ship collisions.

To obtain the torsion angle response data of the platform, Mahony Complementary Filtering was used to solve the torsion angle change of platforms by integrating the information of accelerometer and gyroscope. As can be seen from the torsion angle calculation results of Figure 26, the pitch angle (Y-axis rotation) changed significantly with a maximum angle of 1.2°, followed by the roll angle (X-axis rotation) with a maximum angle of 0.15°. The heading angle (Z-axis rotation) did not change significantly with the impact event, but it deviated about 1.1° from the X direction (north direction). Therefore, Conjecture 2 could be verified. In other words, the misalignment between the installation axes of MIMU and the geographical coordinate system leads to the displacement component of the Y-axis and the rotation of the Z-axis. All the above analyses show that ship collisions will cause displacement responses and make the platform tilt or even twist. In addition, the relationship between the accelerometer and the gyroscope is also revealed: The impact on the X direction caused the Y-axis to rotate, that is, the pitch angle changes. The impact direction had a component on the Y-axis to cause the X-axis rotation, which is, the roll angle changes. The Z-axis rotation was caused by a small-angle oblique impact, which was the change of heading angle.

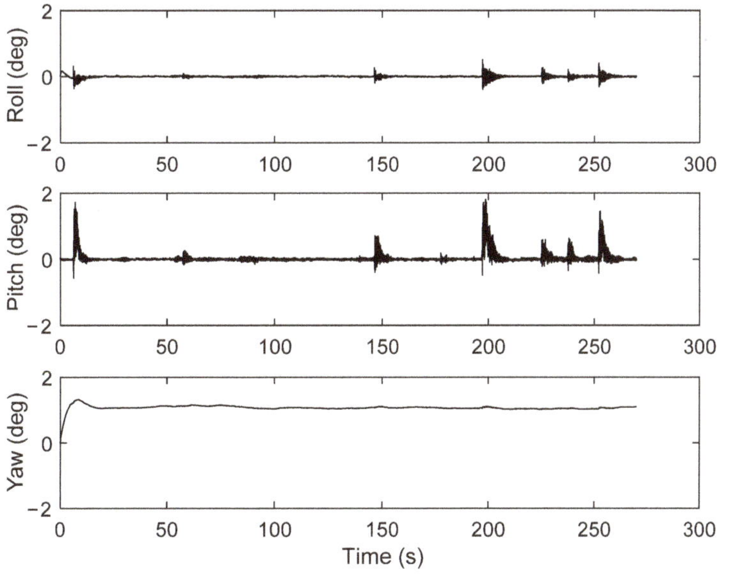

Figure 26. Three-axis torsion angle by Mahony complementary filter using MIMU.

6. Conclusions

In this paper, a model combined by HHT and VMD is proposed to extract multidimensional dynamic responses characteristics of time, frequency, and energy of offshore oil platforms. A series of shaking-table tests and field tests in an offshore oil platform in Shengli Oilfield at Dongying City were performed using a single GPS receiver and a single accelerometer. By comparing analysis results between PSD and VMD–HHT, we found that dynamic response characteristics of ship collisions cannot be extracted entirely relying only on PSD. In contrast, VMD–HHT could effectively extract the characteristics of time, frequency, and energy simultaneously. Based on the analysis results of VMD–HHT, the dynamic displacements were reconstructed using the accelerometer in the MIMU, and the maximum displacement calculated by VMD–HHT based FDIA was 6 mm. Moreover, acceleration and angular rate obtained by MIMU had a high similarity and clear axial correlation, and such integration could also be used to calculate the torsion angle of offshore oil platforms; the maximum torsion angle is up to 1.1°.

This novel proposed method could aid in the avoidance of collisions of offshore oil platforms and be further applied in oil exploitation and platform maintenance and monitoring. However, VMD–HHT model will not be able to effectively extract feature information when the noise frequency is close to the frequency of dynamic response events.

Author Contributions: Conceptualization, J.W. and X.L.; methodology, J.W.; validation, J.W., X.L. and W.L.; investigation, J.W.; resources, J.W. and C.H.; data curation, F.L.; writing—original draft preparation, X.L.; writing—review and editing, J.W., X.L., W.L., F.L. and C.H.; funding acquisition, J.W. All authors have read and agreed to the published version of the manuscript.

Funding: This research was funded by The National Natural Science Foundation of China, Grant Number 41874029, and The National Key Research and Development Program of China, Grant Number 2020YFD1100201.

Data Availability Statement: Not applicable.

Conflicts of Interest: The authors declare no conflict of interest.

References

1. Yang, X.F.; Wang, N. Comprehensive safety assessment method for offshore platform in different application scenarios. *Energy Energy Conserv.* **2021**, *1*, 198–200.
2. Meng, X.; Dodson, A.H.; Roberts, G.W. Detecting bridge dynamics with GPS and triaxial accelerometers. *Eng. Struct.* **2007**, *29*, 3178–3184. [CrossRef]
3. Hu, G.S. *Digital Signal Processing: Theory, Algorithm and Implementation*; Tsinghua University Press: Beijing, China, 2012.
4. Mateo, C.; Talavera, J. Short-time Fourier transform with the window size fixed in the frequency domain. *Digit. Signal Process.* **2018**, *77*, 13–21. [CrossRef]
5. Debnath, L.; Shah, F.A. *Wavelet Transforms and Their Applications*; Birkhäuser: Boston, MA, USA, 2015.
6. Wang, D.J.; Xiong, Y.L. A precise kinematic single epoch positioning algorithm using moving window wavelet denoising. *Geomat. Inf. Sci. Wuhan Univ.* **2015**, *40*, 779–785.
7. Yi, C.; Lv, Y.; Xiao, H. Multi-sensor signal denoising based on matching synchro squeezing wavelet transform for mechanical fault condition assessment. *Meas. Sci. Technol.* **2018**, *29*, 045104. [CrossRef]
8. Luo, Y.; Huang, C.; Zhang, J. Denoising method of deformation monitoring data based on variational mode decomposition. *Geomat. Inf. Sci. Wuhan Univ.* **2020**, *45*, 784–791.
9. Huang, N.E.; Shen, Z.; Long, S.R. The empirical mode decomposition and the Hilbert spectrum for nonlinear and non-stationary time series analysis. Proceedings of the Royal Society A: Mathematical. *Phys. Eng. Sci.* **1998**, *454*, 903–995. [CrossRef]
10. Chan, W.S.; Xu, Y.L.; Ding, X.L. An integrated GPS–accelerometer data processing technique for structural deformation monitoring. *J. Geod.* **2006**, *80*, 705–719. [CrossRef]
11. Barbosh, M.; Singh, P.; Sadhu, A. Empirical mode decomposition and its variants: A review with applications in structural health monitoring. *Smart Mater. Struct.* **2020**, *29*, 093001. [CrossRef]
12. Rehman, N.; Park, C.; Huang, N.E. EMD via MEMD: Multivariate noise-aided computation of standard EMD. *Adv. Adapt. Data Anal.* **2013**, *5*, 1350007. [CrossRef]
13. Wang, T.; Zhang, M.; Yu, Q. Comparing the applications of EMD and EEMD on time–frequency analysis of seismic signal. *J. Appl. Geophys.* **2012**, *83*, 29–34. [CrossRef]
14. Shrivastava, Y.; Singh, B. A comparative study of EMD and EEMD approaches for identifying chatter frequency in CNC turning. *Eur. J. Mech.* **2019**, *73*, 381–393. [CrossRef]

15. Yeh, J.R.; Shieh, J.S.; Huang, N.E. Complementary ensemble empirical mode decomposition: A novel noise enhanced data analysis method. *Adv. Adapt. Data Anal.* **2010**, *2*, 135–156. [CrossRef]
16. Niu, Y.; Xiong, C. Analysis of the dynamic characteristics of a suspension bridge based on RTK-GNSS measurement combining EEMD and a wavelet packet technique. *Meas. Sci. Technol.* **2018**, *29*, 85–103. [CrossRef]
17. Wang, J.; Li, Z.; Gao, J. EMD-wavelet based stochastic error reducing model for GPS/INS integrated navigation. *J. Southeast Univ.* **2012**, *42*, 406–412.
18. Song, J.G.; Zhao, C.X.; Lin, S.H. Decomposition of seismic signal based on Hilbert-Huang transform. In Proceedings of the 2011 International Conference on Business Management and Electronic Information, Guangzhou, China, 13 May 2011; pp. 840–843.
19. Zhou, Y.; Chen, W.; Gao, J. Application of Hilbert–Huang transform based instantaneous frequency to seismic reflection data. *J. Appl. Geophys.* **2012**, *82*, 68–74. [CrossRef]
20. Huang, N.E. *Hilbert-Huang Transform and Its Applications*, 2nd ed.; World Scientific: Singapore, 2017.
21. Shi, C.; Niu, Y.J.; Wei, N. Application of the HHT-EEMD approach in analysis of GPS height time series. *J. Geod. Geodyn.* **2018**, *38*, 661–667.
22. Mou, Z.; Niu, X.; Wang, C. A precise feature extraction method for shock wave signal with improved CEEMD-HHT. *J. Ambient Intell. Humaniz. Comput.* **2020**, *3*, 1–12. [CrossRef]
23. Luo, Y.Y.; Yao, Y.B.; Huang, C. Deformation feature extraction and analysis based on improved variational mode decomposition. *Geomat. Inf. Sci. Wuhan Univ.* **2020**, *45*, 612–619.
24. Dragomiretskiy, K.; Zosso, D. Variational mode decomposition. *IEEE Trans. Signal Process.* **2014**, *62*, 531–544. [CrossRef]
25. Wang, M.; Guan, L.; Gao, Y. UAV torsion angle measurement based on enhanced mahony complementary filter. *IEEE Int. Conf. Mechatron. Autom.* **2018**, 545–550.

symmetry MDPI

Article

Data Anomaly Detection of Bridge Structures Using Convolutional Neural Network Based on Structural Vibration Signals

Yixiao Zhang and Ying Lei *

School of Architecture and Civil Engineering, Xiamen University, Xiamen 361005, China; 25320181152869@stu.xmu.edu.cn
* Correspondence: ylei@xmu.edu.cn

Abstract: Structural monitoring provides valuable information on the state of structural health, which is helpful for structural damage detection and structural state assessment. However, when the sensors are exposed to harsh environmental conditions, various anomalies caused by sensor failure or damage lead to abnormalities of the monitoring data. It is inefficient to remove abnormal data by manual elimination because of the massive number of data obtained by monitoring systems. In this paper, a data anomaly detection method based on structural vibration signals and a convolutional neural network (CNN) is proposed, which can automatically identify and eliminate abnormal data. First, the anomaly detection problem is modeled as a time series classification problem. Data preprocessing and data augmentation, including data expansion and down-sampling to construct new samples, are employed to process the original time series. For a small number of samples in the data set, randomly increase outliers, symmetrical flipping, and noise addition methods are used for data expansion, and samples with the same label are added without increasing the original samples. The down-sampling method of symmetrically extracting the maximum value and the minimum value at the same time can effectively reduce the dimensionality of the input sample, while retaining the characteristics of the data to the greatest extent. Using hyperparameter tuning of the classification weights, CNN is more effective in dealing with unbalanced training sets. Finally, the effectiveness of the proposed method is proved by the anomaly detection of acceleration data on a long-span bridge. For the anomaly detection problem modeled as a time series classification problem, the proposed method can effectively identify various abnormal patterns.

Keywords: structural health monitoring; deep learning; data anomaly detection; convolutional neural network

Citation: Zhang, Y.; Lei, Y. Data Anomaly Detection of Bridge Structures Using Convolutional Neural Network Based on Structural Vibration Signals. *Symmetry* **2021**, *13*, 1186. https://doi.org/10.3390/sym13071186

Academic Editor: Sergei Alexandrov

Received: 5 June 2021
Accepted: 25 June 2021
Published: 30 June 2021

1. Introduction

In the field of structural health monitoring, the problem of data accumulation has been paid more and more attention. During the real-time monitoring of bridges, a large number of data are generated every day. These data containing the damage information of the bridge structure are the basis of bridge state assessment and long-term performance prediction. However, the installed sensors are exposed to harsh environments. As the working time increases, the performance of the sensor will decrease, which may cause sensor failure or data anomalies [1]. In the absence of an effective data processing mechanism, anomalies not only increase the cost of storage but also fail to guide the formulation of bridge maintenance strategies.

The existing data anomaly detection methods can generally be divided into model-based methods and data-driven methods. Basically, model-based methods rely on finite element models to reflect inherent structural characteristics. A series of statistical and mechanical models have been established to predict the output of the measurement [2–5]. Model-based methods can achieve better detection accuracy. However, when dealing with large numbers of SHM data, it is difficult to create a reliable explicit finite element model to describe the structural behavior of the structure in service [6].

Data-driven methods include statistical process control and machine learning methods. They do not rely on finite element models and directly analyze measured time series data, which hopefully alleviates the shortcomings of model-based methods [7]. Among data-driven methods, deep-learning-based methods have the potential to learn from big data containing abnormal data to automatically diagnose various abnormal data. Recently, deep learning has been increasingly applied to solve time-series-related tasks [8–10], including time series classification, time series prediction, and time series anomaly detection. Bao [11] et al. proposed a data anomaly detection method based on computer vision and deep learning. The original time series measurement values are first converted into image vectors, and then these image vectors are input to a deep neural network (DNN) to identify various anomalies. Tang et al. [12] proposed a new anomaly detection method using computer vision and deep learning methods. This method first converts the original time series data into images, imitating human-vision-based data collection, and then trains CNN for abnormal classification. Mao et al. [13] combined the generative adversarial network with an autoencoder to improve the performance of existing unsupervised learning methods and used two data sets from full-scale bridges to verify the proposed method.

Supervised deep learning relies heavily on a large number of labeled training data to train the network. However, many abnormal data patterns in actual projects do not have enough labeled data. Therefore, how to efficiently generate a large number of labeled synthetic data with fewer samples is a problem worthy of attention. As an effective tool to improve the quantity and quality of training data, data augmentation is essential for the successful application of deep learning models. The basic idea of data augmentation is to allow limited data to generate more value when new data are not added substantially while maintaining correct labels. Data augmentation has achieved good results in many application scenarios [14]. Sun et al. [15] proposed a simple but effective data augmentation method for generating multi-view 2D pose annotations. Liu et al. [16] proposed an image generation technique to enhance the robustness of the convolutional neural network model. Time-domain transformation is the most direct data augmentation method for time series data. Most of them directly process the original input time series. Cui et al. [17] proposed a sliding window method combined with a Multi-scale Convolutional Neural Network (MCNN) to solve the time series classification problem and achieved good results on a large number of benchmark data sets. Fawaz et al. [18] proposed a new method for generating new time series with DTW and ensembled them by a weighted version of the DBA algorithm. Wen et al. [19] used data augmentation methods such as random mutation and adding random trends in different data sets and proposed a time series segmentation approach based on convolutional neural networks (CNN) and transfer learning. Gao et al. [20] proposed a label expansion method to change those data points near the labeled anomalies and their labels as anomalies, which brings performance improvement for time series anomaly detection.

For the time series classification problem, most studies model the problem as a classification problem based on computer vision, while the classification method directly based on vibration signals is rarely studied. In addition, less research uses time series data augmentation to obtain a more balanced sample set. However, one-dimensional convolutional networks, which are faster for time series problems, are also used in rare cases. In this paper, a data anomaly identification method using one-dimensional CNN is proposed based on bridge monitoring acceleration data, in which data augmentation is employed to process the samples.

2. Data Anomaly Classification Method Based on 1D-CNN

2.1. Bridge Overview and Data Set Composition

This research uses the health monitoring data set of a large-span cable-stayed bridge in China. The main span of the bridge is 1088 m long, and the two side spans are 300 m each, including two 306 m-high towers. The structural health monitoring system of the bridge consists of 38 sensors. The position on the bridge is shown in Figure 1. Sensors include accelerometers, anemometers, strain gauges, global positioning systems (GPS), and thermometers. For this research, one-month (1 January–31 January 2012) acceleration data from all 38 sensors of the SHM system were used for data anomaly detection. The sampling frequency of the accelerometer is 20 Hz. The original continuous measurement data are divided into one-hour time periods, and in a one-month time period, through the method of non-overlapping windows, 744 time series measurement data of each sensor are obtained so as to obtain a total of 28,272 (744 × 38) data. The dimensions of a single data point are 1 × 72,000. Figure 2a–g shows an example of each type of data pattern. Table 1 describes the quantity and characteristics of normal data and six types of abnormal data. Each data point has a real category label. The normal time series measurement data are marked as 1, and the other six abnormal data patterns are marked as 2–7. It can be seen that nearly 52% of the data are abnormal. "Trend" is the main abnormal pattern that constitutes 20% of the data set, followed by "missing" and "square", each accounting for about 10%. On the other hand, the "outlier" only accounts for 1.9% of the data set, followed by "drift", which accounts for 2.4% of the data.

2.2. Data Preprocessing

Since there may be missing values or calculation errors in the process of data acquisition, data cleaning is performed on all data to remove missing or calculated incorrect values, which is reflected in MATLAB as "NAN". In order to keep the data length unchanged as the input of the neural network, all "NAN" values are replaced with 0.

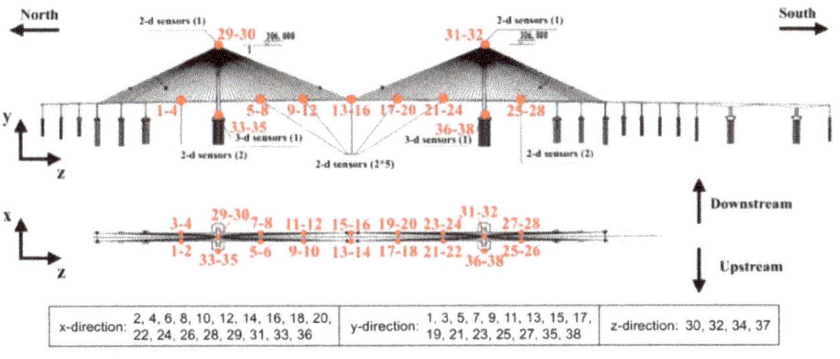

Figure 1. The monitored bridge and the position of the accelerometer on the bridge deck and tower. Source: The 1st International Project Competition for Structural Health Monitoring, July 2020.

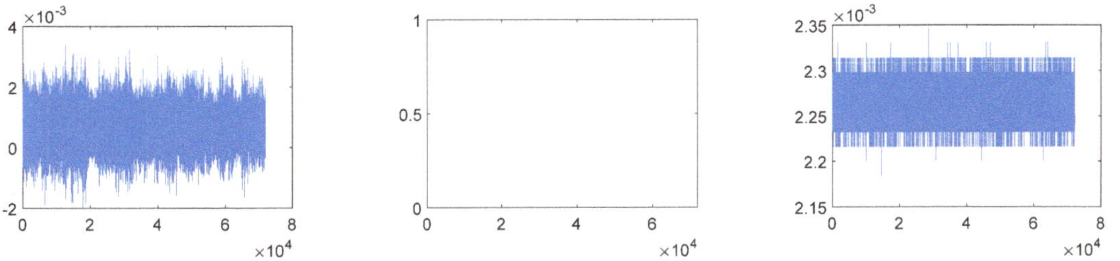

(a) (b) (c)

Figure 2. *Cont.*

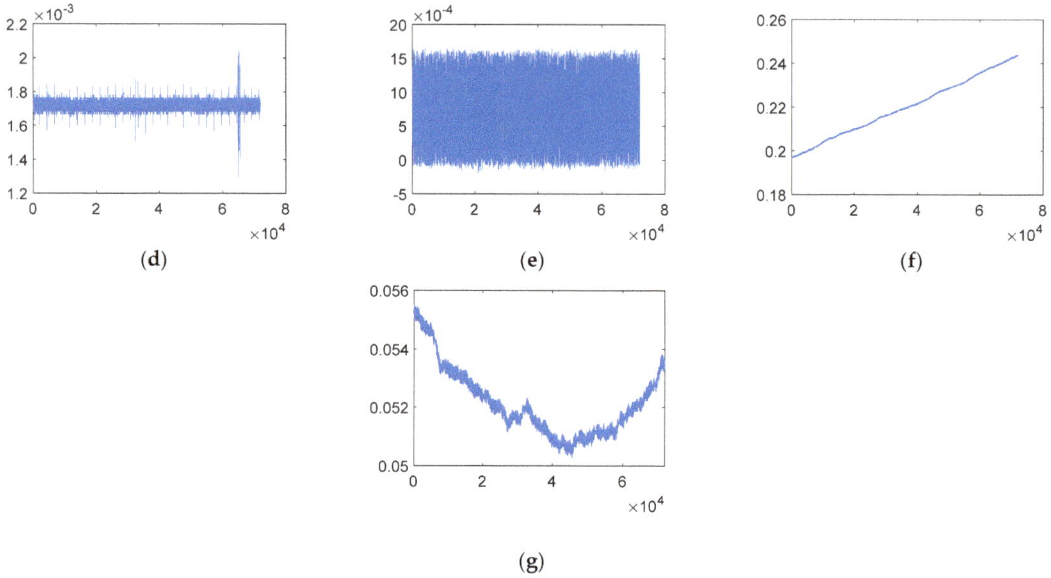

Figure 2. The example for each data pattern. (**a**) Normal data; (**b**) Missing data; (**c**) Minor data; (**d**) Outlier data; (**e**) Square data; (**f**) Trend data; (**g**) Drift data.

Table 1. Description of each type of data pattern.

No.	Anomaly Patterns	Description	Quantity
1	Normal	The time response is normal oscillation curve; frequency response is peak-like (may differ between bridges)	13575 (48%)
2	Missing	Most/all of the time response is missing, which makes the time and frequency response zero	2942 (10.4%)
3	Minor	Relative to normal sensor data, the amplitude is very small in the time domain	1775 (6.3%)
4	Outlier	One or more outliers appear in the time response	527 (1.9%)
5	Square	The time response is like a square wave	2996 (10.6%)
6	Trend	The data has an obvious trend in the time domain and has an obvious peak value in the frequency domain	5778 (20.4%)
7	Drift	The vibration response is non-stationary, with random drift	679 (2.4%)

Zero-mean normalization is used to process the data for one hour so that the normalized data are normally distributed; that is, the mean is zero, and the standard deviation is one. This method can eliminate errors caused by self-variation or large differences in values, making the data more beneficial for subsequent steps. As shown in Equation (1),

$$x^* = \frac{x-\mu}{\sigma} \tag{1}$$

Where x is a one-hour time series $\{x_1, x_2, \ldots, x_N\}$, μ is the mean of all sampling points, σ is the standard deviation of all sampling points, and x^* is the normalized time series.

2.3. Data Augmentation

Augmentation methods should always be selected appropriately for the case under consideration [14]. For example, when applied to a time series containing outliers, the sliding window may not be able to capture the mutation features. Therefore, this research deals with every hour, that is, the full length of the sample.

Data enhancement includes two steps: data expansion of a small number of data samples and down-sampling of all samples.

Data expansion is applied to a small number of samples, namely outlier and drift, in the numerical simulation. Not all abnormal samples need to be expanded.

Outlier data can be defined as individual points of the normal data whose amplitude greatly exceeds the normal range. Therefore, a data expansion method that magnifies individual points is used for outlier samples. x is a normal sample $\{x_1, x_2, \ldots, x_N\}$, and the proposed method is shown in Equation (2),

$$x(p) = mean + \beta \times range \tag{2}$$

where p is a random number between 10 and 60, *mean* is the mean value of x, β is a random number between -2 and 2, and *range* is the difference between the maximum and minimum values in x.

The method of symmetrical flipping and noise addition is used to expand the data of drift samples. The drift data has a random drift upwards or downwards. Therefore, the method of up-and-down symmetrical flipping can construct an effective sample. For the time series $\{x_1, x_2, \ldots, x_N\}$, symmetrical flipping can generate a new time series $\{x'_1, x'_2, \ldots, x'_N\}$ with the same anomaly labels where $x'_i = -x_i$. Different degrees of Gaussian white noise are added to the original sequence to generate more samples with the same anomaly labels. Two examples of data expansion are shown in Figure 3. The horizontal axis represents the number of sampling points, and the vertical axis represents the acceleration amplitude in m/s^2.

The sample dimension of a single hour is 1 × 72,000, which is relatively large as the input of the neural network. Therefore, down-sampling is used to reduce the dimensionality of the sample while retaining the characteristics of the sample as much as possible to increase the efficiency of the neural network. The upper and lower contours of a sample are both useful features. Therefore, a down-sampling method that uses a sliding window to symmetrically extract the maxi-mum and minimum values is used. All 1 × 72,000 samples are down-sampled over the entire sample length. A step size is selected, which is 20 in this article, and the maximum and minimum values in the sequence are taken out for every sampling point of the step size. Therefore, after processing each of the 72,000 samples, a 2 × 3600 sample size will be obtained. The comparison chart of some examples before and after down-sampling is shown in Figure 4a,b. The horizontal axis represents the number of sampling points, and the vertical axis represents the acceleration amplitude in m/s^2.

2.3.1. 1D-CNN

A convolutional neural network (CNN) usually consists of an input layer, convolutional layer (Conv), pooling layer (Pooling), dense layer (Dense), and output layer. In the CNN architecture, the first few layers usually alternate between convolutional layers and pooling layers, and the last few layers close to the output layer are composed of dense layers. CNN is an end-to-end learning method model, which can use the existing supervised gradient descent algorithm to train the model. For time-series-processing problems, the effect of a one-dimensional convolutional neural network (1D-CNN) can be comparable to a recurrent neural network (RNN), and the computational cost is much smaller. For simple tasks such as time series classification, a small one-dimensional convolutional network can completely replace the RNN, and it runs faster [21].

Regardless of whether one-dimensional or two-dimensional convolution is used, convolutional neural networks have a similar structure. The structure starts with a stack of convolutional and pooling layers, and then connected to a flatten layer to convert two-dimensional features into one-dimensional output, and then multiple dense layers can be added for classification or regression. However, there is a little difference between them: one-dimensional convolutional neural networks can use larger convolution kernels [21]. For example, for a two-dimensional convolution layer, a 3 × 3 convolution kernel contains 3 × 3 = 9 convolution vectors; however, for a one-dimensional convolution layer, a convolution kernel of size 3 only contains 3 convolution vectors. Therefore, a one-dimensional convolution kernel greater than or equal to 9 can be easily used.

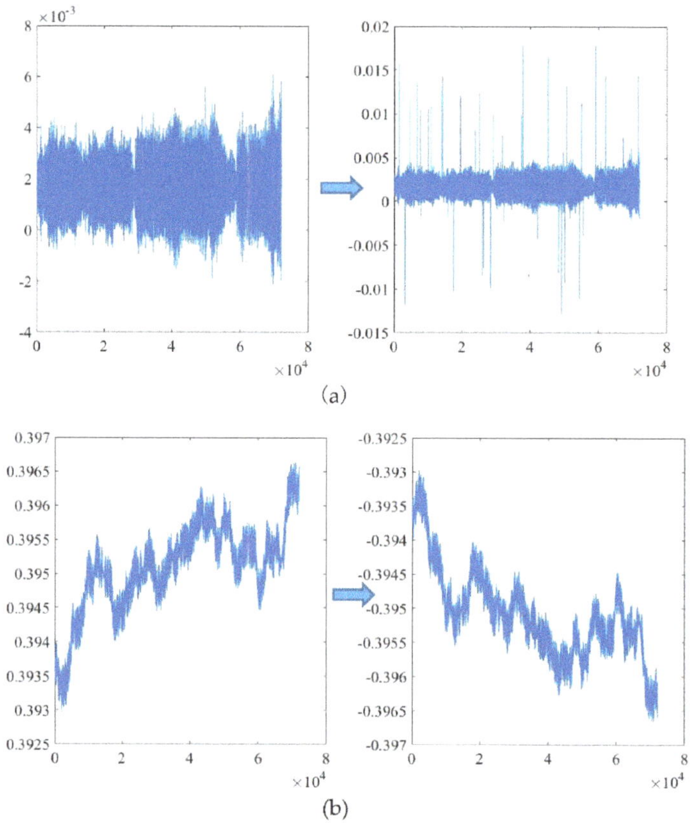

Figure 3. Comparison chart before and after data expansion. (**a**) Data expansion for outlier data; (**b**) Data expansion for drift data.

The Python Science Suite, Tensorflow, and Keras are used to build a neural network architecture with GPU acceleration. The processor and graphics card types of the hardware platform are Inter Core i5-9400F and Nvidia GeForce RTX 2070. The object function in CNN is set to categorical cross-entropy to estimate the difference between the actual data category and the predicted data category. The metric is set to the accuracy to evaluate the performance of the model. In order to minimize the output of the object function, an adaptation of the mini-batch stochastic gradient descent algorithm called Adam is used as an optimizer.

In the classification, the imbalanced training set needs to be considered; that is, the number of normal samples is much larger than the abnormal samples. If an imbalanced training set is used to train the network, all abnormal samples will be predicted as normal samples during the test, and there will still be a high accuracy, but this is meaningless. Therefore, we choose to use the class weight technique [22], which can make important categories of samples contribute more to the object function during training. Batch Normalization (BN) [23,24] is a method that has been widely used in deep network training. The method of adding BN after the convolutional layer and then adding the activation function can save the operator from adjusting the parameters deliberately and slowly. Figure 5 shows the workflow of the proposed method.

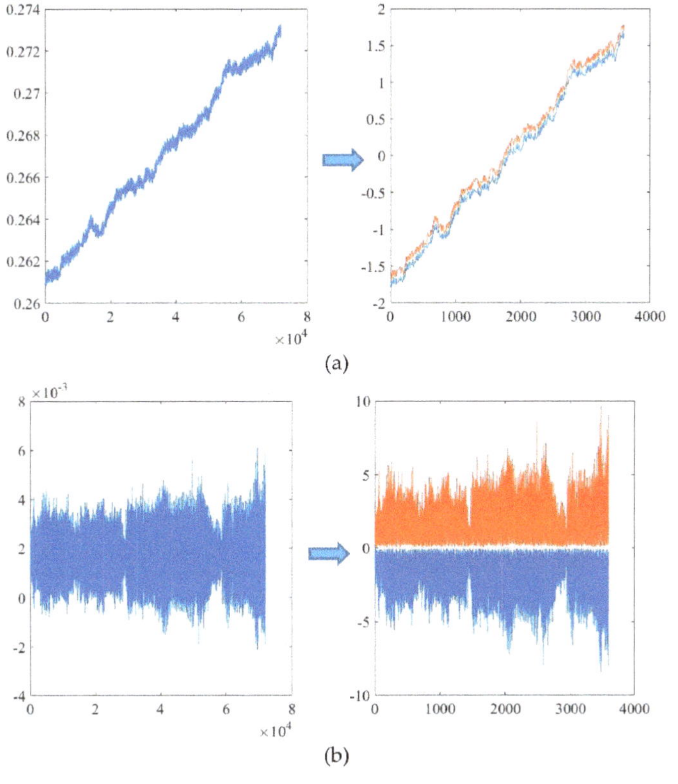

Figure 4. Before and after data preprocessing and down-sampling. (**a**) Preprocessing and down-sampling of trend data; (**b**) Preprocessing and down-sampling of normal data.

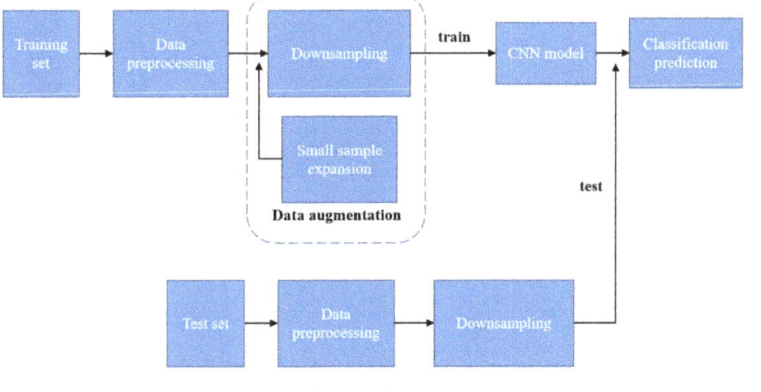

Figure 5. Workflow of the proposed method.

3. Bridge Monitoring Data Verification

According to the proposed method process, anomaly detection is performed on the bridge monitoring data set. First, data preprocessing is performed on all original samples, missing values are deleted, and samples are standardized. In order to test the generalization ability of the model, the data set is divided into training and test sets, and 80% of the samples are randomly selected as the training set. The training set size is 22,616. Twenty percent of the samples are randomly selected as the test set, and the test set size is 5656. In order to simulate real anomalies, the distribution of test samples is unbalanced. Table 2 shows the distribution of selected test samples.

Table 2. Data distribution of the test set.

	1	2	3	4	5	6	7
Anomaly patterns	Normal	Missing	Minor	Outlier	Square	Trend	Drift
Quantity	2688	603	360	106	616	1147	136
Percentage	47.5%	10.7%	6.4%	1.9%	10.9%	20.3%	2.4%

Constructing a balanced training set of various categories is beneficial to the training process. Data expansion is carried out on the small number of anomalies in the training set, namely outlier and drift. The normal samples in all training sets are expanded to outlier samples by magnifying individual points. The Gaussian distributed noise with 2%, 3%, 4%, 5%, 6%, 7%, and 8% standard deviation to the signals are added to each drift sample once, and symmetrical flip it once to obtain 8 times the number of drift samples. Therefore, an additional 10,860 (13,575 × 80%) outlier samples and 4345 (679 × 80% × 8) drift samples were obtained. After adding to the training set, the new training set size is 37,821 (22,616 + 10,860 + 4345).

Down-sampling is implemented on the test set and new training set samples, and the dimensionality of the samples is reduced from 1 × 72,000 to 2 × 3600 while retaining most of their features.

In order to build the 1D-CNN architecture, two one-dimensional convolutional layers are stacked to obtain the deep features of the sample more efficiently, and a flatten layer and two dense layers are connected to convert two-dimensional features into one-dimensional output. The last layer of the network uses the softmax multi-classifier. In short, softmax is the value that maps the output of the previous layer to (0,1) through the softmax function. The sum of these values is 1, which can be understood as a probability. The node with the largest probability is selected as the predicted abnormal data type. The network structure is shown in Figure 6. The detailed structure of 1D-CNN is shown in Table 3. The hyperparameter configuration is shown in Table 4.

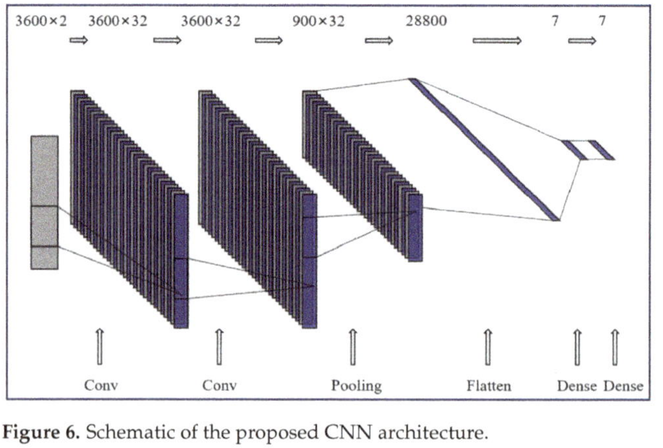

Figure 6. Schematic of the proposed CNN architecture.

Table 3. The detailed architecture of CNN.

Layer	Type	Input Shape	Output Shape	Kernel Num	Kernel Size	Stride	Padding	with BN	Activation
1	Conv	(3600, 2)	(3600, 32)	32	16	1	Same	True	Leaky ReLU
2	Conv	(3600, 32)	(3600, 32)	32	16	1	Same	True	Leaky ReLU
3	Pooling	(3600, 32)	(900, 32)	None	4	4	Valid	False	None
4	Flatten	(900, 32)	(28800)	None	None	None	None	None	None
5	Dense	(28800)	(7)	None	None	None	None	False	Leaky ReLU
6	Dense	(7)	(7)	None	None	None	None	False	Softmax

Table 4. The configurations of training process.

Name	Value	Description
Batch size	128	The size of data batch used in every training iteration
Initial learning rate	10^{-3}	The initial learning rate of Adam algorithm
Patience	40	A parameter of early stopping
α	0.01	A parameter in Leaky RELU function 0.01 in every activation function

Mean Squared Error (MSE) as a loss function for training and validation can be expressed as:

$$MSE = \frac{1}{N}\sum_{i=1}^{N}(Y_i - Y_{0,i})^2 \tag{3}$$

where Y represents the predicted value, and Y_0 represents the true label value. N represents the total number of samples.

In the training process, the training set is divided into 12.5% as the verification set. During the training process, the training loss and the validation loss (MSE) are monitored, and the training accuracy and verification accuracy (Accuracy) are also monitored. The change of the loss function and the change of the accuracy are shown in Figures 7 and 8.

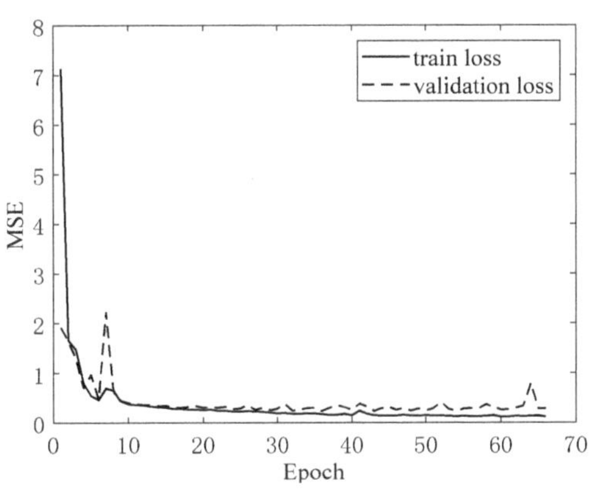

Figure 7. Training and validation loss curve.

It can be seen that the overall loss value shows a downward trend, and the overall accuracy shows an upward trend. The amplitude is large at the beginning of training, indicating that the learning rate is appropriate. There are glitches and oscillations locally, possibly because a large batch size is selected for a large number of samples, and there are a small number of samples with incorrect labels in the real-world data set. After the loss value and accuracy stabilized, the final training and validation accuracy reached more than 95%.

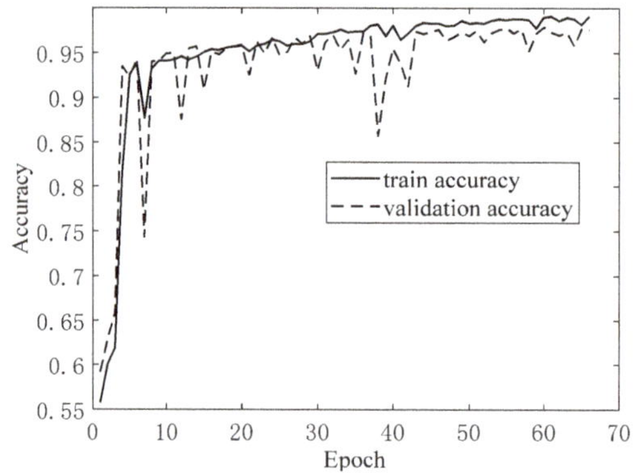

Figure 8. Training and validation accuracy curve.

Table 5 shows the classification results in a statistical way. In the statistical analysis of binary or multiple classifications, precision, recall, and F1 score are measures of the accuracy of the classification results, and the last one is the harmonic average of the first two. Recall is relative to the sample, that is, how many positive samples in the sample are predicted correctly. Take the missing-type samples in Table 5 as an example. There are a total of 603 missing-type samples. If 602 are predicted correctly, the recall is 602/603 = 99.83%. Precision is relative to the prediction result. It indicates how many of the samples whose predictions are positive are correct. Taking the normal-type samples as an example, a total of 2590 samples are predicted to be normal types. If 2542 predictions are correct, the precision is 2542/2590 = 98.15%. Recall and precision indicators are sometimes contradictory. If a comprehensive indicator is used to express the results of recall and precision, the most common method should be the F1 score as follows:

$$F_1 = 2\frac{precision \cdot recall}{precision + recall} \times 100\% \tag{4}$$

Where F_1 represents F1 score, *recall* represents recall, and *precision* represents precision.

Table 5. The prediction result of the test set.

		Predicted Data Pattern								
		1	**2**	**3**	**4**	**5**	**6**	**7**	**Total**	**Recall (%)**
Real data pattern	**1-normal**	2542	0	60	72	14	0	0	2688	94.57
	2-missing	0	602	1	0	0	0	0	603	99.83
	3-minor	26	0	326	8	0	0	0	360	90.56
	4-outlier	21	0	1	84	0	0	0	106	79.24
	5-square	1	0	1	2	612	0	0	616	99.35
	6-trend	0	0	2	0	0	1110	35	1147	96.77
	7-drift	0	0	0	0	0	13	123	136	90.44
	Total	2590	602	391	166	626	1123	158	5656	95.45
	Precision (%)	98.15	100.0	83.38	50.60	97.76	98.84	77.85		
	F1 score	0.96	1.00	0.87	0.62	0.99	0.98	0.84		

It can be seen that the proposed method can effectively identify various data patterns. The recall of normal, missing, minor, square, trend, and drift categories can reach above 90%. Except for the low F1 score of outlier and drift, the other types are all high. A small number of minor samples are classified into the normal category. Some outlier samples are classified into the normal category, and a few are classified into the minor category. The outlier sample may have only a few peaks, and most of the features of the outlier sample are very similar to the normal sample, and the feature that is too small will be lost in the convolution process. Trend and drift are partly confused, probably because they both have slanted features.

4. Conclusions

This paper modeled the anomaly detection problem into a time series classification problem. The original time series undergoes data preprocessing and data augmentation to get a sample set with more uniform distribution, more obvious features, and smaller dimensions. Data augmentation includes data expansion and down-sampling. For small samples, the methods of symmetrical flipping, adding noise, and randomly generating outliers are used for data expansion, and samples with the same label are added without increasing the original samples. The down-sampling method of symmetrically extracting the maximum and minimum values can effectively reduce the dimensionality of the input sample and retain its features. Build a one-dimensional convolutional neural network model that is faster for time series classification problems. Adding the hyperparameter tuning of class weights makes the network more effective in dealing with an unbalanced training set. The method is verified with the acceleration data of a long-span cable-stayed bridge for one month. For the anomaly detection problem modeled as a time series classification problem, the results show that the proposed method can automatically detect a variety of data anomaly categories with high precision.

The proposed method can accurately identify most types of abnormal data, but for abnormal types with very inconspicuous features, such as outlier data, there is still much room for improvement in recognition accuracy. In future work, time series augmentation will not only be carried out in the time domain, but will be expanded to the frequency domain, or more advanced methods (such as GAN) will be used to expand samples.

Author Contributions: Conceptualization, Y.Z. and Y.L.; methodology, Y.Z.; validation, Y.Z. and Y.L.; investigation, Y.Z.; data curation, Y.Z.; writing—original draft preparation, Y.Z.; writing—review and editing, Y.L.; funding acquisition, Y.L. All authors have read and agreed to the published version of the manuscript.

Funding: This research was funded by the National Key R&D Program of China via Grant No. 2018YFC0705606.

Institutional Review Board Statement: Not applicable.

Informed Consent Statement: Not applicable.

Data Availability Statement: Restrictions apply to the availability of these data. Data was obtained from The 1st International Project Competition for Structural Health Monitoring with the permission of The 1st International Project Competition for Structural Health Monitoring organizing committee.

Acknowledgments: The raw data involved in this study were obtained from the organizing committee of the 1st International Project Competition for Structural Health Monitoring (IPC-SHM (2020)), and the authors thank IPC-SHM (2020) organizing committee for its valuable data resources.

Conflicts of Interest: The authors declare no conflict of interest.

References

1. Yi, T.H.; Huang, H.B.; Li, H.N. Development of sensor validation methodologies for structural health monitoring: A comprehensive review. *Measurement* **2017**, *109*, 200–214. [CrossRef]
2. Chang, C.M.; Chou, J.Y. Damage detection of seismically excited buildings based on prediction errors. *J. Aerosp. Eng.* **2018**, *31*, 04018032.1–04018032.9. [CrossRef]
3. Kang, F.; Li, J.; Dai, J. Prediction of long-term temperature effect in structural health monitoring of concrete dams using support vector machines with jaya optimizer and salp swarm algorithms. *Adv. Eng. Softw.* **2019**, *131*, 60–76. [CrossRef]

4. Moriot, J.; Quaegebeur, N.; Le Duff, A.; Masson, P. A model-based approach for statistical assessment of detection and localization performance of guided wave-based imaging techniques. *Struct. Health Monit.* **2017**, *17*, 1460–1472. [CrossRef]
5. Wang, H.; Zhang, Y.M.; Mao, J.X.; Wan, H.P.; Tao, T.Y.; Zhu, Q.X. Modeling and forecasting of temperature-induced strain of a long-span bridge using an improved bayesian dynamic linear model. *Eng. Struct.* **2019**, *192*, 220–232. [CrossRef]
6. Chandola, V.; Banerjee, A.; Kumar, V. Anomaly detection: A survey. *ACM Comput. Surv.* **2009**, *41*, 15. [CrossRef]
7. Xu, B.; He, J.; Masri, S.F. Data-based model-free hysteretic restoring force and mass identification for dynamic systems. *Comput. Aided Civ. Infrastruct. Eng.* **2015**, *30*, 2–18. [CrossRef]
8. Fawaz, H.I.; Forestier, G.; Weber, J.; Idoumghar, L.; Muller, P.A. Deep learning for time series classification: A review. *Data Min. Knowl. Discov.* **2019**, *33*, 917–963. [CrossRef]
9. Han, Z.; Zhao, J.; Leung, H.; Ma, K.F.; Wang, W. A review of deep learning models for time series prediction. *IEEE Sens. J.* **2019**, *21*, 7833–7848. [CrossRef]
10. Chalapathy, R.; Chawla, S. Deep learning for anomaly detection: A survey. *arXiv* **2019**, arXiv:1901.03407.
11. Bao, Y.; Tang, Z.; Li, H.; Zhang, Y. Computer vision and deep learning-based data anomaly detection method for structural health monitoring. *Struct. Health Monit.* **2019**, *18*, 401–421. [CrossRef]
12. Tang, Z.; Chen, Z.; Bao, Y.; Li, H. Convolutional neural network-based data anomaly detection method using multiple information for structural health monitoring. *Struct. Control. Health Monit.* **2018**, *26*, e2296. [CrossRef]
13. Mao, J.; Wang, H.; Spencer, B.F. Toward data anomaly detection for automated structural health monitoring: Exploiting generative adversarial nets and autoencoders. *Struct. Health Monit.* **2020**. [CrossRef]
14. Wen, Q.; Sun, L.; Song, X.; Gao, J.; Wang, X.; Xu, H. Time series data augmentation for deep learning: A survey. *arXiv* **2020**, arXiv:2002.12478v1.
15. Sun, J.; Wang, M.T.; Zhao, X.; Zhang, D.J. Multi-View Pose Generator Based on Deep Learning for Monocular 3D Human Pose Estimation. *Symmetry* **2020**, *12*, 1116. [CrossRef]
16. Liu, B.; Zhang, Y.; He, D.J.; Li, Y.X. Identification of Apple Leaf Diseases Based on Deep Convolutional Neural Networks. *Symmetry* **2018**, *10*, 11. [CrossRef]
17. Cui, Z.; Chen, W.; Chen, Y. Multi-scale convolutional neural networks for time series classification. *arXiv* **2016**, arXiv:1603.06995.
18. Fawaz, H.I.; Forestier, G.; Weber, J.; Idoumghar, L.; Muller, P.A. Data augmentation using synthetic data for time series classification with deep residual networks. *arXiv* **2018**, arXiv:1808.02455v1 [cs.CV].
19. Wen, T.; Keyes, R. Time series anomaly detection using convolutional neural networks and transfer learning. *arXiv* **2019**, arXiv:1905.13628v1 [cs.LG].
20. Gao, J.; Song, X.; Wen, Q.; Wang, P.; Sun, L.; Xu, H. Robusttad: Robust time series anomaly detection via decomposition and convolutional neural networks. *arXiv* **2020**, arXiv:2002.09535.
21. Francois Chollet. *Deep Learning with Python*; Manning: New York, NY, USA, 2018; pp. 188–190.
22. Keras. Available online: https://keras.io/ (accessed on 31 October 2019).
23. Alom, M.Z.; Taha, T.M.; Yakopcic, C. A State-of-the-Art Survey on Deep Learning Theory and Architectures. *Electronics* **2019**, *8*, 292. [CrossRef]
24. Lin, Y.; Nie, Z.; Ma, H. Structural Damage Detection with Automatic Feature-Extraction through Deep Learning. *Comput. Aided Civ. Infrastruct. Eng.* **2017**, *32*, 1025–1046. [CrossRef]

symmetry

MDPI

Article

Mixed Sensitivity-Based Robust H_∞ Control Method for Real-Time Hybrid Simulation

Xizhan Ning [1,2]

[1] College of Civil Engineering, Huaqiao University, Xiamen 361021, China; xzning@hqu.edu.cn
[2] Key Laboratory for Intelligent Infrastructure and Monitoring of Fujian Province, Huaqiao University, Xiamen 361021, China

Abstract: Real-time hybrid simulation (RTHS), dividing the emulated structure into numerical substructures (NS) and physical substructures (PS), is a powerful technique to obtain responses and then to assess the seismic performance of civil engineering structures. A transfer system, a servo-hydraulic actuator or shaking table, is used to apply boundary conditions between the two substructures. However, the servo-hydraulic actuator is inherently a complex system with nonlinearities and may introduce time delays into the RTHS, which will decrease the accuracy and stability of the RTHS. Moreover, there are various uncertainties in RTHS. An accurate and robust actuator control strategy is necessary to guarantee reliable simulation results. Therefore, a mixed sensitivity-based H_∞ control method was proposed for RTHS. In H_∞ control, the dynamics and robustness of the closed-loop transfer system are realized by performance weighting functions. A form of weighting function was given considering the requirement in RTHS. The influence of the weighting functions on the dynamics was investigated. Numerical simulations and actual RTHSs were carried out under symmetric and asymmetric dynamic loads, namely sinusoidal and earthquake excitation, respectively. Results indicated that the H_∞ control method used for RTHS is feasible, and it exhibits an excellent tracking performance and robustness.

Keywords: real-time hybrid simulation; H_∞ control; time delay; mixed sensitivity

Citation: Ning, X. Mixed Sensitivity-Based Robust H_∞ Control Method for Real-Time Hybrid Simulation. *Symmetry* **2021**, *13*, 840. https://doi.org/10.3390/sym13050840

Academic Editor: Jan Awrejcewicz

Received: 16 April 2021
Accepted: 7 May 2021
Published: 10 May 2021

1. Introduction

Real-time hybrid simulation (RTHS) [1], or the real-time substructure pseudo-dynamic test, is a cost-effective and versatile experimental technique to evaluate structural performance under dynamic excitations. It originated from the pseudo-dynamic test [2] first proposed by a Japanese researcher in the 1970s, which is known as hybrid simulation (HS) nowadays. HS takes advantage of numerical analysis and physical experiments, in which the emulated structure is divided into several substructures: the part that cannot be simulated exactly is experimentally tested in the laboratory, which is denoted as the physical substructure (PS), and the rest is simulated by a computer program, which is denoted as the numerical substructure (NS) [3,4]. In RTHS, boundary conditions between the two substructures are imposed on the PS by a transfer system, a servo-hydraulic actuator or shaking table, in a real-time manner. This allows RTHS the ability to test rate-dependent components, such as TMD, AMD, and MR dampers. In recent decades, much progress has been achieved [5–8].

However, due to the inherent nonlinear dynamics of the actuator–specimen system, the desired displacement cannot be realized at the end of the time integration step. This is often called the time delay, which will decrease the accuracy and may lead to RTHS instabilities [9]. Therefore, to carry out a successful RTHS, the detrimental effect of time delay must be mitigated. Horiuchi et al. [9] assumed a constant time delay and proposed a polynomial extrapolation method. Subsequently, more accurate strategies have been investigated to consider the variation in time delay [10–13].

Modern control theory was also used to deal with time delay, where the control plant includes the servo-hydraulic and the specimen. The inverse control technique was introduced by Chen and Ricles [14], where the servo-hydraulic actuator is modeled by a first-order transfer function. Carrion and Spencer [15] proposed a model-based control approach, where a low-pass filter is combined with the inverted actuator system plant. In this method, a higher-order control plant can be used. Ning et al. [16] proposed an adaptive feedforward control method, where the Kalman filter is used to estimate the adjustable parameters. Xu et al. proposed a frequency evaluation index-based compensation for RTHS [17]. A two-stage delay compensation method, combining the feedforward and polynomial extrapolation, was proposed by Wang et al. [18]. A polynomial-based feedforward prediction algorithm was combined with a robust linear-quadratic-gaussian controller that was proposed by Zhou et al. [19] to deal with the adverse effects of time delay. Ning et al. [20] proposed an adaptive feedforward and feedback control method based on a discrete control plant, of which the model order is not restrained.

However, there are differences between the control plant model and the actuator–specimen system. Hence, preliminary discussions have been made on the robustness of the control strategy to model uncertainties [21–23]. In this study, a mixed sensitivity-based H_∞ control method was introduced to deal with the time delay and uncertainties in real-time hybrid simulation. The H_∞ control theory is overviewed in Section 2. The selection of a performance weighting function is presented in Section 3, where the influence of the weighting function on the system dynamics is discussed. Subsequently, the proposed method is validated through numerical simulations and actual RTHS in Sections 4 and 5, respectively.

2. Overview of H_∞ Control Theory

The standard setup of H_∞ control is shown in Figure 1. In this figure, w, u, z, and y are vector-valued signals: w is the exogenous input, typically consisting of command signals, disturbances, and sensor noises; u is the control signal; z is the performance output that is to be minimized; y is the measured output. G is the generalized plant and K is the controller to be designed.

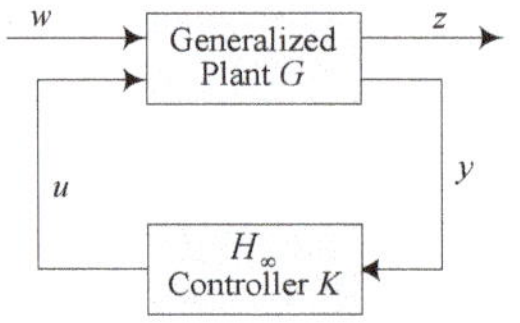

Figure 1. Block diagram of the standard H_∞ control.

In mixed sensitivity-based H_∞ control, the generalized control plant, or augment control plant, can be formulated from feedback control. For a typical feedback control diagram shown in Figure 2a, three functions: sensitivity function S, complementary sensitivity function T, and controller sensitivity R, are defined, and they are calculated by

$$S = (1 + PK)^{-1}, T = I - S = PK(1 + PK)^{-1}, R = K(1 + PK)^{-1}. \tag{1}$$

where P is the transfer function of the control plant, and K is the to-be-designed controller. The sensitivity function S is the transfer function between the reference input r and tracking error e, or between the disturbance and measurement output y. The complementary sensitivity function T is the transfer function between the reference input r and measurement output y. The controller sensitivity R reflects the control effort, which is the transfer function between the reference r and controller output u.

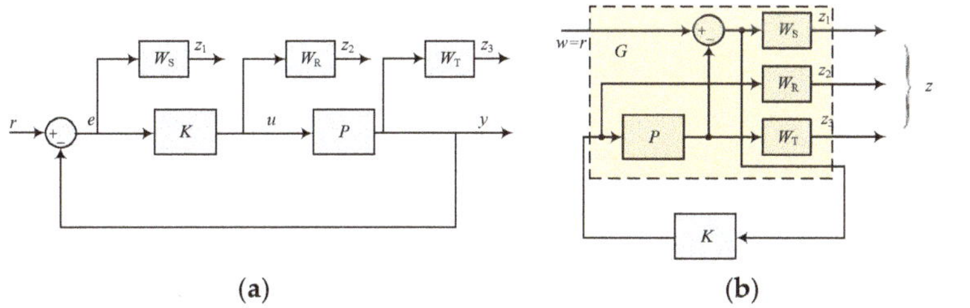

Figure 2. Formulation of the standard H_∞ problem. (**a**) Weighted feedback control system; (**b**) equivalent standard H_∞ problem.

In RTHS, it is expected that the loading system can accurately reproduce the command signal and is less sensitive to external disturbances, i.e., $S \rightarrow 0$ and $T \rightarrow 1$ are demanded. To meet this requirement and consider the robustness index of additive and multiplicative uncertainties [24], performance weighting functions, namely W_S, W_R, and W_T, are introduced to the feedback control loop, as shown in Figure 2a. Thus, the equivalent standard H_∞ block diagram can be reached, which is shown in Figure 2b. Hence, the generalized plant G, from (w, u) to (z, y), is given as follows:

$$\mathbf{G}(s) = \begin{bmatrix} G_{11} G_{12} \\ G_{21} G_{22} \end{bmatrix} = \begin{bmatrix} W_S & -PW_s \\ 0 & W_R \\ 0 & PW_T \\ 1 & -P \end{bmatrix}. \tag{2}$$

In H_∞ control theory [24], the controller is synthesized by optimizing the H_∞-norm of the cost-function, a transfer function from the exogenous input w to the performance output z, which is calculated by

$$\mathbf{T}_{wz} = \mathbf{G}_{11} + \mathbf{G}_{12}\mathbf{K}(\mathbf{I} - \mathbf{G}_{22}\mathbf{K})^{-1}\mathbf{G}_{21} = \begin{bmatrix} W_S(s)S(s) \\ W_R(s)R(s) \\ W_T(s)T(s) \end{bmatrix}, \tag{3}$$

Thus, the H_∞ control problem can be formulated as follows: find a controller K that makes the closed-loop system internally stable, and make the H_∞ norm of Equation (3) the least (optimal), or less than a given positive constant (suboptimal) [24]. However, it is often not necessary to design an optimal controller in practice, and it is usually much cheaper to obtain controllers that are very close in the norm sense to the optimal ones. Hence, the H_∞ suboptimal controller was used in this study, in which the cost-function satisfies

$$\|\mathbf{T}_{wz}\|_\infty < \gamma, \tag{4}$$

where γ is a positive number, and the minimum value is in relation to the generalized plant G [24].

Let a possible state-space realization for the generalized plant G be calculated by

$$\begin{aligned} \dot{x} &= Ax + B_1 w + B_2 u \\ z &= C_1 x + D_{11} w + D_{12} u \\ y &= C_2 x + D_{21} w + D_{22} u \end{aligned} \tag{5}$$

where x is the state vector, and the dimensions of w, u, z, and y are compatible with that of x.

Suppose G satisfies the following assumptions [24]:

- (A, B_2) is stabilizable and (C_2, A) is detectable;
- $D_{12} = [0; I]^T$ and $D_{21} = [0\ I]$;

- $\begin{bmatrix} A - j\omega I & B_2 \\ C_1 & D_{12} \end{bmatrix}$ has full column rank for all ω;
- $\begin{bmatrix} A - j\omega I & B_1 \\ C_2 & D_{21} \end{bmatrix}$ has full row rank for all ω.

Then, a controller can be designed employing the DGKF method by solving two Riccati equations [24].

It should be noted that the internally stable controller K is not unique in the suboptimal problem, and the central controller is used in general.

3. Weighting Function and Its Influence

Section 2 shows that the performance weighting function plays an important role in H_∞ controller design. In this section, the weighting function with adjustable parameters is proposed. Moreover, the influence of the parameters on the system dynamics is discussed.

3.1. Selection of Performance Weighting Function

3.1.1. Weighting W_S

W_S is the weighting function of the sensitivity function S. In the RTHS used for earthquake engineering, the frequency of the commanded signals and disturbances is generally very low. Hence, to track the reference signal with high accuracy and to suppress the external disturbances, the sensitivity function S should be small enough over the low-frequency range, while in the high-frequency band beyond the command signal, there are no additional restraints for the sensitivity function S. Therefore, the weighting function W_S should have a high gain over the low-frequency range. Hence, W_S can be selected as

$$W_S = \frac{as + b\omega_d}{s + c} \tag{6}$$

where a, b, and c are adjustable parameters, and ω_d is the desired bandwidth of the system. The singular value of ${W_S}^{-1}$ is shown in Figure 3. It is seen from the figure that, if c is small enough, that the sensitivity function S will tend to zero when the frequency is not very high, which means that the controlled system can realize the command perfectly. Typically, –60 dB is small enough for the sensitivity function S when the frequency is low; hence, it is recommended that c is less than 0.01.

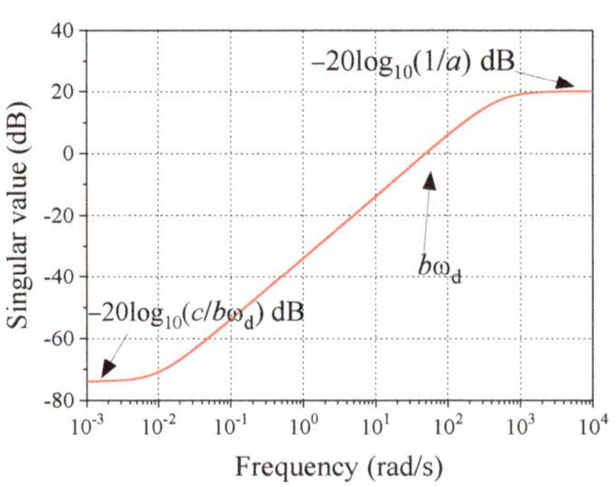

Figure 3. Singular value of ${W_S}^{-1}$.

3.1.2. Weighting W_T

W_T is the weighting function of the complementary sensitivity function T, which represents the characteristic of the multiplicative uncertainties [24]. Hence, the weighting W_T can be selected as

$$\sigma\left(\frac{G(j\omega)}{G_0(j\omega)} - 1\right) < \sigma(W_T(j\omega)), \forall\omega \tag{7}$$

where G is the actual plant, G_0 is the nominal or analytical plant, and σ is the singular value. Once W_T is determined, it can remain unchanged, because it is related to the model of the control plant.

In RTHS, it is expected that T should be close to 1 to achieve the reference command, especially over the concerned frequency band. Afterward, T should be small enough to suppress the modeling errors over the high-frequency range, which will also diminish the effect of measurement noise. Furthermore, if the gain of W_T decreases quickly over the high-frequency range, the measurement noise will be suppressed effectively. Hence, for practical purposes, the recommended form of function W_T is given by

$$W_T = hs^2 + ms + n \tag{8}$$

where h, m, and n are adjustable parameters. It should be noted that $W_T P$ should be a rational function. If not, the form in Equation (8) should be modified. Examples can be found in the subsequent sections.

3.1.3. Weighting W_R

The weighting W_R is associated with the additive uncertainties [24]. Typically, the weighting W_R is introduced to satisfy the premises of the H_∞ control theory. While it is seen in Section 2, W_R is the weighting of the controller sensitivity R, which is related to the command sent to the servo valve in RTHS. A high controller gain will destabilize the system due to measurement noise. Hence, to suppress the maximum magnitude of the controller, the weighting W_R can be selected as a small constant. The weighting W_R can be eliminated.

3.2. *Influence on the System Dynamics*

In RTHS, it is expected that the servo-hydraulic actuator can realize the desired displacement command quickly and precisely. Hence, to give guidance on weighting function selection, the influence of the parameters in the weighting function on the dynamics, namely the settling time, overshoot, and steady-state error, were investigated in this subsection. The control plant, from the controller output u to measurement displacement y, was taken as

$$P = \frac{20.99}{s + 1.418}. \tag{9}$$

By several trials, the weighting functions were selected as follows:

$$W_S = \frac{0.0475s + 170.1}{s + 0.001}, W_T = \frac{2500\,(s+250)\,(s+14.14)}{(s + 1.571 \times 10^4)^2}, W_R = 0.001. \tag{10}$$

It should be noted that as the control plant is a first-order transfer function, the form of the weighting function W_T is modified to guarantee that $W_T P$ is a rational function. The step response was obtained and is shown in Figure 4. It is seen that the controlled system could respond to the reference command quickly, and there was no steady-state error, indicating a perfect tracking performance.

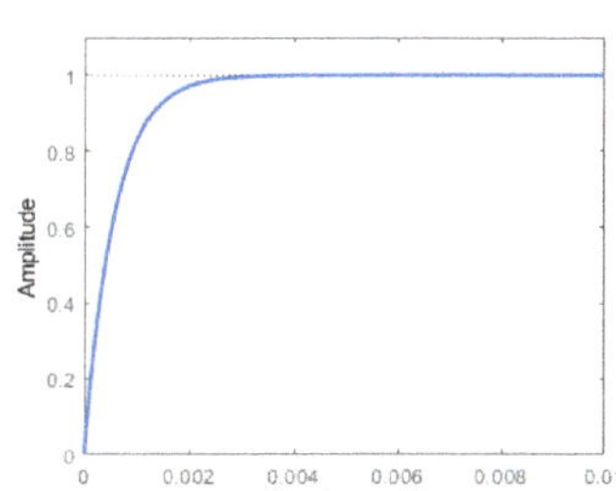

Figure 4. Step response of the H_∞-controlled system.

3.2.1. Influence of W_S

1. Parameter a in W_S

To analyze the influence of the parameter a on weighting function W_S, the other parameters in the weightings were kept unchanged. By changing a from 0 to 3.5, the dynamics were calculated via the step response, which is shown in Figure 5. It is seen from the figure that with the increase in parameter a, the settling time grew rapidly, while the overshoot decreased quickly. The steady-state error almost remained the same. Hence, it is recommended that a small positive value should be used for parameter a.

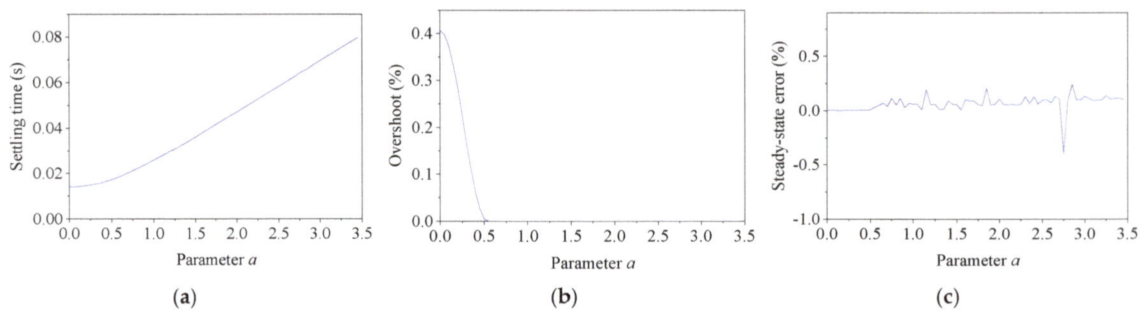

Figure 5. Effect of parameter a on W_S. (**a**) Settling time; (**b**) overshoot; (**c**) steady-state error.

2. Parameter b in W_S

When designing the weightings in Equation (10), the desired system band was 20 Hz and b was 1.35. To analyze the influence of the parameter b on weighting function W_S, the other parameters in the weightings were kept unchanged. By changing b from 0 to 1.5, the dynamics were calculated by the step response, which is shown in Figure 6. It is seen in Figure 6 that when parameter b was less than 0.1, with the increase in parameter b, the settling time decreased rapidly, while the overshoot almost remained unchanged. When parameter b was larger than 0.1, with the increase in parameter b, the settling time decreased very slowly, while the overshoot increased quickly. The steady-state error almost remained unchanged. It is seen in Equation (6) and Figure 3 that with the increase in b, the system bandwidth widened, resulting in the response speed accelerating and the overshoot increasing. Hence, it is recommended that a larger b should be used if one expects a faster response speed.

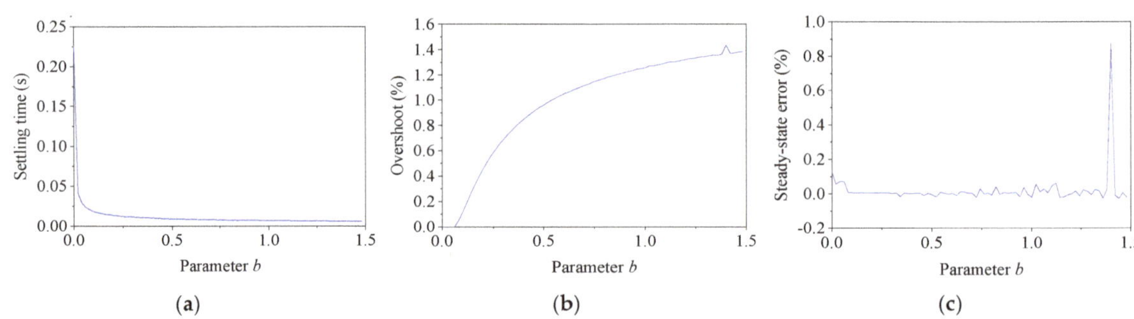

Figure 6. Effect of parameter b on W_S. (**a**) Settling time; (**b**) overshoot; (**c**) steady-state error.

3. Parameter c in W_S

The parameter c was changed from 0 to 1.5 to investigate its influence, while the other parameters in the weightings were kept unchanged. The dynamics were calculated by the step response, which is shown in Figure 7. It is seen that with the increase in parameter c, the settling time and the overshoot almost remained unchanged, while the steady-state error increased. Hence, it is recommended that the parameter c should be small enough to eliminate the steady-state error, which is very important to RTHS.

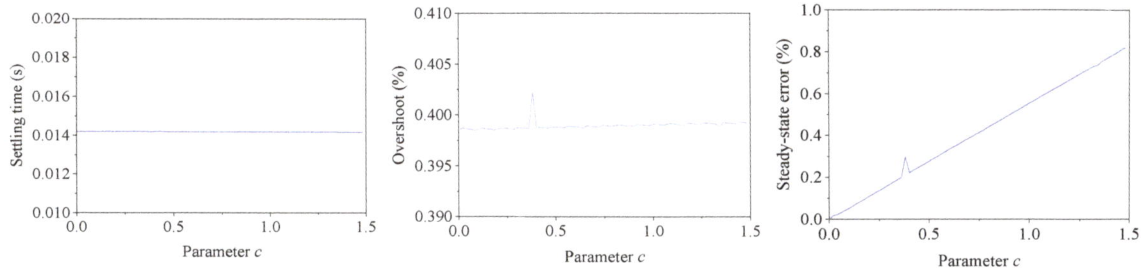

(a) (b) (c)

Figure 7. Effect of parameter c on W_S. (**a**) Settling time; (**b**) overshoot; (**c**) steady-state error.

3.2.2. Influence of W_T

For convenience of analysis, the weighting function W_T in Equation (10) can be rewritten as

$$W_T = \frac{0.0225s^2 + 5.9431s + 79.5216}{(0.003s + 47.1239)^2}. \quad (11)$$

1. Parameter h in W_T

Keeping the other parameters in the weightings unchanged, the parameter h was varied from 0 to 0.3 to examine its influence. The dynamics were calculated via the step response, which is shown in Figure 8. With the increase in parameter h, the settling time decreased first and then increased rapidly. The overshoot almost remained unchanged at the beginning and then increased quickly, while the steady-state error decreased quickly and then remained unchanged.

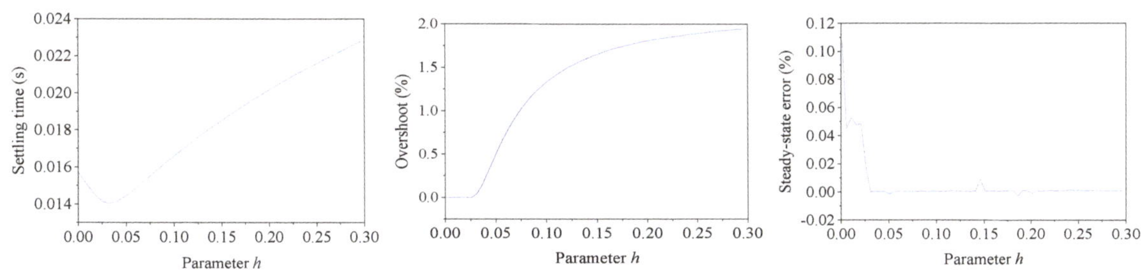

(a) (b) (c)

Figure 8. Effect of parameter h on W_T. (**a**) Settling time; (**b**) overshoot; (**c**) steady-state error.

2. Parameter m in W_T

To analyze the influence of the parameter m on weighting function W_T, the other parameters in the weightings were kept unchanged. By changing m from 0 to 30, the dynamics were calculated by the step response, which is shown in Figure 9. With the increase in parameter m, the settling time almost remained unchanged first and then decreased quickly. Then, the settling time increased with the increase in parameter m. The overshoot decreased with the increase in parameter m, and then almost remained unchanged regardless of how the parameter m increased. Meanwhile, the steady-state error remained unchanged first and then fluctuated at a relatively high value.

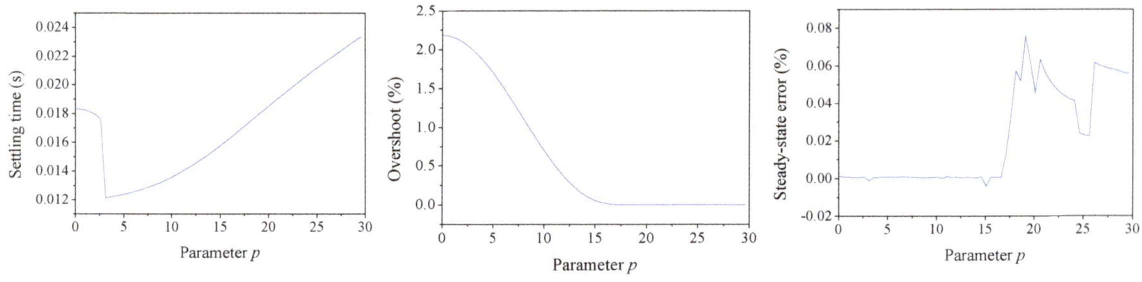

(a) (b) (c)

Figure 9. Effect of parameter m on W_T. (**a**) Settling time; (**b**) overshoot; (**c**) steady-state error.

3. Parameter n in W_T

By varying the parameter n from 0 to 450, the dynamics were calculated via the step response correspondingly, which are shown in Figure 10. It is seen that with the increase in parameter n, the settling time and steady-state error varied in a very small range, which could be viewed as unchanged. Meanwhile, the overshoot increased with the parameter n.

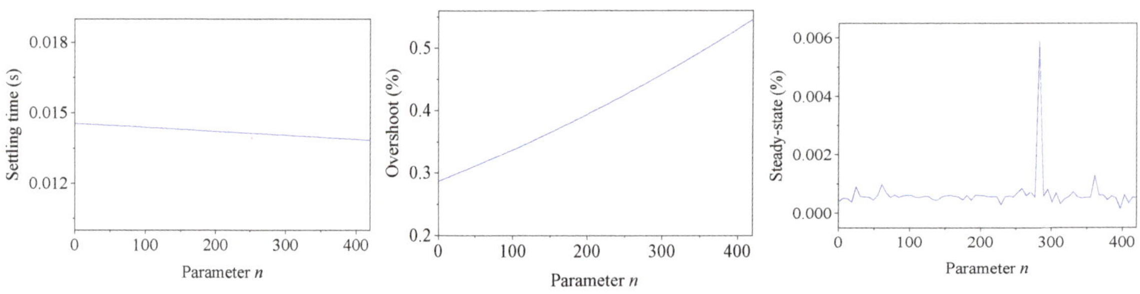

(a) (b) (c)

Figure 10. Effect of parameter n on WT. (**a**) Settling time; (**b**) overshoot; (**c**) steady-state error.

3.2.3. Influence of W_R

The weightings W_S and W_T were kept unchanged, and the weighting W_R was varied from 0 to 1.5. The dynamics were calculated via the step response, which is shown in Figure 11. With the increase in weighting W_R, the settling time increased. When the weighting W_R increased in a very small range near zero, the overshoot decreased quickly. Then, the overshoot almost remained at zero with the increase in W_R. The steady-state error fluctuated at a relatively small value. Hence, it is recommended that a small positive number is used for weighting W_R.

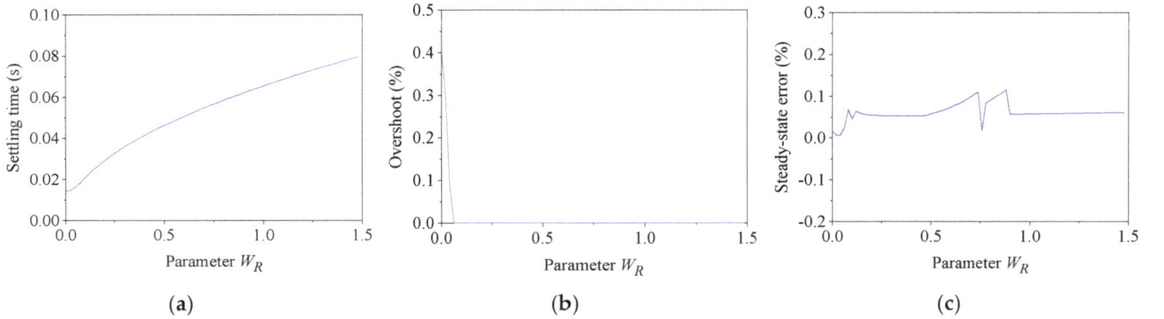

Figure 11. Effect of W_R. (**a**) Settling time; (**b**) overshoot; (**c**) steady-state error.

4. Numerical Validation

In this section, a nonlinear model of the servo-hydraulic actuator [25] was employed to validate the effectiveness of the proposed H_∞ control method used for RTHS.

Typically, the H_∞ controller is designed by employing a nominal analytical model of the physical testing system, in which the PS is included. Hence, a linear model of the servo-hydraulic actuator, or nominal plant, is obtained from the nonlinear model for design convenience, which is given by

$$P(s) = \frac{2\beta A_p k_0}{M_E V s^3 + (C_E V + 2\beta k M_E k_0)s^2 + (K_E V + 4\beta A_p^2 + 2pA_p^2 + 2k\beta C_E k_0)s + 2k\beta K_E k_0}. \quad (12)$$

where the symbols and their values are listed in Table 1. The damping coefficient C_E is calculated by the damping ratio and natural frequency of the specimen. Then, the transfer function of the nominal model is

$$P(s) = \frac{7.748 \times 10^6}{s^3 + 165.7s^2 + 3.706 \times 10^5 s + 5.235 \times 10^5}. \quad (13)$$

Table 1. Values of system parameters for simulation [25].

Item	Parameter	Value	Parameter	Value
Servovalve	Natural frequency	816.81 rad/s	Damping ratio	0.7
	Servo-valve gain k_0	1.0674 m^2/s	Supply pressure p	19.995
Actuator	Piston area A_p	0.0248 m^2	Volume V	0.0069 m^3
	Effective bulk modulus of oil β	677.8 MPa	Pressure difference feedback gain k	0.0002
Specimen	Mass M_E	56.289 kg	Damping ratio	0.05
	Stiffness K_E	2276.3 kN/m		

It should be noted that the units have been transformed into the international system of units.

4.1. Robustness Investigation

In a real application, there are differences between the nominal model and the physical testing system, and the characteristics of the PS will vary during RTHS. Hence, the robustness of the H_∞ controller was investigated numerically in this section.

4.1.1. Modeling Uncertainties

- Controller design

When designing the H_∞ controller, the modeling uncertainties and measurement noise should be considered first, followed by the dynamics of the controlled system. Hence, it is recommended that the designed order of the weighting function is W_T, W_S, and W_R.

To determine the modeling uncertainties, a series of linear numerical models of the nonlinear model were calculated. Then, the uncertainties were obtained, whose singular values are shown in Figure 12. Later, the weighting W_T can be designed. It is seen in the figure that the modeling error was very small when the frequency was not very high; then, the modeling uncertainties increased with the frequency. Hence, the weighting W_T should cover the uncertainties over all the frequency ranges. Through several trials, the expression for W_T was given by

$$W_T(s) = 2 \times 10^{-5}s^2 + 0.005s + 0.25 \tag{14}$$

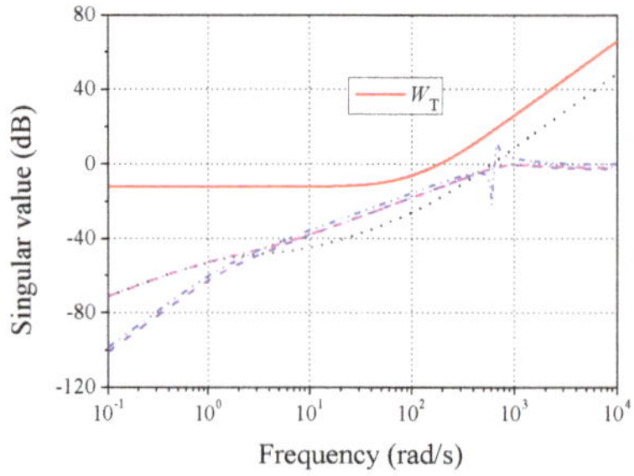

Figure 12. Modeling uncertainties and weighting function W_T.

Subsequently, the dynamics of the controlled system were considered. It is expected that the controlled system should respond to the command quickly and without a steady-state error. Hence, the other two weightings were determined after several trials, which were

$$W_S(s) = \frac{0.01s + 63}{s + 0.001}, W_R = 0.1 \tag{15}$$

Eventually, a feasible solution was reached, and the H_∞ central controller was

$$K(s) = \frac{7,731,236.8628(s + 1.413)(s^2 + 164.3s + 3.704 \times 10^5)}{(s + 1.021 \times 10^6)(s + 1171)(s + 568.5)(s + 0.001)} \tag{16}$$

Afterward, the performance of the controlled system was examined through performance curves and the step response, which are given in Figure 13. A fast response speed is observed in Figure 13a, and there is no steady-state error. It can also be seen from Figure 13b that the complementary sensitivity function T was almost identical to 1 when the frequency was less than 10 Hz, indicating that the controlled system could track the command very well. It is seen in the figure that the curves of the sensitivity function S and complementary sensitivity function T were below the weighting functions ${W_S}^{-1}$ and ${W_T}^{-1}$, respectively, indicating that the selected weighting function could meet the robust performance [24]. Moreover, over the concerned frequency, the sensitivity function S was far less than 1, indicating that the controlled system exhibited a strong robustness considering the disturbance.

- Modeling errors

To investigate the robustness of the H_∞ controller to modeling uncertainties, two cases were considered here. For the first case, the multiplicative uncertainties of 50% were considered, while for the second case, the controller designed employing a linear model was used for the nonlinear model. Step responses for the two cases are shown in Figure 14, and the response without any uncertainties is also given in the figure for convenience of comparison.

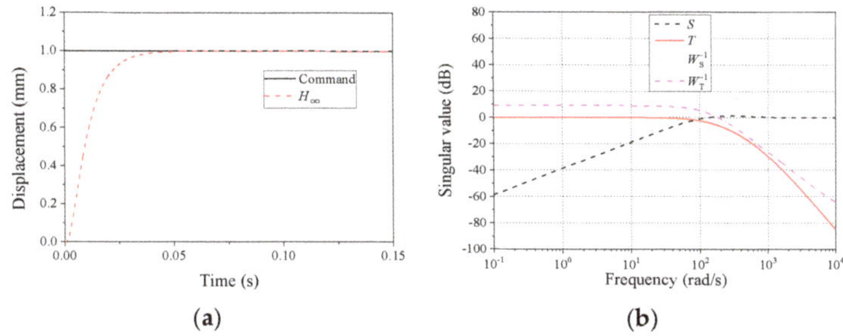

Figure 13. Performance of the controlled system. (**a**) Step response; (**b**) performance curve.

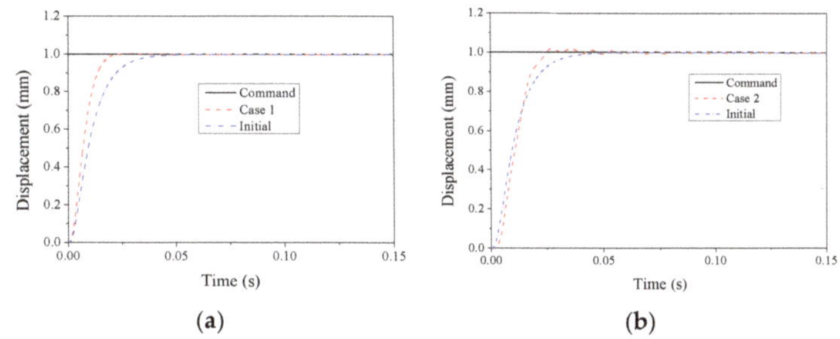

Figure 14. Step responses. (**a**) Case 1: multiplicative uncertainties; (**b**) case 2: nonlinear model.

It is seen in the figure that when the multiplicative uncertainties were considered, steady-state errors and overshoots were still not observed, while the settling time was less than the initial state. When the H_∞ controller was used for the nonlinear model, it was found that there were obvious fluctuations, and overshoot and steady-state errors occurred. However, the overshoot was very small (less than 0.5%), and the steady-state error tended to zero. Hence, the H_∞ controller exhibited strong robustness to modeling uncertainties.

4.1.2. Variation of the Specimen

- Stiffness

Two different stiffness coefficients were considered for the PS, 1.5 and 0.1 times the initial one, respectively. The time histories of the change ratio under the step response are given in Figure 15. It is seen from the figure when the stiffness coefficient varied, the steady-state error occurred. The change ratio for the stiffness decrease was smaller than that for the stiffness increase, and they were within the acceptable range. Hence, the H_∞ controller was robust to the variation in stiffness.

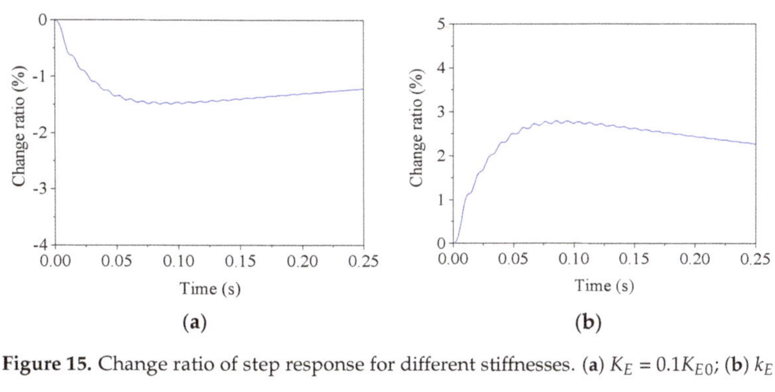

Figure 15. Change ratio of step response for different stiffnesses. (**a**) $K_E = 0.1K_{E0}$; (**b**) $k_E = 1.5K_{E0}$.

- Damping

To assess the robustness of the H_∞ controller, two different damping ratios, 0.1 and 1.5 times the initial one, respectively, were considered for the PS. The time histories of the change ratio under the step response are given in Figure 16. It is found that the change ratios were negligible compared to those in Figure 15. With time, the change ratio tended to zero, which means that the steady-state error was zero. This indicates that the damping characteristics had limited influences on the step responses, and the H_∞ controller exhibited a strong robustness.

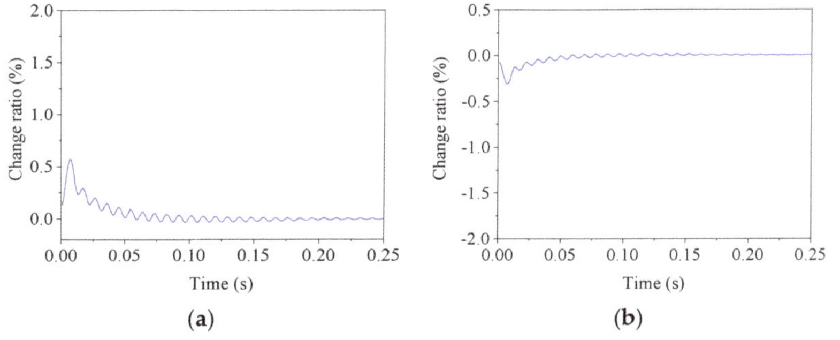

Figure 16. Change ratio of step response for different stiffnesses. (**a**) $C_E = 0.1C_{E0}$; (**b**) $C_E = 1.5C_{E0}$.

4.2. Virtual RTHS

In this section, virtual RTHSs were conducted on a single-degree-of-freedom (SDOF) structure. The mass was 1200.8 kg, the structural stiffness was 2276.3 kN/m, and the damping ratio was 5%. A nonlinear PS was employed, of which the mass and stiffness of the PS were 56.6384 kg and 2276.3 kN/m, respectively, the yield displacement was 7.6 mm, and the stiffness coefficient after yield was 0.12. The earthquake excitation was El Centro (NS, 1940), whose peak ground acceleration (PGA) was scaled to 0.16g. In the virtual hybrid simulation, the α method [26] was adopted to solve the equation of motion, and the time integral was 0.01 s. A pure dynamic analysis of the whole structure was performed to serve as an exact reference solution, where the stepwise integral method was CDM-RST [25], and the time step was 0.01 s. Due to the PS being changed, the controller was re-designed, and the controller was:

$$K = \frac{14,893,750.8909(s+2.094\times10^4)(s+1.047\times10^4)(s+1.413)(s^2+164.3s+3.704\times10^5)}{(s+1.305\times10^6)\,(s+2.096\times10^4)\,(s+1.031\times10^4)\,(s+1196)\,(s+669.9)\,(s+0.001004)} \quad (17)$$

The time histories of displacement obtained by virtual RTHS were compared with those of the reference solution, which is shown in Figure 17. It is seen in the figure that a shifting was observed for the displacement due to the yielding of the structure. However, the responses were in good agreement with those of the reference solution under the earthquake excitation.

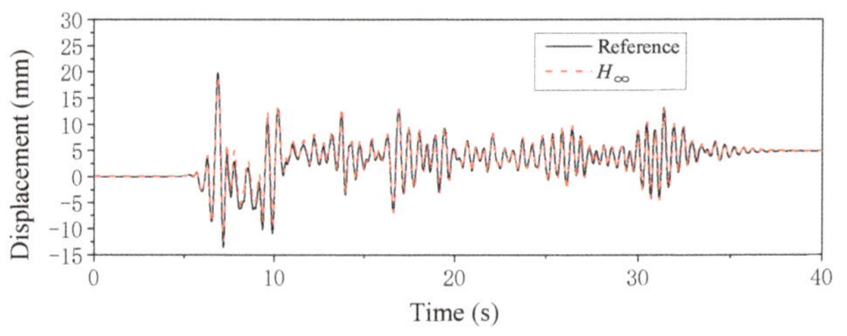

Figure 17. Displacement time histories for virtual RTHS.

5. Experimental Validation

5.1. Overview of the Test

5.1.1. Experimental Setup

To validate the effectiveness of the proposed H_∞ control method, actual real-time hybrid simulations were conducted on an SDOF structure. The loading system at the Structural and Seismic Testing Center of Harbin Institute of Technology was employed, which consists of an FTS system and a dSPACE 1104 board. The FTS system comprises a servo-hydraulic actuator, displacement sensor, force sensor, and servovalve. The displacement sensor is an LWH-200 with a maximum stroke of 200 mm, which is produced by Novotechnik. The force sensor is PSD-5tSJTT with a capacity of 5 T, which is produced by Vishay. The servovalve is MOOG-761-3005, produced by Moog. The experimental setup is shown in Figure 18.

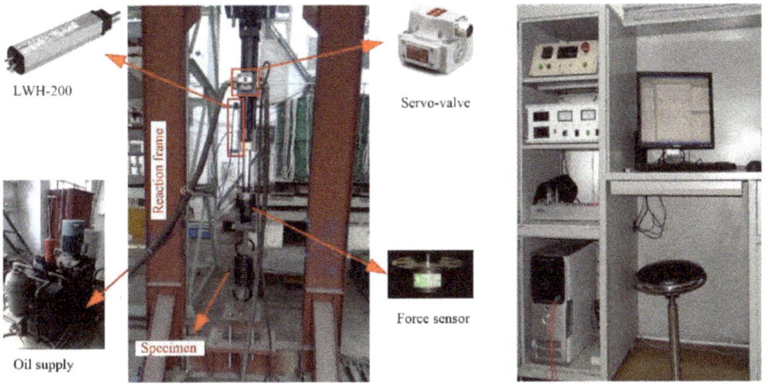

Figure 18. Loading system setup. (**a**) FTS system; (**b**) control hardware and system.

Considering the restraints of the loading system, a spring was taken as the specimen and was loaded in the axial direction. In RTHS, the central difference method was adopted to solve the equation of motion, and the time integral was 0.02 s. The equation of motion solved by the LSIM command in MATLAB served as the exact reference solution, where the time step was 0.02 s.

To diminish the effect of measurement noise, an elliptic filter was designed by the digital filter design module, with stop frequencies of 100 and 30 Hz for displacement and force, respectively. The sampling time for the filter was 1000 Hz.

5.1.2. Controller Design

Similar to that used in Section 4, the analytical model was established theoretically using the technical index of the servo-hydraulic actuator, which is provided by the manufacturer. The transfer function from u to y is

$$P = \frac{4\beta k_0 A_p}{\left(4\beta A_p^2 + 4pA_p^2 + VK_e\right)s + 4\beta k k_0 K_E} \tag{18}$$

where the parameters are listed in Table 2. Hence, the numerical model is given by

$$P(s) = \frac{33.74}{s + 0.5739} \tag{19}$$

Table 2. Values of system parameters.

Parameter	Value	Parameter	Value
Servo-valve gain k_0	0.0512 m^2/s	Supply pressure p	19.995
Piston area A_p	0.0015 m^2	Volume V	0.0043 m^3
Effective bulk modulus of oil β	677.8 MPa	Pressure difference feedback gain k	0.0002

The mathematical model in between the input and output in the frequency domain is shown in Figure 19. It is seen in Figure 19a that within the frequency range of interest, the model was in good agreement with the test data, while for the phase, differences were observed when the frequency became large. However, the analytical model can reflect the major dynamics of the actual testing system.

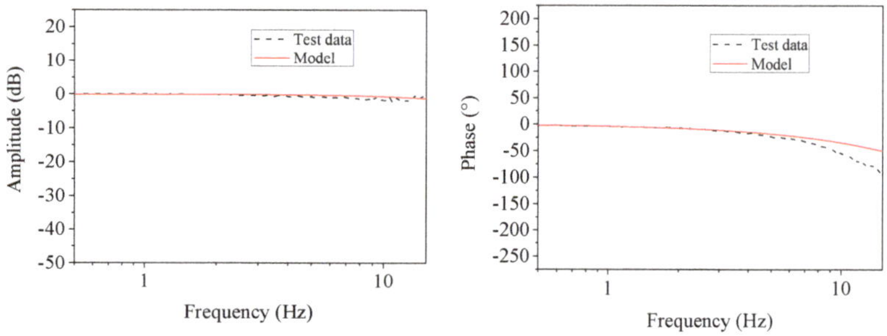

(**a**) (**b**)

Figure 19. Model validation. (**a**) Magnitude; (**b**) phase.

The weighting functions were determined using the recommended method given in Section 3. After several trials, they were selected as

$$W_S = \frac{0.05s + 62.83}{s + 0.01}, W_T = \frac{833.3333(s+805)\ (s+19.51)}{(s+2.09 \times 10^4)\ (s+1.05 \times 10^4)}, W_R = 1 \times 10^{-6} \tag{20}$$

Eventually, a feasible solution was achieved and the controller was

$$K(s) = \frac{7.6251 \times 10^8\ (s + 2.094\ \times 10^4)\ (s + 1.047\ \times 10^4)\ (s + 0.5739)}{(s + 2.812 \times 10^{10})(s + 0.01048)(s^2 + 1892s + 1.086 \times 10^2)} \tag{21}$$

It should be noted that the H_∞ controller was designed in the continuous-time domain, while the dSPACE is a digital sampling system. Hence, the H_∞ controller was converted to discrete form by the c2d command in MATLAB. The discrete method was Tustin, and the sampling time was identical to that in dSPACE, namely 1000 Hz.

5.2. Loading System Verification

The feasibility of the designed H_∞ controller was verified first. The step response is given in Figure 20. It is seen in the figure that there was a pure delay in the system and no steady-state error. The rise and settling times were 0.018 and 0.048 s, respectively. The time of first achieving the command was 0.025 s. It seems that the H_∞ controller could not complete the RTHS. However, the step response is very rigorous, so the additional test was carried out to further verify the dynamic performance.

A sinusoidal signal was discretized with a sampling time of 0.02 s, which was used as the command signal. The response is shown in Figure 21. It is seen in the figure that at the end of each step, the measurement displacements were in good agreement with the command, indicating that the H_∞ controller was suitable for RTHS.

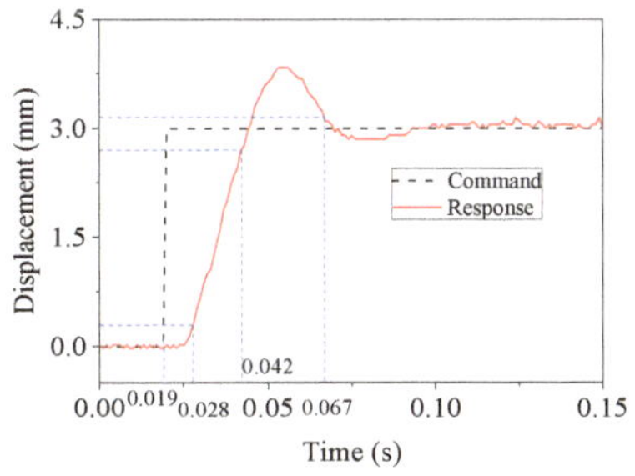

Figure 20. Step response with H_∞ controller.

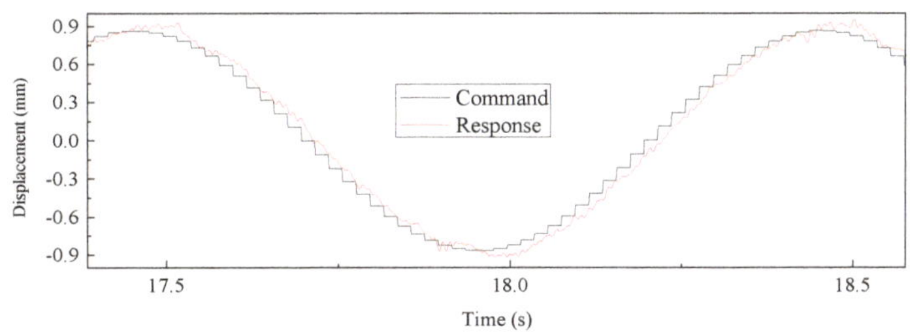

Figure 21. Tracking performance under sinusoidal signal.

5.3. RTHS

5.3.1. Sinusoidal Excitation

RTHS under sinusoidal excitation was first carried out by employing the H_∞ controller. The frequency of the excitation was 2 Hz, and the amplitude was increased slowly from 0 to 344.492 N. The period and damping ratios of the simulated structure were 1 s and 0.02, respectively. The stiffness of the NS was the same as that of the PS, which was 34 kN/m. The tracking performance of the actuator with the H_∞ controller was examined first, and the commanded and measured displacements are shown in Figure 22. It is seen in the figure that at the end of each time integration step, the displacement measurement was almost identical to that of the command. Moreover, the reference solution is also given in Figure 22. It is seen that the measured displacements were in good agreement with the reference ones.

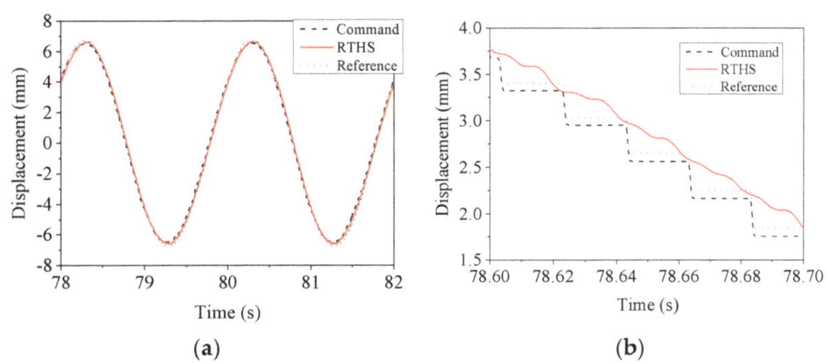

Figure 22. Steady-state response under sinusoidal excitation. (**a**) Overall view; (**b**) enlarged view.

5.3.2. Earthquake

A series of RTHSs were conducted on a spring specimen with a stiffness of 35 kN/m. Three different numerical substructures, whose stiffnesses were 35, 17.5, and 0 kN/m, were considered. The damping ratio was 0.1. For the NS with a stiffness of 35 kN/m, the period was 2 s. The earthquake excitation was El Centro (NS, 1940), and the peak ground acceleration (PGA) was scaled to 50 gal. The displacement time histories, overall and enlarged view, are shown in Figure 23, together with the reference solution.

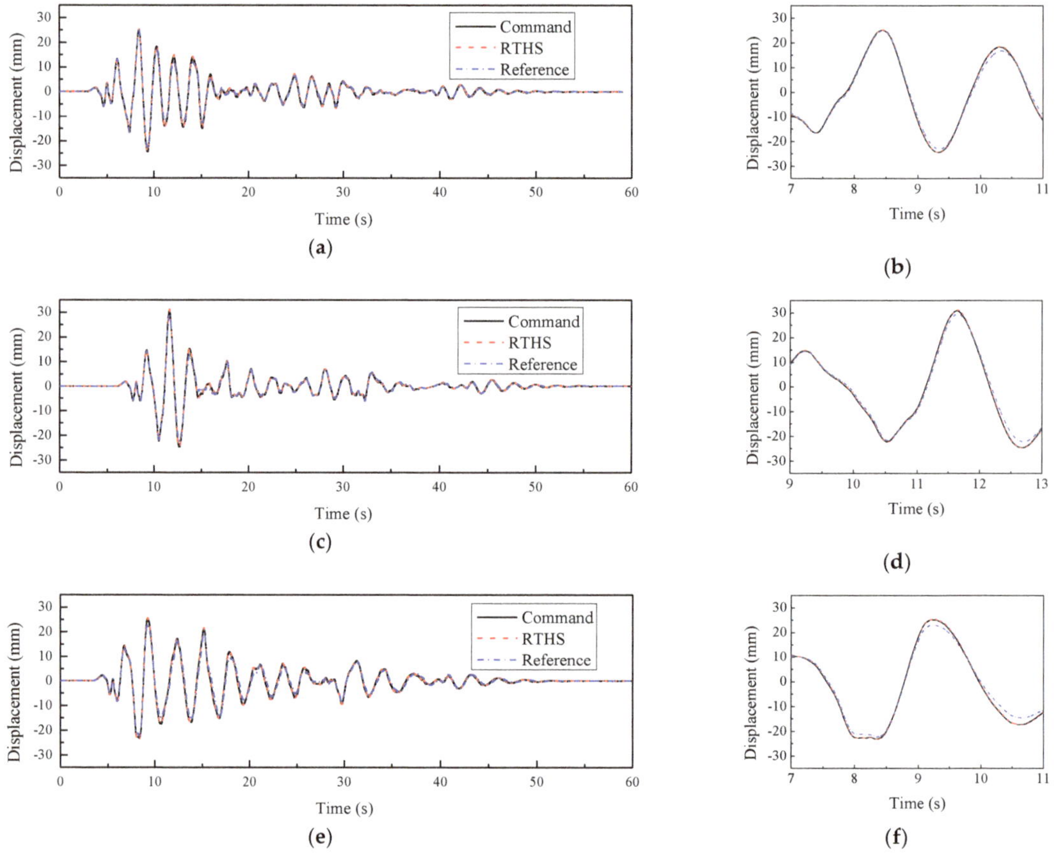

Figure 23. Displacement time histories under earthquake excitation. (**a**) Overall view for Case 1; (**b**) enlarged view for Case 1; (**c**) overall view for Case 2; (**d**) enlarged view for Case 2; (**e**) overall view for Case 3; (**f**) enlarged view for Case 3.

It is seen in the figure that the measured displacements were almost identical to the command for the three cases, indicating a perfect tracking performance of the H_∞-controlled system. They were further verified by the RMSE collected in Table 3.

Table 3. Stiffness of NS and RMS error of hybrid simulation.

Case	Stiffness of NS	*RMSE* (%)	
		Tracking	RTHS
1	35 kN/m	2.77	12.47
2	17.5 kN/m	3.04	13.07
3	0 kN/m	2.31	17.96

If focusing on the displacement response and reference solution, one will find that the displacements of RTHS matched the reference well before the first displacement peak. Subsequently, differences emerged, especially at each positive or negative peak.

It is seen in the figure that the tracking performance was perfect, but the simulation results were barely satisfactory. The reason is that the measured force was not synchronized with the desired displacement. There were two different filters, which may introduce additional time delays differently.

6. Conclusions

Real-time hybrid simulation is a powerful technique to evaluate the structural performance under dynamic loads, especially for structures with velocity- or acceleration-dependent components. Due to the inherent nonlinearities of the servo-hydraulic actuator and the uncertainties in the systems, the boundary conditions between the two substructures could be realized completely. Hence, a mixed sensitivity-based H_∞ control method was proposed for RTHS. The main conclusions that could be reached are as follows.

1. The principle of the H_∞ control theory was presented briefly. Theoretically, the H_∞ control strategy is an optimization problem. By introducing the performance weighting function to the feedback control system, the standard H_∞ control problem can be formulated.
2. The weighting function selection method was proposed, and the influences of the weighting function on the system dynamics were discussed. Typically, W_S should be close to an integrator to eliminate the steady-state error, and a large numerator will generate a fast response speed. W_T should be determined by evaluating the model uncertainties in advance, and it should have a slope of approximately 40 dB/dec over the high-frequency range to suppress the unmodeled dynamics and measurement noise. A small positive constant value is usually used for W_R.
3. The robustness of the H_∞ controller was investigated numerically. When considering the model uncertainties and characteristic variation in the specimen, the overshoot and steady-state error varied in an acceptable range, indicating the strong robustness of the H_∞ controller.
4. The effectiveness and feasibility of the proposed method were validated via numerical simulations and actual RTSHs. When considering the nonlinear characteristics of the specimen, the actual modeling uncertainties, or the measurement noise, the H_∞-controlled system showed an excellent tracking performance, indicating that it is suitable to use the H_∞ controller for RTHS.

However, it should be noted that the proposed method was only validated via a linear elastic specimen, and a nonlinear physical substructure should be used to further validate the feasibility. Furthermore, as the H_∞ controller is still a feedback controller in essence, the time delay cannot be eliminated. Hence, it is necessary to combine the H_∞ controller with other strategies.

Funding: The Scientific Research Fund of the Institute of Engineering Mechanics, China Earthquake Administration (Grant No. 2020D14), the National Natural Science Foundation of China (Grant No. 51908231), and the Natural Science Foundation of Fujian Province of China (Grant No. 2020J01058).

Institutional Review Board Statement: Not applicable.

Informed Consent Statement: Not applicable.

Data Availability Statement: The data presented in this study are available on request from the corresponding author.

Conflicts of Interest: The author declares no conflict of interest.

Nomenclature

a, b, and c	Adjustable parameters in weighting function W_S
e	Tracking error
h, m, and n	Adjustable parameters in weighting function W_T
j	Imaginary unit
k	Pressure difference feedback gain
k_0	Servo-valve gain
p	Supply pressure
r	Reference input
u	Control signal or controller output
w	Exogenous input
x	State vector
y	Measured output
z	Performance output
A, B, C, D	Coefficient matrix
A_P	Piston area
G	Generalized plant
G_0	Nominal or analytical plant
K	Controller
K_E	Stiffness of specimen
M_E	Mass of specimen
P	Transfer function of the control plant
R	Controller sensitivity
S	Sensitivity function
T	Complementary sensitivity function
T_{wz}	Transfer function from input w to output z
V	Volume
W_S, W_R, and W_T	Weighting function
β	Effective bulk modulus of oil
γ	Positive number
ω	Frequency
ω_d	Desired bandwidth
σ	Singular value

References

1. Nakashima, M.; Kato, H.; Takaoka, E. Development of real-time pseudo dynamic testing. *Earthq. Eng. Struct. Dyn.* **1992**, *21*, 79–92. [CrossRef]
2. Hakuno, M.; Shidawara, M.; Hara, T. Dynamic destructive test of a cantilever beam, controlled by an analog-computer. *Pro. Jpn. Soc. Civ. Eng.* **1969**, *171*, 1–9. [CrossRef]
3. Nakashima, M.; Takai, H. Computer-actuator online testing using substructure and mixed integration techniques. In Proceedings of the 7th Symposium on the Use of Computers in Building Structures, Architectural Institute of Japan, Tokyo, Japan, 10–11 December 1985; pp. 205–210.
4. Dermitzakis, S.N.; Mahin, S.A. *Development of Substructuring Techniques for on-Line Computer Controlled Seismic Performance Testing*; University of California: Berkeley, CA, USA, 1985.
5. Wu, B.; Bao, H.; Ou, J.; Tian, S. Stability and accuracy analysis of the central difference method for real-time substructure testing. *Earthq. Eng. Struct. Dyn.* **2005**, *34*, 705–718. [CrossRef]
6. Wu, B.; Xu, G.; Wang, Q.; Williams, M.S. Operator-splitting method for real-time substructure testing. *Earthq. Eng. Struct. Dyn.* **2006**, *35*, 293–314. [CrossRef]
7. Chen, C.; Ricles, J.M. Analysis of implicit HHT-α integration algorithm for real-time hybrid simulation. *Earthq. Eng. Struct. Dyn.* **2012**, *41*, 1021–1041. [CrossRef]
8. Huang, L.; Chen, C.; Guo, T.; Chen, M. Stability Analysis of Real-Time Hybrid Simulation for Time-Varying Actuator Delay Using the Lyapunov-Krasovskii Functional Approach. *J. Eng. Mech.* **2019**, *145*, 04018124. [CrossRef]
9. Horiuchi, T.; Inoue, M.; Konno, T.; Namita, Y. Real-time hybrid experimental system with actuator delay compensation and its application to a piping system with energy absorber. *Earthq. Eng. Struct. Dyn.* **1999**, *28*, 1121–1141. [CrossRef]
10. Darby, A.; Williams, M.; Blakeborough, A. Stability and delay compensation for real-time substructure testing. *J. Eng. Mech.* **2002**, *128*, 1276–1284. [CrossRef]
11. Ahmadizadeh, M.; Mosqueda, G.; Reinhorn, A. Compensation of actuator delay and dynamics for real-time hybrid structural simulation. *Earthq. Eng. Struct. Dyn.* **2008**, *37*, 21–42. [CrossRef]

MDPI
St. Alban-Anlage 66
4052 Basel
Switzerland
Tel. +41 61 683 77 34
Fax +41 61 302 89 18
www.mdpi.com

Symmetry Editorial Office
E-mail: symmetry@mdpi.com
www.mdpi.com/journal/symmetry

www.ingramcontent.com/pod-product-compliance
Ingram Content Group UK Ltd.
Pitfield, Milton Keynes, MK11 3LW, UK
UKHW060123010826
14089UKWH00006B/111

9 783036 542379